THE Sweeps

THE Sweeps

Behind the Scenes in Network TV

Mark Christensen & Cameron Stauth

William Morrow and Company, Inc., New York

Library of Congress Cataloging in Publication Data

Christensen, Mark.
  The sweeps.

  1. National Broadcasting Company, Inc.   I. Stauth,
Cameron.   II. Title.
PN1992.92.N37C49   1984   384.55′4′0973   84-9021
ISBN 0-688-03912-X

Printed in the United States of America

First Edition

1 2 3 4 5 6 7 8 9 10

BOOK DESIGN BY NANCY DALE

To our parents

Acknowledgments

We would like to thank our editor, Jim Landis, and our tireless and greathearted representative, Richard Pine, for the considerable efforts they both made stitching this project together.

David Kelly provided absolutely vital editing, writing, and manuscript planning from which the book benefited greatly.

We would also like to thank Deborah Wenner and Shari Levine for their love and support, Richard Meeker for his counsel, David Stauth for help with research, and Faye Caswell, Joan Aschim, and Jill Wilson for their efforts in typing and transcription.

During the course of a year we interviewed over two hundred people in the television industry. Many made extensive contributions that may not be immediately apparent reading this book. We'd like to thank, in particular, author Laurie Burrows Grad, NBC Senior Vice-President of Business Affairs John Agoglia, NBC Vice-President of West Coast Broadcast Standards Maurice Goodman, head of television publicity for MGM Rob Maynor, attorney Paul Ziffren, actor George Wendt, news anchor Kirk Matthews, casting director Vickie Rosenberg, and producer Marshall Herskovitz.

Too, we'd like to express our considerable appreciation to comic genius–producer–magician Mark Maxwell-Smith, who spent days' worth of his nonexistent spare time giving us vital background material, and to Gary Goldberg and the entire staff

7

of *Family Ties*, the nicest and sanest people we met in Hollywood. Thanks likewise to NBC Vice-President, Press, West Coast, Gene Walsh and his extremely able associate Dorothy Burns for their year's worth of help, and to *Cheers* producer Jim Burrows for introducing us to the real realm of quality television.

Finally, we'd like to donate our firstborn children to John Case, without whose support this entire escapade would have been severely compromised.

Contents

Cast of Characters

THE ACTORS

Harry Anderson—star of NBC's *Night Court*, guest star on NBC's *Cheers* and *Saturday Night Live*

Ed Asner—president of the Screen Actors Guild, former star of CBS's *Lou Grant* and *The Mary Tyler Moore Show*

Teri Copley—star of NBC's *We Got It Made*

Ted Danson—star of NBC's *Cheers* and the 1983 ABC-TV movie *Something About Amelia*

Corky Hubbert—guest star on CBS's *Magnum, P.I.*, star of the 1984 feature film *Not for Publication*

Shelley Long—star of NBC's *Cheers* and the 1984 feature film *Irreconcilable Differences*

John Ratzenberger—co-star of NBC's *Cheers*

Karen Salkin—star of *Karen's Restaurant Revue*, Group W Cable, Los Angeles

James P. Sikking—co-star of NBC's *Hill Street Blues*

Tim Thomerson—co-star of CBS's *The Apple Dumpling Gang*

THE PRODUCERS

Steven Bochco—executive producer of NBC's *Hill Street Blues* and *Bay City Blues*

Lynn Farr Brao—co-executive producer of NBC's *We Got It Made*

Steve Cannell—executive producer of NBC's *The A-Team*, ABC's *Hardcastle and McCormick*, and NBC's *The Rousters*
Glen and Les Charles—co-executive producers of NBC's *Cheers*
Dick Ebersol—executive producer of NBC's *Saturday Night Live*
Gordon Farr—co-executive producer of NBC's *We Got It Made*
Earl Hamner—executive producer of NBC's *Boone*, creator of CBS's *The Waltons* and CBS's *Falcon Crest*
Allan Katz—executive producer of NBC's *The National Snoop*
Glen Larson—executive producer of NBC's *Manimal*, ABC's *The Fall Guy*, and ABC's *Trauma Center*
Jeff Lewis—co-executive producer of NBC's *Bay City Blues*
Garry Marshall—executive producer of ABC's *Happy Days* and *Laverne and Shirley*
Larry Rosen—co-executive producer of NBC's *Jennifer Slept Here*
Fred Silverman—co-executive producer of NBC's *We Got It Made*, former head of programming for all three networks
Ziggy Steinberg—producer of NBC's *Another Jerk*
Mark Tinker—supervising producer of NBC's *St. Elsewhere*
Larry Tucker—co-executive producer of NBC's *Jennifer Slept Here*
Reinhold Weege—executive producer of NBC's *Night Court*

THE DIRECTORS

James Burrows—director and co-executive producer of NBC's *Cheers*
Tom Chamberlin—director of the regional/cable *Pillars of Portland*
Bob Radler—producer-director of MTV videos
Alan Rafkin—director of NBC's *We Got It Made*
Jay Sandrich—director of NBC's *Night Court*

THE WRITERS

Tony Colvin—writer for ABC's *Just Our Luck*
Scott Gorden—writer for ABC's *Just Our Luck*

THE NETWORK: NBC

Hamilton Cloud—head of on-air comedy programs

Warren Littlefield—head of comedy development
Jeff Sagansky—senior vice-president for series programming
Steve Sohmer—senior vice-president for promotion, specials, daytime, and Saturday morning
Brandon Tartikoff—president of NBC Entertainment
Joel Thurm—vice-president for talent (in charge of casting)
Grant Tinker—chairman of the board

THE NETWORK AFFILIATES

Steve Currie—program director for CBS's KOIN-TV, Portland, Oregon
Ancil Payne—president of King Broadcasting, Seattle
Irwin Starr—general manager of NBC's KGW-TV, Portland, Oregon

CABLE AND NONNETWORK

Larry Colton—writer-producer of the regional/cable *Pillars of Portland*
James Devaney—president of JPD Television Network, a "low-power television" network
Paul Klein—president of the Playboy Channel
Norm Smith—president of the Pleasure Channel
Ted Turner—chairman of Turner Broadcasting Systems

THE STUDIOS

Peter Grad—head of TV development for Twentieth Century-Fox
Gary Nardino—president of Paramount Television

THE JOURNALISTS

Ben Brown—reporter for *USA Today*
William Donaldson—writer for the *Tulsa Tribune*
Donna Rosenthal—reporter for the *National Enquirer*
Ben Stein—columnist for the *Los Angeles Herald-Examiner*

THE AGENTS AND PUBLICISTS

Jim Canchola—agent for Tony Colvin and Scott Gorden
Beth Herman—Rogers and Cowan publicist for Karen Salkin
Danny Robinson—agent for Harry Anderson
Cynthia Snyder—publicist for several *Hill Street Blues* cast
    members

"Freddie Silverman, at his most salesmanlike, convinced us [Rhoda] should be married during sweeps week, which was much more quickly than we were really comfortable with. Freddie felt everybody should be married in sweeps week. He probably did it himself."—ALLAN BURNS in *TV Guide*

Four times every year, the A. C. Nielsen Company measures the audience of every network affiliate in America. These crucial rating periods, called the sweeps, determine local advertising rates. Strong sweeps mean a strong network. Weak sweeps mean trouble.

Prologue

This was the part Lynn Farr Brao liked best: *pitching*. Selling ideas. Writing was fun; so was producing and casting and rehearsing and editing. So much of TV work was just hard play—you told stories, you acted them out, people clapped and cheered, and when it was all over you were rich and famous. It was a delight. But this was the hard nub of it, the gravitational center of the entire industry—*pitching*. My *God*, it was exciting.

Lynn's cohort of the moment, Fred Silverman, once programming head for each of the three networks and now an independent producer, was pitching to Brandon Tartikoff, the president of NBC Entertainment, as they gathered in Tartikoff's modest, comfortable Burbank office. "If there's one thing I know, Brandon," Silverman was saying, "it's how to put a new twist on an old format." Tartikoff, thirty-four, boyish in a blue sweater, not wearing a tie—no one in the room had on a tie—smiled and nodded his agreement. Then his assistants smiled and nodded. Who could argue the point? Fred Silverman was the most famous recycler in the business.

Lynn noted the programmers' nodding with satisfaction. It was the oldest of salesmen's ploys—get the buyer agreeing, even if he's only agreeing that it's a nice day, or that Chrysler makes a great car. As good as she was at pitching, she thought, Silverman was better.

They were discussing the script of a prospective pilot for the 1983–84 season. Silverman, who had been fired by NBC three years earlier, had been given a commitment to put a show on the air for ar least thirteen episodes as part of his severance pay. About a month before this meeting, Silverman had called Lynn and her partner, Gordon Farr, to whom she'd once been married, and asked them if they had any ideas for a show. NBC, Silverman said, was looking for a 9:30 P.M. half-hour program that would appeal to adults—in other words, a sexy sitcom. It had taken Lynn and Gordon exactly seven minutes—they timed it—to come up with an idea. Which was: A beautiful girl moves in with two single guys as a live-in maid. The concept was, of course, derivative of *Three's Company,* which featured two girls and a guy, *but that was the whole point.* In TV, nothing pitches like a rip-off.

The script they had written—for a fee of $30,000—had one of the guys dumping his girlfriend and making a play for the luscious live-in.

Tartikoff didn't like it. He and his two equally boyish underlings explained their reasoning to Silverman, Lynn, Gordon, and Alan Rafkin, the show's director. If the guys hit on the maid in the first episode, they said, the poor girl will never be safe. What about this: The guys hire the maid, their girlfriends get jealous, the doorbell rings, and the maid opens the door and jumps into the arms of former football player John Matuszak, pretending that he's her boyfriend, as you break for the first-act commercial.

Lynn and Gordon thought the idea was patently lame, a real quadriplegic, since it took the action away from their main characters, shifting it to a one-time guest star. But they held their tongues. "Not bad," Gordon allowed.

The fact was, if the script was made into a pilot, and the pilot into a successful show, Gordon Farr would make in the neighborhood of $40,000 per weekly episode. And it had not been so awfully long ago that Gordon Farr had been making $250 a week writing game shows. So a bit of artistic compromise was not so hard to swallow.

And what could Silverman say? As a network chief notorious for meddling with his producers' wares, he'd practically *invented* network interference.

Nor was Alan Rafkin going to balk. At fifty-four—practically ancient in the television business—his goal in life was to grab on to one more hit and gracefully finish off a career that had started with *The Donna Reed Show*. Rafkin smiled absently and tried to look as if he thought the idea were brilliant. But he wasn't even thinking about the script. He was thinking about the three network programmers. "My God," he thought to himself, *"my shoes are older than these kids."*

An hour after it started, the meeting was over. They filed out of Tartikoff's office one by one, then rode the elevator a single floor down to ground level. Once past the security guard and into the parking lot, Silverman began to bitch about the network's telling them how to do their jobs.

But Gordon Farr cut him off. "Shit, Fred," he grinned, "you're the guy who *started* that. Before you came along, TV was fun to work in."

Silverman smiled and shut up.

Lynn and Gordon would soon submit a second script, with Matuszak's part in it, that Tartikoff would like, and a third one that he would like even better. Tartikoff would okay the third script and tell them to "go to pilot." He would make only one final decree—get rid of the Matuszak character. So, in the end, they had a script that was essentially the one they had started with.

A new season was beginning.

PART ONE Getting In

1
Affirmative Action

Early March 1983

J. Antony Colvin was young, talented, black, and in trouble. He had tax problems. Bad ones. Worse tax problems than he alone could solve. If worse came to worst, J. Antony Colvin would have to enlist the aid of his attorney, his CPA, and finally his mother and father, who, between them, might be able to advance him enough money to extract the future king of network television from the $20,000 hole he had dug for himself.

It had begun with a letter. Shuffling through his mail, looking for something interesting to read, Tony's eyes had fixed on an envelope return-addressed "Internal Revenue Service." Leaping from the enclosed pages were the words "back taxes" and the figure "$19,400." For a moment, Tony, so tough he had once been a starting wide receiver for the University of Washington, thought he might swoon. Instead, he sat down, calculating. It took him about ten seconds to realize that he wouldn't have $19,400 even if he liquidated his every asset, up to and including his cars, clothes, vacuum cleaner, and waffle iron.

He met with his accountant and together they began to draft a letter to the tax people. But then he got sidetracked by *Cheers*, the NBC comedy on which he served as producer's assistant. The show was in trouble, he found himself lost in it, and the $19,400 was relegated to a nagging, unsettled obligation scratching at the back of his mind.

Then, one muggy Los Angeles day in the early spring of 1983, he arrived home from Paramount Studios, where *Cheers* was filmed, to see a young woman, a gorgeous young woman, exiting from his apartment cluster. He passed her on the sidewalk. When he got to his apartment door, he saw that she had left him a message—a note wedged between the apartment's door and its jamb. The note ordered him to contact her at her IRS office immediately. He called to make an appointment. She wanted him there the next afternoon. When he told her he worked in the afternoon, she replied, in so many words, that that was his problem, not hers.

Standing in line at the tax office the next day, he watched the guy in front of him get worked over by the lovely young agent. It wasn't a pretty sight. She told the poor shmo to have all his records ready in two days. "But I can't do it that fast," he said. Well, she replied, she *could* order him to have the records *immediately*—forget the two days—if she chose. Tony took a deep breath. This was crazy. How could somebody so cute . . . It was like finding Snow White tattooing arms at Auschwitz.

Waiting for the ax to fall, Tony ruminated on his self-sponsored misfortune. Gone were the days when he could have leveraged his way out of this kind of mess by just signing a check. Gone were the days when he was the fastest-rising young non-white exec at First Interstate Bank, where he had vice-presidents working under him and was being paid so well he could afford a fleet of cars that included an Italian De Tomaso Pantera, a Corvette, a four-wheel-drive Blazer, a Dodge Power-Wagon, a Volvo, a Mazda, and, for going to the grocery store, a Chevy Nova.

No, he had given all that up for art. Well, for network television, at any rate. And TV had changed his life. That was for sure.

The guy facing the instant audit finally slunk away from the pretty tax agent's desk, his chest and shoulders so slack inside his jacket that he no longer looked vertebrate.

Tony Colvin—comedy writer par excellence—took the offensive. He made up a joke on the spot. He hit the iron maiden with a one-liner that might not have cut it on prime time, but was enough to make her laugh. And, the next thing he knew, it was "Well, Mr. Colvin, would you like a little extra time to get your records in order?"

"Not really," he allowed. "A week or so would be fine."

"Great," she said. "See you in a week."

When he got back to his office at Cheers, Tony put in a call to a lady friend who was a tax whiz, hoping to enlist her in a seven-day blitzkrieg to outflank the $19,400 deficit. As he was hanging up, a familiar bark of command issued from a producer's office: "Tony! My lunch!"

Instantly, J. Antony Colvin, until recently the twenty-six-year-old regional manager of quality assurance at First Interstate Bank—a man wired into a total compensation package of about $70,000 per annum—was double-timing like a frightened bellhop to the Paramount commissary, where he waited in line to receive the producer's broiled chicken. And then he high-stepped back to the Cheers office, lest the bird, protected only by a thin white bag, get cold.

That night, his tax expert pal sat and stared at him after she had briefly shuffled through his records. "This makes no sense," she said. "It says here the Cheers people are only paying you a thousand dollars a month. Gross, not take-home. I know winos who make more than that. Explain to me how a guy who brings home eight hundred a month, tops, could have a twenty-thousand-dollar tax problem. Explain that to me!"

Words failed Tony, because he would have to begin by admitting that he had used most of the missing twenty grand to pay for the mere chance to trade his affluent banker's life for that of a $250-a-week gofer. That he had fought and schemed and worked his ass off for the very privilege of running warm chicken to the nail-biting producer of a television show that half the people in town thought was on the fast track to oblivion.

To fully explain, Tony knew, he would have to describe a Land of Oz that was real. It was a land where the Munchkins worried about which color Mercedes to drive on Monday and which on Tuesday. It was a land where a fired dishwasher who had been expelled from one of the seediest diploma mills in the Midwest could write his way to six-figure status in one year. A land where the tin men and the straw men and the cowardly lions could land on their butts nine tries out of ten and still end up richer than good thieves. In this land Tony now lived in, the wizard's balloon soared up, up, and all of Kansas watched—all of America watched—an average of seven hours every day! Even though everyone knew the balloon was mostly just a lot of hot air in a pretty wrapper. In this very strange land, Tony knew he could—if only he kept trying, and clicked his heels just right— one day speak his thoughts to every man, woman, and child in the United States, and they would listen, and Tony would become wealthier than any banker in California. And all of this was not over the rainbow. It was here, now, in the cash-and-carry America of 1983.

Hoping it wouldn't take the rest of the evening, Tony sighed and began to tell his friend how he had become an employee in the world of television, a host of gourmet midnight picnics— and, not incidentally, a spy.

Paramount Studios is an Alamo of security. It consists of four square blocks of buildings, sets, and flowered courtyards, surrounded by a bleached-cream masonry wall twelve feet tall and three feet thick, a wall that could withstand a mortar attack, or perhaps an inner-city riot, and is most certainly capable of repelling the unwanted visitations of unemployed actors, carnivorous agents, tabloid gossip-mongers, star-famished sightseers, and all the other hungry hustlers who daily mill around the studio's several entrance gates. These gates, patrolled by a large, suspicious security force, admit employees and the few guests who have managed to obtain dated, signed passes, which can originate only inside the walls. Even the chief executive officers of other studios and the presidents of the television networks can have a hard time busting through the guarded gates without these passes, and sometimes confrontations arise between young guys in baggy, sweat-stained security uniforms and moguls in three-piece suits that invariably culminate, after many threats, in a victory for the men in the less expensive uniforms.

Occasionally there are breaches of the security system, such as the case of the thief, believed to be responsible for the theft of *Cheers* star Shelley Long's wardrobe in the summer of 1983, who fooled security agents by punching in a fake time card at a blue-collar entrance station. For the most part, though, there is a clear distinction at Paramount between who is "in" and who is "out," and a person who is out finds this condition illustrated quite dramatically by the three-foot-thick wall.

Tony got in.

Back in the summer of 1981, Tony had been at the beginning of what seemed an almost limitless career at First Interstate Bank, an elephantine chain of Western banks that had put Tony in charge of quality control of data processing for its southern California region. He had an office in Pasadena and also one in the downtown Los Angeles building, which dominates the city's skyline.

His ascendance in the banking world had not seemed extraordinary to him, because he was already accustomed to affluence, and even more accustomed to the wave-making grit that it takes to forge personal gain out of a largely indifferent world. When Tony Colvin was born, in Houston, shortly after World War II, his father was working three jobs at once, almost twenty hours per day, doing physical labor, driving a cab, and working as a short-order cook. As a child, Tony would accompany his mother when she drove his dad from job to job. Mr. Colvin was saving his money. His dream was to buy a Dairy Queen.

Eventually, he put away enough to buy the ice cream place, which he ran himself, usually staying there for all of its operating hours. The drive-in was so successful that he was able to buy another. Then he used his experience and capital to open a full-service restaurant and, some time later, a nightclub. By the time Tony was a teenager, his father also owned cattle, and the land the cattle grazed on, and the trucks the cattle were shipped in. He was a wealthy Houston businessman. He cut back his workday to about seventy-two hours per week and added a bit of decor to his office . . . which was in the back of the first Dairy Queen.

Tony admired his father's work habits, despite the fact that they had contributed to a divorce from Tony's mother, and Tony put them to work in his own life. He became an excellent student, eventually earning three college degrees, one in computer science, one in administration, and an M.B.A. in management. He also became a superior athlete. As hard and lean as a railroad spike, with just under two hundred pounds molded around a six-foot-four frame, Tony earned a first-string spot catching passes for the University of Washington.

Tony also inherited from his father a sense of rugged individualism, which surfaced not only in his scholastic and athletic ambition, but also in his disdain for most aspects of the civil rights movement. He didn't need anybody trying to help him by getting him welfare or some lame CETA job or making it easy for him to get into or pay for college. And did he spend Saturday afternoons lounging around jazz clubs sipping sweet red wine and listening to Charlie Mingus? No. Tony preferred to kick back at home with a dry martini, Luciano Pavarotti on the stereo, and a crisp new *Barron's* or *Business Week* in his lap. He resented it when people leaned on him to do all that trendy identify-with-black-culture crap. After all, just because you were Norwegian didn't mean you spent your free hours eating sardine sandwiches and hanging around the Grieg Hall.

All of this—the family success story, the three college degrees, the tough-jock reputation, and the creed of Republican individualism—opened the world of banking for him like a magical incantation. But he wanted out.

He wanted to be in television. Television was . . . everywhere. It was watched by . . . everyone. Tony himself watched TV all the time. And when he watched a show he paid attention to its writing, and he often found himself saying, sometimes out loud, "I could do that."

Tony had been writing stories for years. To him, it was the only kind of work that made him feel more free, instead of less free. Once, he'd gone down to his garage to get something out of his car. While he was there, he got a story idea and sat in his car while he wrote it out. When he got back to his apartment, he saw that four hours had passed. Magically. Instantaneously. At the bank, four hours was eternity. To get paid for writing— for doing something that made time stand still—that seemed to Tony Colvin like heaven.

But it wouldn't be enough to publish some obscure book that would end up among a billion others in the Library of Congress.

Or, for that matter, a best seller that might reach a hundred thousand people and pay his bills for two and a half years. Tony wanted to blaze in the public mind like neon. He wanted to write a prime-time TV show. And to be compensated commensurately with the other young multimillionaires in the field.

More specifically, Tony wanted to write prime time for NBC. NBC was, in his mind, the *elegant* network, where ideas got a hearing. It was the class act, like his De Tomaso Pantera, almost hypnotic in its allure.

So he applied for a job at Paramount, which produced many NBC shows, in the data-processing department. That was the cover he needed to breach those three-foot-thick walls. In order to land the job, Tony had to pare down his résumé drastically, claiming one degree instead of three, because he was ridiculously overqualified for the position. The data-processing job was on the graveyard shift, which would enable Tony to keep his lucrative banking position. From personal observation, Tony had learned that working two shifts was far from impossible.

When he got the Paramount job, he began his day at about the time Johnny Carson came on the air, which was, if nothing else, at least a time of light freeway traffic. His Paramount job ended during the 9 A.M. rush hour, though, so he had to fight traffic to arrive, red-eyed, at his banking job. After eight hours of banking, he grabbed enough sleep to power through another shift at Paramount.

Almost as soon as he arrived at Paramount, he began to face a second challenge. In order to win a "creative" job, he had to get through another wall, the one that separates the clerical, business, maintenance, and construction workers from the writers, directors, producers, and actors. This invisible barrier can be even tougher than the one outside. Still, Tony was confident.

All over town young men and women were holding down jobs as pages, carpenters, or secretaries while they waited for a chance to act, design, direct. It was an old showbiz cliché in action—Judy Garland hangs around backstage waiting for the leading lady to land on her keister—but, a fat percentage of the time, it worked. TV, Tony knew, was essentially a meritocracy, in which an unconnected person could still make it on the basis of brains and balls. This was more true than ever in the 1980s, because cable TV was opening up vast new fields to plow.

For Tony, now ensconced in Paramount data processing, the next step was simple. All he needed, really, was the help of the janitors.

So, the picnics. Gourmet picnics. Every night. For the custodians and guards. Every evening, just about the witching hour, they would come clinking up the stairs to Tony's office, their noisy, crammed key rings riding their hips like six-shooters, in order to feast on the cold cuts, thin-sliced turkey, potato salad, French bread, rice pudding, baked beans, and other deli delights that would emerge from Tony's picnic baskets. And always there was the Heineken. A case of the expensive, imported beer, so light and icy, appeared in Tony's office each night, with Tony, the gracious host, popping the tops off the frosty green bottles and passing them around. In order to keep a sense of surprise, Tony would bring different edible treats each night, but the brand of beer remained a constant; it was a brand not generally affordable by the people who so gratefully guzzled it. The picnics were all the more delicious for the guards and janitors in coming from a person who was a notch or two above them in the Hollywood caste structure, and their appreciation abounded. Tony became a graveyard legend.

Tony asked for a return favor. The guards and janitors had been waiting for it. But when it came, it wasn't as much of a favor as they had feared it might be. Could Tony borrow their keys, to look around in some of the offices, in order to better learn what was going on at Paramount? He would be careful, of course, not to displace any papers or to take anything out of the offices.

Well, hell yes, old buddy, replied the janitors and guards—just don't steal anything, don't try to jimmy any locked files, don't ever let on where you got the keys, and, last but not least, don't forget to start bringing some ice to throw on top of the Heineken, since it tends to get a touch warm by four in the morning.

Tony became a white-collar prowler. The Paramount executives kept most of their sensitive materials under lock and key, but some of the less important papers were left on top of desks overnight. Using a plastic card that operated the photocopy machines—a gift from a janitor—Tony duplicated the more interesting-looking documents. Then, late one night, he found it—his ticket into television.

It was a thin sheaf of papers describing a program designed to train and maintain a staff of producer's assistants. The idea of the program, said the papers, was to find a certain type of young person, bright and aggressive, willing to work long hours, interested in the postproduction work that must be done after a show is shot, but also other-directed enough to be a happy gofer, sufficiently limited in horizon to be content with the job. They did not want someone who would run off the first time a job one rung up the ladder appeared.

Tony took his copy of the program description home, studied it, and then got in contact with the sponsor of the program. He spelled out his desires and qualifications—remarkably like those sought for the as yet unannounced program. He was informed that yes, there might indeed be something for a young man like him at Paramount. Tony was told to get in touch with John Barber, director of current programming at Paramount. Barber handed Tony down to his assistant, a woman named Lenore French. She was polite but noncommittal. Tony began a series of phone calls and letters to both Barber and Emmanuella Upchurch, director of personnel, painting a picture of himself as a bright, humble kind of guy, willing to work long hours to learn postproduction, but desiring a stable position that would not change for several years—in short, the perfect round peg for their round hole. Even so, nothing of consequence happened. Tony was not the only young guy in Hollywood who could psyche out the desires of the people holding the keys to the kingdom. Competition was stiff.

Tony kept up his barrage of letters, phone calls, and resumés, each letter referring to the one that had preceded it. He kept them up even after he'd been forced to quit his Paramount graveyard-shift job. His boss at Paramount had been looking for another job and had applied for one at First Interstate Bank, where he was directed to go see one of the bigwigs, a certain Mr. Colvin, who bore a striking resemblance to the kid he'd recently put on the graveyard shift.

But by then it didn't matter, because Tony was closing in on the producer's assistant program. In February of 1982, he finally was told that something might be opening up in March. He called in March and sent in another résumé. He called in April and sent in another résumé. And again in May. At the end of May, they called him, told him something actually was opening up, and said that they believed they already had his résumé. "You damn well ought to have it," he thought.

He got an interview. He told them exactly what they wanted to hear, and he could tell he'd made a sale. The job was his. Now to place him with a particular show. . . . What did he think, the interviewer asked him, about Garry Marshall's show, *Joanie Loves Chachi?* The show was a spin-off from *Happy Days*, which had also begotten *Mork & Mindy*, as well as *Laverne and Shirley*, a show so successful that the president of Paramount Television had equated its earnings with those of the movie *Star Wars*. In fact, practically everything that Garry Marshall touched turned to gold, to the degree that he had been able to make his sister, Penny, a comedic superstar by casting her in *Laverne and Shirley*. Garry Marshall was by far the most important producer at Paramount, and was considered an absolutely vital component in the success of ABC, which used his comedies as early-evening magnets to attract the youthful audience that the network so coveted. If there was a single preeminent comedy producer in television, Garry Marshall was it.

"*Joanie Loves Chachi?*" Tony stammered, uncharacteristically tongue-tied. In truth, he'd rather have stayed in his three-piece banker's straitjacket than work on Garry Marshall's *Joanie Loves Chachi*, which he considered the type of mindless brain-candy that had given the word "sitcom" a bad name. Garry Marshall might indeed be the Croesus of comedy, but what Tony wanted was to work for one of the hot, young talents that Paramount Television President Gary Nardino had been breaking his back to lure to the studio, like Gary Goldberg or Lloyd Garver, who were putting together a new show. Or, better yet, Ed Weinberger, or James Burrows, or Glen and Les Charles, all of whom had worked on *Taxi*, the flagship of the intelligent comedies that Paramount was producing. Tony knew that these men were looking for something to do beyond *Taxi*, which they had taken artistically about as far as they could, and which ABC was making noises about dumping. If any of these men were working up something with Paramount, Tony wanted to be part of it.

"Well," said the interviewer, during Tony's long pause, "there is another project that's going to get a chance on NBC. It's by the Charles brothers and Jimmy Burrows. It's about a bar. Right now, they're calling it *Cheers*."

Tony tried not to reply too quickly or boldly, in order not to violate his carefully constructed psycho-graph as the perfect slave. "That might be good," he said.

He was sent to be interviewed by the producers of *Cheers*.

The three producers teased him about his three-piece suit. Arriving at the interview directly from his First Interstate job, he stuck out in the *Cheers* offices like Richard Nixon at Woodstock. Glen Charles, his brother Les, and Jim Burrows, all in their late thirties or early forties, favored golf clothes, beach clothes, and anything made by Levi Strauss or Nike, and had acquired sufficient fame and fortune within the industry to wear whatever they pleased, whenever they pleased.

They told Tony what a horrible job being a producer's assistant was, and how much he would grow to loathe it. They told him about the long hours, the relative impossibility of rising through the hierarchy, the humiliation of being little more than an errand boy, and the $1,000 monthly salary, which did not get a person very far in Los Angeles, where a modest two-bedroom apartment rented in the $600-per-month range. Tony, who had been having a bit of trouble maintaining his lifestyle on $70,000 a year, said nothing. The producers also reminded him that NBC had only agreed to buy their first thirteen shows. Going beyond even that limited run would be difficult, and achieving long-lasting success would be a real long shot.

After all, 75 percent of all new shows are canceled after only one year.

Tony nodded, undaunted. They asked him when he could start.

"Right now," he said.

When Tony started work, *Cheers* was nothing more than a script. But it was a very valuable script, since NBC had just agreed to buy thirteen episodes of the show solely on its strength. Selling a series without first making a premiere episode, or pilot, is practically unheard of in television, since it is a quintessential example of buying a pig in a poke. It was testimony to the reputations of the Charles brothers and Jim Burrows—and to the quality of their script—that NBC had chosen to purchase their unseen show, particularly in light of the fact that each of the thirteen episodes would cost over $300,000.

Although it would seem that a television show might be accurately envisioned before it is filmed—by reading its script, knowing its cast members, and talking to the people who will produce the show—it is usually considered impossible to tell if a show will work until after it's actually shot. Even then, network executives rarely have the slightest notion if a particular pilot episode will turn into a series that will strike the fancy of America. Probably the most often repeated adage in the television business is "If I knew what would work, I'd be the richest man in town."

The Charles brothers and Jim Burrows, therefore, were very pleased that NBC had allowed them to "go to series" after having seen only their first script, but they were not inordinately surprised. After all, previously, just after they had formed a partnership and shopped themselves to each of the three networks, they had signed a deal with NBC guaranteeing that they would be able to put at least one program on the air for at least thirteen weeks. Also, they knew that the new chairman of NBC, Grant Tinker, former producer of The Mary Tyler Moore Show, which had starred his ex-wife, and former head of the independent production company MTM, was highly appreciative of their individual talents, and of the type of cerebral "character comedies"—as opposed to situation comedies—that they were able to produce.

Tinker, in fact, had been Jim Burrows's first sponsor in Hollywood. While Burrows was still a stage manager in New York, he had written a letter to Mary Tyler Moore, with whom he had worked during her Broadway appearance in Breakfast at Tiffany's, telling her how much he liked her new TV show. Grant Tinker wrote back to Burrows, inviting him to come out to L.A., at MTM's expense, to study directing. Burrows jumped at the offer. He was anxious to escape the New York theater scene, where the massive shadow of his Pulitzer Prize–winning father,

Abe Burrows, loomed over his every effort. He'd long existed as *the son* of the author-producer of *Guys and Dolls*. *The son* of the creator of *How to Succeed in Business Without Really Trying*. After five months of studying TV directing, mostly by observing the MTM directors, Burrows was asked to direct *The Mary Tyler Moore Show* and MTM's *Rhoda* and *Phyllis*, where he met the Charles brothers. The three of them later went to work together at Paramount, on *Taxi*, where Burrows won two Emmy Awards and gradually gained acclaim as the best director in television comedy. By the early 1980s, Burrows, who looked almost like a monk from the shoulders up due to a shiny pate ringed by a tangle of black hair and a thicket of a beard, was able to say, "Now people will say to me, 'Gee, I didn't know your dad was Abe Burrows.' "

As a TV director, Burrows's enjoyment of his personal accomplishments was matched by his appreciation of television's financial rewards, which far surpass those of Broadway. A successful TV director can earn in the range of $15,000 to $20,000 per weekly episode, and will receive a hefty royalty payment each time one of his episodes is shown as a syndicated rerun. And as a producer of *Cheers*, as well as director, Burrows was able to assign himself a weekly salary range that was about *double* what most directors earned. Also as a producer, he was co-owner of a possible syndicated-rerun sale. If the show was to last four or five years, it would have enough episodes to sell to stations all across the country as a syndicated rerun; about a hundred shows are needed for syndication, since the stations gobble them up at the rate of five shows per week. And this syndication sale could amount to $100 million, or even more. Burrows's share of that sale would make him rich.

Burrows, therefore, was absolutely determined to make the most of *Cheers*.

As Tony began his duties at the fledgling *Cheers* operation, duties as demanding and demeaning as the producers had promised, the biggest problem the company faced was finding actors to fill the show's roles. The starring roles of Sam, the womanizing bartender who was a former pro baseball player, and Diane, the attractive, intellectual waitress, were the ones that had the producers most worried. The male and female leads needed to exude a natural attraction for one another, but also an abrasive friction that would spark the "heat," or conflict, that is considered the vital spine of every television comedy. Approximately three hundred actors and actresses were interviewed, most in L.A., some in New York.

Eventually, the search centered on three pairs for the two roles. One pair consisted of William Devane, the film actor who had appeared briefly on TV in the series *From Here to Eternity*—and in the miniseries that had spawned the show—coupled with Lisa Eichhorn, a star of *Yanks* and other movies. This pair was the "name" duo, who could be expected to bring an instant audience to the show through their reputations. The second team was Fred Dryer and Julia Duffy, and the third was Ted Danson and Shelley Long.

Ted Danson, a tall, good-looking thirty-three-year-old with swept-back Jack Kennedy hair and a proven store of talent for comedy and character roles, had not been in any NBC projects since the latter part of the 1970s, when he had landed on the blacklist of former NBC President Fred Silverman, the only man in TV history ever to control programming at all three networks. Silverman, once called the man with the golden gut because of his uncanny ability to spot hit shows, had fallen from grace at NBC, the last of his network roosts, when his golden gut suddenly seemed to sour. One of the pilots he'd commissioned during the darkening days of his final network presidency was called *Allison Sidney Harrison*, which starred Ted Danson as a father who enlists his young daughter in private-eye perils.

The pilot fared poorly, and Silverman blamed Danson, partially because testing surveys showed that audiences didn't find Danson sympathetic. "How *could* the audience like him?" said another NBC executive. "He was sending his little girl out to chase guys with guns!" Nevertheless, the pilot grabbed Danson's career momentum like a hydraulic brake.

By 1981, though, Silverman and his blacklist were gone, and Ted Danson was the secret favorite of the *Cheers* producers in the contest for the show's male lead. Danson had been approached by Jim Burrows one day while he was at Paramount, doing a guest role on *Taxi*, and had huddled with the three producers in their offices upstairs and across the narrow studio street from the sound stage where *Cheers* was to be filmed. He wanted the role badly.

When Shelley Long was approached about the project, she was shooting *Night Shift* with Paramount alum Henry Winkler, who, along with Robin Williams, had become a major film star after a few years of cranking out sitcom next door to the Paramount commissary. Shelley Long was not sure she wanted to become involved in a series, which could stretch on indefinitely and could virtually devour an actress's most glamorous, vital years. She had made another movie, the as yet unreleased *Losin' It*, in which she would garner very good reviews, and she feared that she might be trading the money and stature of a film career for a series that, even if it did defy the odds and achieve success, would probably become an arduous daily grind. On the other hand, though, were the all too obvious examples of Winkler and Williams.

After several individual tryouts, the three leading pairs for the parts were brought in for what amounted to a read-off. Each of them would read a scene, and a pair would be chosen. A host of NBC programming executives and Paramount executives joined the producers and casting agents. The actors and actresses took the stage. This kind of contest—raw, obvious, and unforgiving—is common in the TV decision-making process.

Danson and Long won the parts, though Danson came close to losing the role to Devane, who played it with a tougher air. In the end, though, Danson, because he seemed more vulnerable and approachable than Devane, cashed in on the television adage that "people want to fuck movie stars and hug TV stars."

Now, the *real* problem. Survival. On the face of it, *Cheers* had as much chance of being popular as the Ayatollah Khomeini had of being elected governor of Iowa. *Cheers* had problems. For starters: Danson and Long. While they might be highly respected within the industry, Ted Danson and Shelley Long were nameless nobodies to the viewing public, and TV viewers like nothing better than familiarity. A similar lack of familiarity would also be present in the *Cheers* story lines, because the producers were determined to be innovative. Jim Burrows summed it up: "In *Cheers*, nobody slips on the banana peel."

And the episodes would require a cursory understanding of

the characters, since the program was a character comedy rather than a sitcom. Needing a knowledge of the characters to fully appreciate the show was, of course, a Catch-22, since it required the viewer to watch the show a couple of times *before* all the jokes would seem funny.

It seemed certain that the show's audience would build slowly. In modern network TV, though, a show that is slow to grow is almost always a show that is canceled. The networks tend to shoot first and ask questions later. The mind-numbing financial outlay that is needed to put a program on the air these days legislates powerfully against producing anything other than quick winners.

But the producers of *Cheers*, and the executives at Paramount, were betting on a single factor: that NBC, under its current management, was a vastly different entity from any other television network in the history of the medium.

NBC was, in fact, in chaos. The network's prime-time ratings had fallen from a 1975 high of 19.8 percent of television homes to 15.5 percent at the time of Grant Tinker's arrival in 1981. NBC's audience had slipped even more since then, and was only *half* of ABC's by November 1982. Many of the network's 215 affiliates had begun bumping network fare in favor of their own locally produced programs or those purchased from independent syndicators. Not only did this rob the network of current advertising revenues, but vastly lowered the ratings used to set prices for NBC's commercial airtime in the future. During the month of November 1982 alone, there were 334 instances of network preemption. On a single evening that month, a full quarter of NBC's affiliates chose not to air the network sitcom *Facts of Life*, slashing its ratings by a full 6 points.

In order to come up with new and better prime-time programming, NBC was being forced to spend heavily on development. In 1981, the new chairman, Tinker, found himself having to replace nine hours of prime-time series, an amount nearly equal to CBS and ABC replacements combined. It is an industry rule of thumb that to create one prime-time series suitable for regular airing, four series pilots must be created. Whether successful or not, they all have to be paid for at roughly $750,000 each. NBC had absorbed $32 million in write-offs the year be-fore for programs shown once or never.

Even NBC's long-standing late-night hit, *The Tonight Show,* was in trouble, "skewing old," in the argot of the trade. At least one NBC affiliate was considering dumping Johnny Carson in favor of the much younger—and presumably hipper—Alan Thicke, who was about to debut his own late-evening opus, *Thicke of the Night,* as a syndicated five-day "strip show" in direct competition with Carson.

And, if anything, the network's problems with its daytime schedule were even worse. "A nightmare," conceded Tinker. As of the spring of 1983, NBC daytime programming was capturing less than 5 percent of the potential total television audience, this thanks to a radical churning of NBC's standard daytime fare by the network's previous administration.

But the most serious of NBC's 1983 problems was that twenty-two network affiliates had jumped ship to *other* net-works over the past five years, lured not only by the prospect of higher ratings but cold cash as well. WSB, the NBC affiliate in Atlanta, moved to ABC after being promised a flat $1.5 million increase in the fees it received as a local affiliate for carrying network programs.

Because the three-network balance of power is a delicate one, a major loss of affiliate stations could snowball into a complete network collapse. This possibility was raised by Tony Schwartz in the May 1983 issue of *Playboy.* Citing competition not only from the other networks but from cable television as well, Schwartz wrote that NBC "is likely to die, or shrink severely, within the next decade."

While Schwartz's prediction was heartily discounted by NBC, there was little question that Grant Tinker's chore was going to be a titanic one.

Ironically, much of the reason the network was in so much trouble was the rescue attempt made by Fred Silverman a few years before. At the time, Silverman was considered the ulti-mate programming genius, a man who had been dizzyingly suc-cessful in programming first CBS and then ABC to the top of the ratings heap during the late sixties and early seventies. Early in 1977, President and CEO Edgar Griffiths of NBC's parent company, RCA, made a decisive move to close the $50 million profit gap between NBC and the prime-time leader, CBS. First, he reversed NBC programming chief Paul Klein's move toward specials and movies, on the grounds that these blockbusters were too expensive to produce—$600,000 to $750,000 per program hour, contrasted to $325,000 to $400,000 for an hour's worth of series programming, a cost not offset by the additional revenues they brought to the network.

Second, and more important, Griffiths announced that he was hiring Fred Silverman, whom he called "the greatest," at $1 million per year for three years, with an additional two-year option. The defection from ABC of Silverman, known for his seemingly unerring abilities to commission and position successful prime-time television programming, had an immediate effect on both networks. RCA's stock leaped $1.25 a share and ABC's dropped $1.75.

The son of a TV repairman, Silverman had made his first mark in television programming by way of his master's thesis at Ohio State University—a four-hundred-page-plus assessment of ABC programming from 1953 through 1959. In what has since been recognized as a classic treatise on television programming strategy, Silverman argued in favor of (1) the use of weekly series to build a steady, mass audience, (2) only occasional use of special programs, (3) reliance on audience research in program scheduling and placement, (4) the importance of reaching young viewers, and (5) the value of promotion. These were the basic ideas he would build on for the next twenty years—ideas that, to a significant extent, have shaped contemporary American television.

Silverman put these ideas into action when he was hired as a junior assistant in the programming department of Chicago's WGN, where he convinced the station's manager to let him reedit a bunch of old Tarzan films into a "new" one-hour show he titled *Zim-Bomba*, starring Bomba, Tarzan's son. The effort was an immediate success and he went on to triple the audience for children's shows for his subsequent place of employment, WPIX in New York, by creating a circus of TV games and promotion around a revival of the old Buck Rogers science fiction series. Shortly thereafter, in 1963, CBS plucked up the twenty-five-year-old Silverman for $17,000 a year. Despite gibes from other network execs that he had been hired "because he was still young enough to fly half-fare" and that "all his programming meetings have to begin before 5 P.M. because 7 is Fred's beddy-bye time," Silverman adopted a style that was hands on and direct. He was booted off the set of *Password* after inquiring of a producer, "Who thinks up this crap, anyway?"

Children's programming became his forte. With efforts like the animated shenanigans of *Superman* and *The Lone Ranger*, he was able to draw kids to CBS by the horde. Then came *Scooby-Doo* and *Shazam*. Thanks largely to Silverman, a high-action, high-violence zoo of super-creatures became the rage—*The Herculoids, Aquaman, Birdman*. Soon, all three networks were awash every Saturday morning with Silverman shows and their clones. His identification with his cartoon characters was complete. While programming daytime soap operas, he had often been moved to tears by various episodes, and the heavy-eyed, dark, portly Young Turk would perform all the parts of his children's shows with great emotion and élan before his CBS superiors. "If you have a good mental picture of what Fred looks like, his performance in the conference room was not to be forgotten," recalls an executive often in attendance.

Regardless, Fred Silverman was bringing success to everything he touched at the network, and in the spring of 1970 he was elevated to vice-president for East Coast programming. In June he accepted the job of chief programmer for CBS. He was thirty-two years old.

Gordon Farr, also thirty-two years old, read about Silverman's promotion. Farr was still up in Canada, "up to my ass in slush," making $150 a week. As he scanned the story of Silverman's ascendance, just one word went through his mind: "Shit!"

Silverman immediately moved to "urbanize" CBS programming, carefully repositioning the incipient hit *The Mary Tyler Moore Show* and then, to considerable corporate nail-biting and Maalox swigging, scheduling *All in the Family,* a program that CBS Chairman William Paley loathed, but which would prove to be CBS's hit of the decade.

Silverman's prosperity at CBS was fueled by eighty-hour workweeks. His commitment to individual shows was limitless.

He was known to spend hours mulling over a single page of script, and it was claimed that he would even go so far as to personally rearrange chairs, sofas, and tables on CBS sitcom sets. Silverman alone was largely responsible for ushering in an era of network interference. *All in the Family* and *The Mary Tyler Moore Show* were followed by Silverman's introduction of *M*A*S*H* and *The Waltons*. Then another stroke of genius. Spin-offs. *Maude*, starring Beatrice Arthur as a liberal bigmouth, was a stepchild of *All in the Family*, where Ms. Arthur had first appeared as Edith Bunker's cousin. Then came *The Jeffersons*, another *All in the Family* offspring. *The Mary Tyler Moore Show* spawned *Rhoda* and *Phyllis*.

Despite his success, Silverman began to feel cramped at CBS. For one thing, he didn't fit the network's mold. "He was fat. He wasn't cool. He lost his temper. He didn't know the right lines. He was sloppy," remembers a colleague. In May of 1975, he jumped to ABC, a network then known as the Poland of broadcasting. He moved quickly to spin off *Laverne and Shirley* from the network hit *Happy Days*, then pulled *The Bionic Woman*, like Adam's rib, from *The Six Million Dollar Man*.

Silverman made few bones about his recipe for top-rated programming: Give the people what they want. At ABC, he became known for soliciting not only the opinions of his college-educated $85,000-per-annum lieutenants, but also those of secretaries and sundry other corporate peasants. To Silverman's mind, these plain folks were "the people who watched prime time."

And if there ever was a Fred Silverman show that reflected this nostrum, Silverman's next big hit, *Charlie's Angels*, doubtless was it. A stunning triumph of form over content, *Charlie's Angels* was a tits-and-guns show that followed the exploits of three gorgeous detectives, inaugurating the one television genre accredited to Fred Silverman—Jiggle TV. Fred's Angels introduced nipples to the small screen.

Silverman's programming hand was felt across the board. ABC's vice-president for children's programming, Squire Rushnell, recalled a network meeting in which someone suggested that a cartoon show might be created around the concept of three girls discovering a caveman frozen in a block of ice. Summarily defrosted, he would come alive and the four would go off and have many adventures. Rushnell claimed that the idea seemed so ridiculous that his eyes actually rolled when he was charged with outlining it to Silverman. Silverman, however, was entranced. Pacing the room, hands flying, he spun the caveman whole there on the spot.

"Yeah," Silverman said, "he's about this high and has a big, furry coat. He's a little guy, but"—at this point Silverman began bellowing—"*he has a great big voice!* He eats everything and everything is yum, yum, yum." The little guy appeared on Saturday mornings that fall.

By this time Silverman had himself begun to attract considerable press attention, and was well on his way to becoming modern television's first corporate superstar. He was also attracting scathing criticism, despite his success with superior shows such as *Roots*, which made history by attracting ninety million viewers to its final installment. After launching such hits at ABC as *Three's Company*, a breasty amalgam of pratfalls and ca-ca jokes, and *The Love Boat*, a giggly flirtfest on an ocean liner, Silverman began to be accused of the shlocking of America. Nevertheless, as chief exponent of what *All in the Family* producer Norman Lear called "television's total winner mentality," Fred Silverman was the hottest item in network television when RCA President Griffiths anointed him as NBC president.

A winner to end all winners, Silverman was then, in the spring of 1977, considered by many in the industry a force that could *not* be reckoned with. "Freddy will always beat you," opined Barry Diller, Silverman's predecessor as chief programmer at ABC. "He is maniacal. For every hour you work, Fred Silverman will work ten. For every minute you spend on a single detail, Fred Silverman will spend sixty. He will not be beaten."

His arrival at NBC made the industry shiver. "This is more than a show business or business phenomenon," said Michael Dann, Silverman's boss at CBS. "It will have a profound effect on the American public, because Silverman will determine how most Americans occupy their time." Dann's was only a minor exaggeration. Well over seventy-five million households in the country have television sets—more than have indoor plumbing—and as many as fifty million Americans watch prime-time network television on any given evening.

When Silverman took the reins at NBC, ABC was in the

number one position, CBS was second, and NBC a somewhat distant third, but Silverman waxed unworried in *Time* magazine: "I think it's old news that ABC is No. 1. . . . There's an inevitability that, just the way ABC came up, that network is going to go down. . . . Now I believe it's going to be our turn. The trick will be to make the right moves."

These moves were quite surprising. When he entered the NBC skyscraper offices at "30 Rock" (Rockefeller Center) in Manhattan that June of 1977, it was to deliver to the world a "new Fred Silverman." The first thing he did was to go for "quality," dumping a sequel to Harold Robbins's steamy *79 Park Avenue* and a smarmy *Love Boat* rip-off called *Coast to Coast*. He announced that he hated the whole 1978 prime-time lineup. Silverman promptly took total control of programming, after claiming that his new NBC confreres had broken the company into fragments and fiefdoms. Executives under him began to refer to themselves as the waiters, as Silverman, convinced that NBC had become the repository for all of Hollywood's worst ideas, began ordering up new shows himself.

Silverman dictated the content of the shows and spared no expense in rushing them to readiness. There was only one problem with this admittedly efficient system—it did not result in shows that the public wanted to watch. NBC's ratings dived.

Silverman's waterloo was *Supertrain*. Under his generalship, NBC spent $12 million developing a show variously described as *Love Boat on Rails* and *Murder on the Orient Express of the Future*. Regardless of energy, hype, and expense, no one—least of all the viewers—could figure exactly what *Supertrain* really was. "I told him there was no way to get this show on the air when he wanted it," *Supertrain* producer Dan Curtis recalled, "but I got the message from him that cost was no object—just get it on the air. I felt sorry for him. He found himself in a rapidly deteriorating situation. When things started to go wrong he would try to fix it all himself." *Supertrain* carpenters built sets without finished designs, scripts were often a goulash of cliché and confusion, and the show was eviscerated by critics. On its final night, *Supertrain* snared only 19 percent of the audience.

By this time, Fred was circling the drain. His Friday night

lineup of four new comedies—*Brothers and Sisters; Hello, Larry; Sweepstakes;* and *Turnabout*—fell right through the bottom of the Nielsens. A joke began to circulate at the network—Q: What's the difference between NBC and the *Titanic*? A: The *Titanic* had an orchestra.

Edgar Griffiths tried to maintain a positive public profile, insisting that the NBC game plan called for several years of rebuilding. But at the end of the 1978–79 season, NBC had registered its worst ratings in ten years. Not only did it trail the other networks by hundreds of millions in annual revenues, but Silverman's recent pilot-shopping spree—fifteen new sitcoms alone during the summer of 1978—cut profits in half: from $102 million in 1977 to $51 million a year later. And of the nineteen comedy pilots he commissioned for the following year, not a single one was worthy, in his opinion, of series commitment.

Fred Silverman had apparently turned a new, weird corner in his career. He was now no longer the czar of successful inanity, nor the "new high-quality Fred." He had become what could most charitably be described as flexible. The word around the network was that he'd lost his touch. One NBC executive told *The New York Times* that, unless RCA pulled in his reins, Silverman might "just sit there and preside over the complete dissolution of the network." By this time NBC profits represented approximately *one fifth* the profits of CBS. This simple fact, ultimately, was the stuff of his undoing. In May of 1981, a new chairman, Thornton Bradshaw, was installed at RCA. One week later, Fred Silverman was kicked out.

The Silverman era in network television was over.

The Grant Tinker era at NBC had begun.

No two network bosses could have presented a stronger contrast. Fred Silverman had been hot-tempered, hyperkinetic, alternately generous and vindictive, a disheveled man who had to have his fingers in everything, and a man to whom the ends— ratings—too often seemed to justify the means—crap programming.

Tinker, on the other hand, was cool and patrician. Fifty-six years old, silver-haired, possessed of white teeth, strong in the jaw, and straight in the nose, Grant Tinker was central casting's ultimate U.S. senator, airline pilot, or Harvard lit professor. While married to actress Mary Tyler Moore, he had created MTM Enterprises, which, in the span of barely a decade, had spawned a string of television hits including *The Mary Tyler Moore Show*, *The Bob Newhart Show*, *WKRP in Cincinnati*, *Lou Grant*, and *Hill Street Blues*. In the process, he had become the most prestigious producer of quality television in America—"perhaps the most liked man in Hollywood," Todd Gitlin noted in his book *Inside Prime Time*, "the one important man about whom I heard not a nasty word anywhere." Gitlin asked Tinker, while Tinker was still running MTM, what he would do if he was to be suddenly put in charge of a network. Tinker replied, "I would seek out the best people and I would say, 'Forget pilots. Just tell me what you want to do, and if I think it's okay I'll give you a series commitment.' " Once in charge at NBC Tinker said, "I feel exactly the same way. When we attract the right people—on and off camera, a star or a writer in whom we have confidence—we will frequently just hold our noses and jump."

Unlike Silverman, Tinker was adept at getting other people to work well for him. While at MTM, he had become a Lone Ranger in the creative community. According to producer Ronald Cohen: "Grant Tinker didn't allow his producers to go to the network and get bloody. He goes to the network. 'Look, you want to take a shot? Take a shot at me. Ronald's back at the office working.' " Tinker explained his utility at MTM this way: "I would go to the network meetings because as an old, elder whatever I am, elder statesman in the business, just being there said something for the producers or, obviously, to them. But it also meant that should it get to war I was one more soldier on our side. Some asshole at the drone level will find some fault that you haven't expected, and you have to beat him to his knees. If you feel strongly enough."

Tinker did not come to NBC giddy about the arena. "I think what is probably the biggest sin of the medium as it exists is that so little sticks to your ribs, that so much effort and technology goes into—what? It's like human elimination. It's just waste." He saw TV programmers now "clonishly following each other in sort of a futile game of leapfrog. The appetite is so voracious that we wind up just feeding shit into it and that's what comes out."

If he considered NBC a major nozzle for the dreck, Tinker's

hiring and firing policies on arriving at the network did not reflect it. There was no purge. Tinker retained Brandon Tartikoff—the youngest programming head in NBC history—as president of NBC Entertainment and tendered him privilege to run the programming as he saw fit. While Tartikoff ranked a full two tiers below Tinker in the NBC hierarchy and answered to both Tinker and NBC President Robert Mulholland, his job was crucial to the network. For it was Tartikoff who would determine the shape and content of all NBC prime-time programming.

In any case, while Tinker brought to the network a promise of quality television, he made it immediately clear that his first priority was to ensure "the viability, the profitability, of the business. That is my primary mandate, to make sure NBC is healthy."

The key to which, he felt, was attracting talent.

By hiring Tinker, NBC hoped to attract the best producers in the business, because, for the first time in any network's history, a man with an outstanding production background was running the show. Tinker's chairmanship reversed the phenomenon in the television industry sometimes unpleasantly described as that of the nigger boss. By tradition, all networks are run by men who have come up through the business, legal, promotion, or programming side of television, usually within the networks. Millionaire producers with years of experience must often pitch their ideas to some youngster who is enthusiastic about his programming job largely because the network has a great dental plan. Earl Greenberg, a former NBC program director, put it this way: "I was an attorney who came to the network and went from the legal side to the creative side because I made some friends. After I got into programming I suddenly realized that I was—by choosing to recommend a show for network production or not—determining the futures of people who had ten times the experience I had in television. I was appalled that I had been given this kind of latitude, so I started asking around. I'd talk to other people at the networks and they'd go, 'Hey, get wise. You're an old salt compared to a lot of us. Fifteen minutes ago I was an accountant for Chevy Trucks, now I'm buying prime time.' "

Consequently, television producers tend to have a general

contempt for network people. But with Tinker at the helm, NBC became "the producer's network." For not only was Grant Tinker a producer, he was considered by many people in the industry to be one of the best two or three ever. Tinker stories were the fuel of minor legend: When MTM was first developing *The Mary Tyler Moore Show*, at the request of CBS, Tinker had hired writers Allan Burns and James Brooks to create the series. They decided to conjure Mary Tyler Moore as a career-minded divorcée. No divorcées, career-minded or otherwise, had ever graced the network airwaves before, and the two writers were summoned to the thirty-fourth floor of CBS's New York headquarters, where, after a brief presentation by the writers, CBS executive Marc Golden said, "There are four things America can't stand: Jews, men with moustaches, New Yorkers, and divorced women." Burns and Brooks made no real attempt to reply, and shortly thereafter a CBS executive went to his office and called Tinker about the writers he had chosen. "Get rid of those clowns," he said.

Tinker refused. He also refused to make his wife's character a thirty-year-old virgin. But he never mentioned the network pressure to his writers. "It wasn't until several years later," said James Brooks, writer–director of *Terms of Endearment*, "that we learned that the network had wanted to dump us. That says a lot about Grant Tinker."

Tinker's announced strategy, when coming to NBC, was a simple one: Get the best people in the business to create the best shows. And if those shows are not immediate hits, have patience.

Some of Tinker's favorite and most critically praised programs, however, such as *St. Elsewhere, Fame, Family Ties*, and *Cheers*, were, as preparations for the 1983–84 season began, languishing near the bottom of the ratings, and showed no signs of making significant ascents. The potential failure of those series, coupled with NBC's long-standing image as a loser, gave the new network chief much to consider as he began his second full season at the helm. And one of his biggest disappointments since taking over was said to be that of the American television audience itself. "People are walking around saying there's nothing good on TV, but they're not making the effort to try our shows," said Tinker.

Nevertheless, Tinker had said that if NBC retreated from quality programming, it would be because he was no longer in charge. Several years earlier, while head of MTM, he had said that poor television programming was "a national crime," and that "someone should go to jail for it . . . probably network executives."

Tinker's affinity for quality programming was not necessarily the myopic faith of a naïve white knight, however. The video revolution of the 1980s—evidenced by increased use of cable, up 250 percent since 1976, and independent stations, as well as other video-access components, such as game cartridges, closed-circuit options, and various computer hookups—had influenced the public's relationship with prime-time network TV. It had divided and reduced it. Radically. In fact, in some areas, like Columbus, Ohio, where cable hookups were widely available in well-programmed, many-channeled tiers, the network share of the viewing audience had already slipped to below 50 percent. NBC, because so much of its programming was, like cable's, aimed at the educated and affluent, was most vulnerable to a cable assault.

Therefore, the Tinker and Tartikoff regime was considered by many producers and critics to be the *last, best hope for free television*. It had assumed the trappings of a crusade.

So 1983 promised to be a watershed year for television—and a year of unique opportunity for crusaders far and wide. The way Tinker put it, "I am coming to the end of my patience stage and into my impatience stage, which means we'd better get up to speed pretty quickly." And if no progress was made during the 1983–84 season? "Then I'd shoot right past impatience into abject frustration or whatever the next stage is—possibly unemployment."

With *Cheers* starting every bit as slowly as everyone had feared it would, Tony Colvin had two problems. The first was his continued employment, which would, of course, evaporate if the show was canceled. Being out of a job would chafe, because Tony needed that $1,000 per month more than he had thought a person could *ever* need a lousy grand a month. And the second problem, of course, was his tax problem.

Tony was up several nights in a row with his tax genius

friend, who turned out to know more than his accountant. But they were getting nowhere. Tony's expense receipts just weren't adding up to enough to offset his years of a $70,000 income. It was beginning to look as if he would have to swallow the god-awful $19,400 tax bill, which would be like swallowing a whole cantaloupe on the leukemic salary of a *Cheers* gofer. He was staring in the face the possibility of no longer being a *Cheers* gofer—of abandoning his dream and trudging back into banking. The idea filled him with dread.

It was tough to be working next to people who made $19,400 *every few days*, but then, that was part of what had inspired the dream. Sometimes, during the first few months of *Cheers*'s run, he'd felt close to being able to make that kind of money, because he was writing sample scripts regularly and, in his own mind, he knew his scripts were good. But absolutely nothing was certain in the TV business—including, of course, the continued life of *Cheers*. The show's ratings were abysmal, even lower than the producers had feared, and now even his paltry $1,000 per month seemed to be in jeopardy.

On what they had planned as their last night of digging into Tony's records, Tony hauled out, from a closet, one more shoe-box of records. They were, like most of his other records, a mess— a tornado of receipts, expenditure notations, and check stubs.

"What's this stuff?" asked his friend, rummaging through the box.

"Expenses I can use if I ever become a writer. Plus a whole bunch of receipts for picnics I used to give at Paramount."

"Picnics?"

"Yes, business lunches, you could say. They enabled me to learn about the producer's assistant training program."

"How much did you spend on the picnics?"

"All told, I think it came to about twelve thousand dollars. The Heineken alone was twenty bucks a case. But I can't deduct it, because I haven't sold any scripts yet."

"You don't have to sell any scripts yet. If this was a bona fide attempt to start a new business, then you are a writer, and you can deduct the picnics!"

At that moment in the early spring of 1983, J. Antony Colvin became a professional television writer, at least in the eyes of the U.S. government. Also, he solved his tax problem. The new deductions put him in a lower tax bracket and reduced what he owed, a sum that had been multiplied by penalties and interest.

By the time the IRS had finished studying the new return he sent them, he owed $9.42. With penalties and interest, the bill came to $46.

2
Day Work

Late March 1983

"Goodbye, America," crowed Corky Hubbert, twisting around in the Pontiac's seat to get the best possible glimpse of the WELCOME TO CALIFORNIA sign as the car in which he was a passenger rocketed south on Interstate 5. "Hello, Haw-lee-wood! I'm back!" Corky Hubbert, four feet eight inches tall, thirty years old, with an uncombed explosion of wiry blond hair and bright buggy eyes that helped make him look like a combination of Bob Dylan, a bottom fish, and several of the Marx brothers, gobbled a handful of aspirin and breathed deeply of the crystalline, piny northern California air. "This time, no fuckups," he vowed, his voice a rasp, as if someone had taken a cheese grater to his larynx. "This time it's straight to the top! Hey, I mean it. Hollywood is dying for a new John Belushi. Except, this time, a real *short* one." He eased back in his seat, wincing a little at the pain in his legs, hands tapping in midair to the beat of the radio on which Sheena Easton's elegantly hysterical voice was brushing at registers heard only by dogs.

Abruptly, he reached back and grabbed a Glad sandwich bag from the Safeway double-strength paper sack that was his luggage. From this bag he extracted a joint the size of a man's finger.

52

"Cool it, Cork," advised the driver, an old pal who'd been enlisted to provide interstate taxi service for Hubbert's latest attempt at superstardom. "There's an inspection station right in front of us, man."

"Christ," Corky breathed, genuinely alarmed. "Let me put this spliff in the glove box."

"No way!" said his buddy, "put it . . . put it . . ." Corky's friend scanned the Pontiac's interior as he braked, and saw green and brown caches of pot everywhere. "My God! It's like Little Maui in here."

"Just be cool." Corky swallowed, eyes fixed on the looming booth of the USDA check station. "This is California. They're hip down here. All they want to bust is medflies." His chubby, childlike hands quickly, artfully, tucked away contraband in the backseat. A moment later an agent put his head to the window and said, "Any fresh fruit in here, plants, or shrubs?"

"No way, man," Corky said. "Never touch the stuff."

The agent—a Latin kid who seemed barely in his twenties—looked at Corky quizzically, at this human dollop with crazy blond hair whose chubby stomach was peeking out below the hem of his flaming-red Aloha shirt. "Okay," said the agent after a beat. "See ya." And the Pontiac eased back into the sparse traffic trickling down the freeway, its interior blooming with smoke as Corky held a fat drop of flame from a kitchen match to the joint and, puffing hard, made it glow and crackle. Corky, who had rheumatoid arthritis so severe that he had spent two years of high school in a wheelchair, loved pot's analgesic effects.

"Last time I was making this run," Corky reminisced, intermittently holding his breath, "I got picked up hitchhiking by this Nazi hippie psycho who spent five hundred miles puffing on joints and telling me what a great guy Adolf Hitler was. This was in nineteen seventy . . . eight, before I'd been on TV. I was splitting from home because my girlfriend was a nymphomaniac. She wore python-skin, spike-heel go-go boots and would sleep with anything. So, anyway, this shmuck is telling me about how cherry it was in the Third Reich and at night when we pull over to crash, he goes, 'By the way, you are not by any chance Juuuuuu-ish, are you?' And I go, 'Oh, no no no no! I'd *kill* myself if I was Juuuuuu-ish.' Then he tells me about all this really great 'scientific research' the Germans did on midgets in 'special camps' and tells me he wants me to sleep under the wheels of his car—where it's nice and cozy. So I said the hell with this crap, and beat feet."

Five miles later Corky prevailed on his friend to stop at a rest camp, so he could take a leak. He walked, a slight hitch to his gait from the arthritis, toward the bathrooms. Four or five red-neck, longhair logger types were standing out front and one said, "What's that dwarf's trip?" as Corky pushed his way through the bathroom door and made for one of the stalls. When people stared at Corky, he liked to think it was because they recognized him from a movie or a TV show. He'd been in *Caveman*, with Ringo Starr and Shelley Long, and in the Hunter Thompson movie *Where the Buffalo Roam*, and in *Under the Rainbow*, with Carrie Fisher and Chevy Chase. In *Rainbow*, he had, as a matter of fact, played a role startlingly similar to the one he was playing now in real life: that of a hard-working young Horatio Alger–style midget on his way to L.A., seeking a big break in show business. Situating himself on the throne, he wondered idly if the Grateful Dead–style truckers, or whatever they were out there, were going to ask him for autographs. The last time *Under the Rainbow* had played on cable, when he was living in Santa Monica, he'd gotten asked to sign autographs three times in two hours—first by some school kids, then by a street freak, and finally by a beautiful, pregnant Chicano woman who invited him into her home, where he signed for her daughters in a room with walls covered by a tasteful blizzard of crucifixes and icons surrounding a portrait of a bleeding Jesus.

Corky glanced toward the floor on his right. There was somebody in the adjacent stall; he could see the cowboy boots. Except they weren't like regular cowboy boots, but boots with practically spike heels that were polished a glossy maroon. That was the trouble with goddamned California, he thought—it was just plain full of oddballs and pervos. Christ, you took the wrong Miss Poopsie for a nooner at a Notel Motel on La Cienega in L.A. and, if you weren't careful, she'd give you some new kind of space-age fungus that'd make your fingers fall off. You couldn't be in the state twenty minutes, it seemed, without ending up in a public john lessening your load in a stall next to some cowboy-bootied fruitcake.

Corky wrapped things up and stepped out of the stall toward the sinks, when he heard a shriek. "My God! A little man!" He turned. There was a woman standing behind him. Then another appeared from the stall next to his and he saw maroon cowboy booties on her feet.

"Holy shit!" he said, bolting out the bathroom door, where all these long-haired no-necks and mouth breathers were standing around laughing their asses off, including his buddy, who was coughing and bent double with glee. "Okay, okay," Corky said, his face purpling. "So I made a little mistake." He hitched back out toward the Pontiac.

Fifty miles deeper into California, as April grew drier and hotter by the mile, Corky had recovered his composure and was pumping himself back up a little with his favorite tale of Ringo Starr and John Matuszak. "It was when I was doing *Caveman* with them and Shelley in Mexico a few years back. You know Matuszak, don't you, the giant lineman for the Raiders? He could drink a margarita the size of a TV set, then run up and down ten flights of stairs. I've seen him do it! He's great. Anyway, one time I had these magic pills, Matuszak didn't know what they were, but he goes, 'Gimme some,' and I go, 'Have you ever had any before?' and he goes, 'Pills aren't my trip, but anything you can handle, I can handle, so gimme!' So I gave him one and he goes, 'You're taking two, I want two.' I go, 'One is enough for a start,' and he goes, 'Corky, if you can handle two, I can handle two.' So, down the hatch. We go to this weird Mexican disco full of people that look like animals. Ringo is there and Tooz starts to rush. A bunch of these guys start fooling around, making a joke by asking Tooz to dance, which was, ha! ha!"—Corky wiped at the corner of his mouth with the back of his hand—"el-mistake-o! Cuz Tooz hates it if you even *pretend* to be a faggot. Then he goes, 'Cork, I'm getting scared, and when I get scared people get hurt.' And so I'm thinking, 'Oh, Jesus, what if he wigs out and eats a Mexican or something?' But Ringo is so cool. He realizes what's happening, so we go over to his place. I end up getting a ride with one of the bosses from the movie and he's going, 'Did you give John drugs, Cork?' And I'm going, 'No, no way, he just doesn't like crowds, that's all.' And this guy is going, 'Corky, you can be a *major star*, but you're going to blow it if you don't start acting more responsible!' Which is a rap I really hate, when people start talking to me like I'm a little boy or something."

Corky glanced out the window. Large buildings here and there, a service station or two—hints of Modesto were beginning to well up on either side of the highway. "Anyway, Tooz and I ended up back at Ringo's place. Ringo was great! He put on *Abbey Road*—which is his favorite Beatles album—and he brought Tooz in for a landing like there was nothing else happening in the world. And like after that"—Corky coughed once, then again—"it may sound weird, but I think I had Tooz's respect. He realized that I could handle some things better than he could, even though he's about the toughest guy in the world."

Corky's legs were covered by copies of trade papers—*Variety, Dramalogue, Hollywood Reporter, Casting Call*. Though he fully intended to be famous before the year was out and was— by his own conservative estimate—already the fourth-top midget in the business, he'd been flying a little below the radar of late. Roles hadn't been easy to come by when he'd left L.A. some months before, tired of the hustle and dirt. He was also close to broke. While he'd walked out of *Under the Rainbow* $23,000 the richer, Corky knew how to spend money. In fact, what remained of his grubstake was in his front pocket, and emerged as a tangle of five- and ten-dollar bills, which expanded in his hand like a time-lapse film of a blossoming flower.

Corky was headed to L.A. with a hot prospect in mind. His friend Allan Sacks, the man who had created the sitcom *Welcome Back, Kotter* for Gabe Kaplan and John Travolta, had created for Corky his very own vehicle, a show called *The Amazing Adventures of Sparky O'Connor*, in which Corky was to star as Sparky, a newsboy who worked the bustling streets of "Big Town, U.S.A.," a metropolis located "somewhere between the two coasts." Sacks had a lot of swat in L.A., and Corky, though down to his last couple of dollars and at a serious Y in his burgeoning career, was confident. He considered himself to be a mouth-watering combination of talent, energy, charisma, and shortness.

"My agent says that as soon as I get back into town I should be able to get into somebody's pilot, but I've got my fingers crossed on *Sparky*. No matter what, I'll make the rounds, and be back in the bucks. I'll work the clubs. Maybe do another *Fall Guy.* You know, Lee Majors's TV show. Did I tell you what happened the last time I was on that show? I was playing this irresponsible midget bookie friend of Lee's who disappears. I was on the set and Lee was doing this scene, playing it like I was dead! Going, 'Corky *was* a partner of mine. Corky *was* great.' I screamed, 'Stop! Stop! Can't you say, "Corky *is* a partner of mine"? "Corky *is* great"? So I can maybe come back again?' And Lee—what a great guy! He says, 'Sure, Cork, no problem.' But if I hadn't been there, I'da been—" he drew a finger across his throat.

He then sighed loudly. "I'll tell ya, man, television is completely crazy, completely random for the actor. It's totally a writer's medium. But it's also going bananas, going crazy. It's getting like the colleges in the sixties. Total anarchy. Cable guys are going after the networks guys. And there's a lot of work. A lot!"

Like most members of the American Federation of Television and Radio Artists, Corky was not used to a year of forty-hour weeks. At any given time, only 15 percent of the union members are employed and the average annual income is estimated at $2,000 a year. While Corky could command $500 a day when times were good, he was not above returning for a visit to the Improv in Hollywood. Which was exactly what he did, when he reached L.A. two days later.

This club, frequented by casting directors and agents, is the most important turnstile to television comedy in the world. Corky was even willing to take another shot at Zoo Night, the one night of the week at the Improv that was totally open, when fetal comic geniuses shared the stage with maniacs and doggerel artists. Nevertheless, the Improv was a regular hangout for many of the top agents and casting directors in Hollywood, and a comic with a good new bit could be plucked from the slapstick barnyard of Zoo Night and find himself trading bon mots with Carson in less time than it takes to file an unemployment check. Lily Tomlin, Robert Klein, Bette Midler, Freddie Prinze, Andy Kaufman, Jimmy Walker, and Robin Williams had all developed their acts at either the L.A. Improv or the Improv in New York. Several months before, Corky had almost made it onto the club's syndicated TV show—a break he'd give anything for now.

Corky sauntered in, ordered a beer from the Improv's bar, and slapped $2 on the counter, his gaze jumping around the room. Jay Leno, Byron Allen, Jimmie Brogan, Rich Hall, Dottie Archibald, and Arsenio Hall were there—half the rising comic talent in L.A. was visible at a glance. "Cable's going batshit," Corky said. "If I could get a cable gig, I'd be fat. Jimmie Brogan made enough on *Laff-a-thon* to buy a house."

Two chrome-haired women wearing tights, leg warmers, and spike heels moved past Corky and sat at the Pac-Man game by the Improv's front windows. "Dig it," he breathed, "total L.A. Girls so hyped out that they sweat just sitting still. Can you imagine them getting married, being moms, lactating? It boggles the mind." Corky stood, trying to figure out if he should go ask Budd Friedman, the Improv's renowned owner, about employment. Abruptly, he was eye level with the bust of a dark and sedately lovely young woman with eyes the size of half-dollars.

"Weren't you in *Under the Rainbow*?" she asked.

"Sure." Corky's mouth turned up at the corners. The woman looked timeless and Mediterranean, a combination of Mona Lisa and Marjorie Morningstar. "As a matter of fact, I was one of the co-stars."

"It sucked, but you were terrific," she replied, putting a piece of paper and a pen in his hand. "Be an angel."

Corky grinned into her chest. "Gladly," he said. "Didn't we go to different schools together?"

She laughed and he scrawled, "No dream too big, no dreamer too small," the official epitaph from the film.

She took it to her sternum, then grabbed Corky's hand, before drifting off into the crowd around the bar. Corky swallowed. "Man, do you realize the *stuff* you get when you make it in this town? There's room for a new Belushi! I swear it. Where's Budd? Fuck it, I need a job; I'm gonna go ask him."

The rumor around town was that he was suffering from cancer, but Brandon Tartikoff, despite the wig that covered his head and the edema that often made his tanned features look wan and puffed, faced NBC's 1983 development season with the same buoyant, cynical optimism that had characterized his three-year reign as the network's top programmer and Grant Tinker's right-hand man.

Watching dailies of a detective spoof whose star handled his role with the finesse of a spastic attempting brain surgery, Tartikoff listened as an associate explained the show's producer saw the guy as "an emerging Jacques Tati."

Easing back in his chair, Tartikoff observed, "Did anyone tell him Jacques Tati is a nine share?"

TV production is an almost agricultural process in which shows are sown, nurtured, and harvested by season. Early spring is the major growing season, and now Tartikoff was already beginning to view pilots—newly budded programs vying for a spot on NBC's fall schedule—at his quarters in L.A.

In television, New York is facts, Los Angeles is ideas, and the rest is flyover. All three networks have their corporate headquarters in New York, and most of NBC's business decisions are made in its offices at Rockefeller Center in Manhattan. But most of its programming and promotion decisions are made in its spread of suites and studios in Burbank, a sleepy north-L.A. suburb that sits in front of a wall of smog-trapping foothills.

The California office space is ordinary. Ten thousand American companies maintain fancier accommodations. NBC Burbank looks like a fair-sized suburban factory—it could be the manufacturing site of golf carts, calculators, or diet pills. From the freeway that sits behind it, it looks only like an anonymous light industry that is doing a brisk enough business to occupy twenty acres of L.A. County land.

Brandon Tartikoff's office was on the second floor of a two-story structure, with a view encompassing Juicy Harvey's burger joint and the small homes of a neighborhood that is almost middle class. His digs were no more ornate than those of half the optometrists in town, and his style was equally lacking in showbiz glitz. Tartikoff had no $50,000 Mercedes, no new-money palace in the Hollywood Hills, not even a swimming pool or a coke habit. Despite an annual income in the middle-six-figure range, Tartikoff drove a Ford Mustang and went home at night to his wife and baby daughter in an unlavish three-bedroom home in Coldwater Canyon.

His style was friendly, casual, cutting, and direct. "He's one of those people," said an associate, "that could be equally at home in a pool hall or at a presidential inauguration."

Tartikoff, at thirty-four, was the first of the "baby-boom programmers," the first member of the generation that grew up on *The Mickey Mouse Club, Leave It to Beaver,* and *Mod Squad* to determine network programming. He was the first network mullah from the generation that drank Diet Cokes, ate LSD, grew its hair to its shoulders, went on surfin' safaris, killed VC, joined the SDS, met the Beatles, and subsidized the wishful thinking of Kahlil Gibran, George McGovern, and Carlos Castaneda.

An English major at Yale, Tartikoff was above medium height and, with his drooping eyes and beak of a nose, could have been the son of his original mentor: Fred Silverman.

Tartikoff's chore was, perhaps, the biggest in television history: to sponsor programming that would stem NBC's ten-year slide toward oblivion. Tartikoff sat mulling over the svelte brainstorms and reinvented inspirations of the myriad producers who'd have loved nothing better than to sting him for a six-to-thirteen-show order worth $5 million to $10 million. It was not uncommon for a man in his position to be pitched thirty separate series ideas a week. Most unworkable, but many only preposterous. Television is endlessly reductive, a foreground medium where simplicity is king. According to Bob Shanks, a former programming chief for ABC, "Imagine . . . that all ideas are small chips of tiles and that there are a limited number of these tiles. We are all playing with the same tiles. The programmer has seen them all. . . . His job is simply to rearrange the tiles in a fresh mosaic."

Tartikoff often sounded like his old mentor, Fred Silverman, when assessing his tasks. "All of television boils down to excisable elements that you can put in twenty-second promos. If you can't have somebody pull a gun and fire it fifty times a day on promos, sex becomes your next-best handle." And, like Silverman, Tartikoff spent no small amount of time chasing after current public tastes. Anticipating a Reagan win in the 1980 presidential elections and assuming a consequent public tilt to the right, Tartikoff televised *Walking Tall,* the tale of a southern sheriff who enforced the law with a club twice the size of a baseball bat. When that didn't work he divined, "It's time for a daffy female."

Nevertheless, Tartikoff also said, many times, that he felt there was a deep, untapped demand for quality programming—thus sounding like his *other* mentor, Grant Tinker. And, like Tinker, he was confident the network could be turned around. For starters, Tinker was a magnet for talent like no other in the industry, a man who had overseen creation of some of the most popular and critically successful shows in television history. Already NBC boasted a string of hip, urbane comedies and dramas—*Hill Street Blues, Cheers, St. Elsewhere,* and *Fame,* whose creators had been largely nurtured at MTM under Tinker's reign. The major problem was that three out of four of these shows were in imminent danger of being canceled, because only twenty million or so people watched them every week, and twenty million people are scarcely a crowd where network advertisers are concerned.

Cheers, for example, had a recent share of 20, meaning that 20 percent of all people watching TV had the show on, producing an audience of about eleven million households, or twenty-four million people. While twenty-four million people could start, say, a fair-sized country, they are not enough to make advertisers salivate, since they must pay $100,000 to $200,000 per thirty-second ad to reach those people. To justify the investment, sponsors want a hit show, one that finishes in the Top Twenty.

To be a hit, a series must have a 30 percent share of the total viewing audience. NBC had only two such shows, *Hill Street* and *The A-Team.* Because each overall network rating point translates roughly into $50 million worth of advertising revenues for the year, retention of shows drawing much below this share becomes costly. And because NBC's overall viewership was so weak, there were almost no existing hits to bring an audience to the new shows—it was like trying to build the Parthenon on quicksand.

In the face of this situation, Tartikoff did have his detractors. Paul Klein, his predecessor at NBC, now president of the Playboy Channel, offered, "Brandon knows about the structure of shows but not how to get ratings. He has no plan. First he says the answer is drama series, then he says comedy, then miniseries. I call him Random Tartikoff."

Observed *Newsweek* television critic Harry Waters of Tartikoff in action in the early spring of 1983: "Tartikoff moved into the network's programming 'war room' with two even younger assistants. Gathered before a large magnetized board containing placards with the names of current NBC series and potential replacements, the trio [was] intently trying out the look of different lineups. As the placards slap on and off the board, the talk is of 'audience flows,' of 'Q scores' and 'comp makeups,' of 'mainstream vehicles,' 'breakout concepts,' 'heat,' and 'buttons.'

"Quite understandably, Tartikoff is obsessed with the mechanics of his craft. The man who selects what so much of America watches is, by necessity, a pure tactician. Ask Tartikoff about how television can upgrade its knee-high intellectual standards and he quickly slides into an analysis of the synergistic relationship between time slots and share points. This form of myopia afflicts Anthony Thomopoulos and Bud Grant, his counterparts at ABC and CBS, as well. With their energies so consumed by competitive strategy, they have little left for big think—or for the instinctive, seat-of-the-pants commitment to a story proposal displayed by Hollywood's legendary showmen. That suppression of creative impulse may help explain why network TV seems to be self-destructing. No scheduling ploy or research printout will ever discover another Archie Bunker or a new M*A*S*H."—the kind of fresh hit the network could sorely use.

Despite large advertising-rate hikes, almost three quarters of NBC net profits were coming not from network programming, but from NBC's five company-owned affiliate stations. Consequently, the NBC profit to parent company RCA was less than what RCA would have made if it had, in the beginning of 1982, put its $1.5 billion network operating budget in a passbook savings account. To no one's surprise, it appeared that Grant Tinker—through Tartikoff—was already beginning to hedge his bets. While NBC shows like Cheers were pulling in most of the network's critical acclaim, it was the more standard buffoonery like The A-Team and Knight Rider that was paying the bills. And among NBC pilots for the upcoming season were shows like Manimal and Yazoo. The first was a proposed series about a man who can turn himself into a whole zoo of animals in order to "fight crime," and the second, about a "former newscaster who finds himself marooned in a land of puppet creatures"—this giving rise to speculation that Tartikoff and Tinker were giving up the ghost on quality programming. Others, however, felt such pilots were anomalies, and predicted that Tinker and Tartikoff would go for quality programming regardless of cost—on the simple grounds that when you're rich, validated, and afflicted with a fatal disease, why not? What in the world do you have to lose?

Road weary and not quite horrified enough to be adrenaline fired, Harry Anderson gazed from backstage at the quickly filling open-air auditorium, shivering slightly in the chill late-March evening air, which was considerably cooler here in New Jersey than it had been in L.A. the previous night. The Great American amusement park's amphitheater was a weekend collecting point for teenagers unable or unwilling to drive through the tunnel to Manhattan, and eight thousand of them were filtering in to watch Harry Anderson's second performance, which was just about to begin. Harry really didn't care how many kids showed up, because he was getting a flat $2,000 fee for the night, with no cut of the gate. As showbiz thrills went, doing his comedy and magic act for a herd of numb teenyboppers was low on the list. These kids. Jesus. Scraggly long hair. Tattoos. Cro-Magnon features. Had he ever looked like that? They were like the mutant offspring of the Ripple and Reds Generation, progeny of Janis Joplin and Fred Flintstone. Did most of them even have opposable thumbs? He could tell from two hundred yards away that almost half of them, bumbling like corks in choppy water, were stoned out of their gourds. They were feeling no pain. They were thinking no thoughts. It was strictly a GTFM performance. Get the fucking money.

Harry was the opening act for the Robert Hazzard Band, a Philadelphia group that specialized in the kind of brain-damage rock that has long been the staple of stadium concerts. The first show had gone well enough. Harry had fed the crowd some of his lowest-common-denominator material, such as Harry the Harelipped Geek, who swallowed snakes ("If you thwallow a thnake, you've really got to oil that thucker up!"), and a gag where he hammered a nail up his nose. Nothing cerebral, just good, old-fashioned stadium humor—the kind that keeps intra-audience bludgeonings to a minimum.

Just before he went onstage, as the mob began to grow restive, Harry was reminded of the time his friend, comedian Albert Brooks, had opened a Richie Havens concert. Backstage, as the crowd chanted "Rich-chee! Rich-chee!" a burly security guard said to Brooks, "Is your name Richie?"

"No."

"Too bad, man. They're gonna killll you."

Then Harry heard his name being blared out over the PA system, took a deep breath, and sauntered onstage, to the coolest of receptions. He began to do a trick, and then stopped and glared at the crowd, as if they were the rudest of intruders. "Look," he snapped, "I didn't expect to see you here, either." The remark won many of them over—they clearly respected aggression, in any form.

As soon as he'd done a couple of bits, though, Harry knew something was wrong. People in the audience were shouting out his punch lines before he got to them. "Christ alive," he thought to himself, "this is one of those places where you pay to get into the park, and can go see the shows as many times as you want." Half these people had already seen the first one.

But he forged grimly ahead, though laying on a bit more sardonic humor than usual. Then he got to a place in the act where he yelled, "Okay, now, everybody throw your money onstage!"

They were waiting for it.

Suddenly there was a hailstorm of metal. Pennies, quarters, nickels, halves—they rained down on him like stinging fire, and then: *whing!* A Kennedy half-dollar skimmed from a hundred yards away squarely into his forehead. Harry felt the welt rising before the half-dollar had even chattered to a spinning halt at his feet.

"All right, you assholes!" Harry yelled at them. "My wife is going to be on this stage in a minute, and if anybody throws so much as one piece of popcorn at her, I'm going to personally come off this stage and kick the living shit out of him!" The crowd cheered, again appreciative of naked aggression, but somehow the magic had left the evening.

Harry wanted out of stadium vaudeville. He wanted to be a TV star. It was, he honestly believed, the only way he would ever be able to lead a normal life.

*　*　*

The next night he was a TV star, if only briefly. He appeared on the (slightly) more civilized side of the Hudson River, in Manhattan, on *Saturday Night Live*, in one of the later shows of the 1982–83 season. He was scheduled to do a spot with his wife and occasional co-performer, Leslie, a delicate, pretty woman who by the spring of 1983 had pretty much abandoned the rambling, peripatetic lifestyle of a performer in order to take care of their three-year-old, Eva, a curly little blonde who had already been categorized at her school as gifted. These days, Harry had to go on the road by himself—another reason he wanted to be a TV star. He thought of being a TV star as having a day job.

The *Saturday Night Live* performance was the seventh that Harry had done. For the past couple of years, he'd been one of the most often recurring nonregulars on the show, a gig he cherished as much as the character role he had played three times on *Cheers* over the past year.

Dick Ebersol, the producer of *SNL*, had Harry on so often because the thirty-year-old comedian brought to the show some of the off-the-wall attitude that had powered *SNL* in its early days. Harry, a lanky, yellow-haired sunflower of a young guy with a perpetual grin, loved to dig into the bizarre, arcane, grotesque side of life in his humor. He was acknowledged as one of the better magicians in the country, but what intrigued him most was not the glossy magic of the center-ring spotlight, but the gritty tricksterism of the carnival sideshow—the Dark Circus, he called it. In fact, part of his act was to puncture the notion that magic can even be performed. After he'd perform an illusion, Harry would find somebody in the audience who looked as if he thought he'd just crossed paths with the supernatural and yell at him, "Hey, c'mon, it's a trick! If your cat has kittens in the oven, you don't call them biscuits, do you?"

The last time he'd been on *SNL*, he'd put a hatpin through what looked like the vulnerable, fleshy part of his forearm, and then glowered at the audience, daring them to abandon the idea that he really was mutilating himself. "Come on!" he shouted. "It's a trick. If I really had a hatpin in my arm, would I do this"— he sawed the needle back and forth like a violin bow. The "flesh" of his arm heaved and churned, as the audience gasped with sickened delight. "Come on," he cried, "would I really cut my-self to make you laugh?" A drop of blood oozed down the hat-pin and splashed to the floor. "Look. If this was blood, would I do this?" He dabbed a finger into the red goo and tasted it. "Ummm. Strawberry!" The audience was stunned. Harry stared at them. "Well, gee," he penitently said, "what kind of tricks do you like?"

A friend, planted in the audience, shouted, "Rabbit tricks!"

"Well, okay," said Harry, pulling out of a box what looked like a live bunny rabbit. "I don't think it's nearly as dramatic," he said, yanking the hatpin out of his arm, "but if that's what you want . . ." He shoved the hatpin through the rabbit's head.

He had been afraid the NBC censor, who spent an inordi-nate amount of time at SNL, would quash the entire act. But all he'd said at dress rehearsal was "Put button eyes on the rabbit, so everyone will know it's fake." While Harry was rehearsing the act, Dick Ebersol, as "blood" splattered at his feet, was eat-ing a bowl of minestrone soup; perhaps the censor reasoned that if Ebersol could keep his soup down, then the act must not be that bad.

The other part of the act that Harry had worried the censor might object to was when he said, as he touched the flesh of his arm while pulling the hatpin out of it, "This part is the fore-skin." But the censor, at the rehearsal, laughed along with everyone else. At the evening shooting, though, the censor's son, after the foreskin joke, whispered into his dad's ear.

Later the censor came up to Harry and said, "You really shouldn't have said 'foreskin.' "

"All right," Harry said amiably. "Next time I won't."

The dress rehearsal for tonight's show, which would be Leslie's first time on network TV, stunk. Ebersol complained that the act contained too much "dead air" in it, too many moments when nothing was being said. The act consisted of a contest be-tween Harry and Leslie, to see if Harry could get out of a strait-jacket before Leslie could escape from being bound hand and foot to a chair. Both escape acts were genuine; Harry really could wriggle out of a straitjacket, by popping his shoulder out of its socket. It was a stunt he'd done in a brief role in a Francis Ford Coppola movie, *The Escape Artist*, which had never been released to theaters, but which Harry had seen on an American Airlines flight—and hated so much that he'd asked for his $3 back. Leslie also had mastered her escape routine; she could squirm out of practically any set of ropes that held her.

The gag of the act was for Leslie, playing the conventionally docile magician's assistant in a sexy dress and fishnet stockings, to win the contest, beating the magician at his own game—but not before Harry, in his straitjacket, had, in a monstrous display of jealous showbiz pettiness, kicked Leslie—whom he had already introduced as his own wife—to the floor, still bound to the chair, and stamped at her writhing attempts to free herself. Whenever they did the act, the crowd always cheered for Leslie.

But Ebersol said the whole thing was too flat. Harry made up a joke for one of the dead spaces, which came while he was being laced into the straitjacket. He looked at the audience and proclaimed, "You are looking at a man who does not *know* the meaning of the word 'glutamate.' " Not a bad joke, but still the act was moved to later in the show, after the audience dropped off, a sign that Ebersol didn't have much faith in it. When Harry first appeared on *SNL*, he had been so hot they'd pushed Miles Davis back and moved him up. Could he be on the downhill slide *already*?

The dress rehearsal at *SNL* was taped, just in case a performer had an accident between the rehearsal and the live show, a few hours later. This precaution had been instituted several years previously, when the former cast had been a more raucous, drug-using crew. But it had proven necessary only once, when Buck Henry had suffered a minor injury and the dress tape had been used as the show.

Harry had not seen the regular cast members—Eddie Murphy, Joe Piscopo, Tim Kazurinsky, and the others—until they'd all been made up together for the evening shoot, partially because he had only been required to visit the stage, and the *SNL* offices, a couple of times prior to the shoot, and also because all of the cast members had their own offices and tended to hang out in them. The *SNL* offices were on the seventeenth floor of 30 Rock. *SNL* occupied about two thirds of the seventeenth floor, while the NBC election unit worked in the remaining one third.

The seventeenth floor, despite being the hip headquarters of modern network TV, was as ordinary a collection of cubicles and lounges as any other floor in the building, generally devoid of even a New Wave poster or an abandoned roach in an ashtray. Even the floor's kitchen, which sat in the middle of the row of offices, and had been used a few years back by some of the more adventurous of the initial SNL cast members to cook free-base cocaine, was unadorned by any signs of anticorporate activity, past or present.

The election unit also shared Studio 8-H, where SNL was shot, and was the only production entity that had priority access to 8-H. The studio is a large one, compared to most TV stages, which generally are no more imposing than the auditoriums of small, rural high schools. Studio 8-H is larger because it was built in another era—originally, for Arturo Toscanini and the NBC Orchestra. The studio had also been home base to Milton Berle and Sid Caesar, the giants of TV's "golden age," who had set the tone of television comedy for years to come. The studio's legacy of pioneering programming had been, to no small degree, honored by SNL, which had enlarged the boundaries of acceptable TV humor in the mid-1970s and, in doing so, had spawned a number of entertainment stars: John Belushi, Chevy Chase, Bill Murray, Dan Aykroyd, and, most recently, Eddie Murphy, who by 1983 had signed a movie deal with Paramount worth about $15 million. Murphy, obviously, could have afforded to leave SNL, but chose to remain, enjoying the deferential treatment granted by Ebersol, who, only a few years earlier, had loomed over him mighty as God.

Ebersol, a longtime NBC programming executive, married to actress Susan Saint James, was a rising power at the network, due mostly to his rescue of SNL after the mass exodus of the original cast following the 1979–80 season. For a year the show had floundered and was nearly canceled, but then Tartikoff had brought back Ebersol, who'd left the network to do independent production. Ebersol had been responsible for SNL's being on the air in the first place, in the mid-seventies, when he'd been head of NBC late-night programming. Tartikoff gave Ebersol a mandate to hire an all-new cast and writing staff. For his actors, Ebersol had drawn primarily on Chicago's Second City comedy group, which had proven to be the effective minor leagues for the initial *SNL* cast, and had found Eddie Murphy in New York comedy clubs, where the comedian, barely old enough to legally buy a beer, was an underground hit.

Harry was disappointed that Ebersol wasn't wild about the escape-race routine, because at this point in his career, Ebersol's *SNL* was vital to him. Harry's nightclub fee had escalated appreciably after he had begun appearing on *SNL* in the 1980–81 season. And nightclubs, for Harry, still paid the rent. He had been performing in clubs for about five years, but had only risen to his current $50,000-or-so-per-year income as a result of being on TV. Ironically, *SNL*, which only paid about $500, was now a loss leader.

Harry had made his first big upwardly mobile leap in club appearances in the late 1970s, when an agent who needed a fill-in for a Las Vegas act saw Harry at a small club in suburban L.A. The Vegas act had been as an opener for Kenny Rogers, who had recently attained superstar status. Harry had eventually prospered on the Vegas-Reno circuit, even though, during his first regular engagement, he and ventriloquist buddy Jay Johnson had tricked the head of security at the Circus Circus casino into wearing a pair of handcuffs. They had then refused to release the man, and had ended up on the casino's aerial trapeze, both drunk, while the security chief watched from below, mirthlessly. Jay Johnson had gone on to four starring years on the series *Soap*, while Harry plied his trade during the graveyard shift, sometimes in side lounges where keno winners would be announced over his act.

Harry had made eleven appearances on Merv Griffin's show during this time, and twelve on John Davidson's, which had helped jack up his fee. But he could never get on Johnny Carson's *Tonight Show*, even though he'd submitted tapes many times.

Once, at a party, he'd run into Jim McCawley, the talent coordinator for *The Tonight Show*, the man who held the keys to that particular kingdom. Before Harry could even say hi, McCawley had said, "I saw your tapes. And you're just not funny. Doing magic acts that end up not working—it's just not funny."

Sometimes the nightclub circuit seemed like a dead end, but

Harry was not at all sure that he would ever be able to get beyond it. He didn't consider himself an actor. Gradually, he was growing more restless with life on the road. After watching firemen cart body after body out of the MGM Grand Hotel in Las Vegas during that establishment's disastrous fire, he began insisting that he not be stationed higher than the tenth floor at any hotel he stayed in. And he took to asking for airplane seats in the far rear of the plane, as close as possible to the "black box" of flight information that is built to withstand a crash.

Worse than the presence of the irritating elements of road life, though, was the absence of his wife and daughter.

This *SNL* show was a reprieve, of sorts, because both Leslie and Eva were with him. Earlier in the day, he'd taken Eva shopping, then they'd all gone to the Statue of Liberty.

After the show, he and Leslie went to a brief wrap party, then returned to the Berkshire Place, where Eva, watched by a friend, was sleeping.

They packed and boarded a red-eye flight back to L.A. They indulged in no celebration. *Saturday Night Live,* Harry thought, for all its pomp and circumstance, was turning into just one more case of GTFM: Get the fucking money.

The next day, home from New York, Harry, Leslie, and Eva took a shuttle bus from the L.A. airport to Van Nuys, in the San Fernando Valley, which wasn't too awfully far from where they lived, just beneath the enormous tract of ridges, canyons, and green lawns that is Griffith Park. Harry and Leslie had lived for six years in rural Oregon and still didn't relish the urban feat of driving into L.A. International Airport, where motorists rushing for planes slide through red lights and accelerate through stop signs.

When they got to their house, a pleasant, neo-Spanish stucco high enough in the foothills to be above the marine layer of fog, a message was waiting on their answering machine. It was from Harry's agent, Danny Robinson, who asked for a call-back and promised good news. Danny had found a pilot, he said, that was going to make Harry a TV star.

In fact, though he didn't know it yet, Harry was about to enlist in a game of strategy, being played on a big magnetic board not far away, and for keeps, by Brandon Tartikoff.

* * *

It was some small cheer to Tartikoff's hard-pressed team that ABC, still riding a wave of cartoonish programs Fred Silverman brought to the network in the mid-1970s, seemed to be running out of steam by spring of 1983. The only real commercial vitality in the ABC schedule came from the six shows of independent producer Aaron Spelling, including *Dynasty, The Love Boat,* and *Fantasy Island.* Industry people said that ABC stood for Aaron's Broadcasting Company. But even the Spelling shows were starting to show signs of wear and tear. Tartikoff thought that *Fantasy Island* and *The Love Boat* both could be knocked off by shows with a little more brain appeal, or action appeal, or even sex appeal—anything that offered more than totally predictable plots and laugh-track humor.

ABC had just won the most recent ratings period with its miniseries *The Thorn Birds,* but its regular series, its nuts-and-bolts programming, were woefully lacking in distinction. For years ABC had mined the youth vein with sitcoms like *Happy Days, Laverne and Shirley, Mork & Mindy,* and *Joanie Loves Chachi,* but the kids who provided the backbone audience for these shows were growing up and switching to other shows, and were not being systematically replaced by a younger generation. And the breasts-and-fights dramas provided by Spelling also appeared to be nearing the end of their natural life-spans, mostly because every possible situation that could be concocted in these one-dimensional offerings had, over the years, been used, then used again. Even the ratings victory provided by *The Thorn Birds* had a bittersweet taste to it, since the show had been used in April rather than in the May sweeps, when audiences of local affiliates are measured in order to determine ad rates; many ABC affiliates were incensed about the scheduling. Their reasoning was "If you network guys aren't smart enough to hype the sweeps, then what the hell are you doing running the company?"

Adding to ABC's problems was the recently announced report of its second-quarter earnings, almost 100 percent below the previous year's quarterly profit. Network officials blamed it on a lack of tax credits and spending on new video enterprises.

The mood at ABC was defensive and apprehensive. A producer who had spent the spring trying to develop a project at

ABC called the network's programmers "the biggest bunch of incompetent assholes I've ever tried to work with. All anybody cares about is covering their own ass with research statistics."

It seemed possible, at this juncture in TV history, that ABC might decline into the third-place position that it had once occupied for so many years. There had been a joke that went: "Know how to end the Korean War? Put it on ABC, and it'll be over in thirteen weeks." Then there was the joke that went: "Know how to end the Vietnam War? Put it on ABC, and it'll . . ." Finally the joke "Why can't the FBI find Patty Hearst? She's hiding on ABC." It had seemed for so long that ABC, which got started sixteen years after its competitors, would never catch up. Then Fred Silverman had made the ABC programs the dumbest and most successful on TV. When jiggle and car chases and Monday night football had first come to television on ABC, they had seemed at least contemporary, if not mentally stimulating. Now these elements just seemed tired.

More encouraging to Tartikoff than the apparent vulnerability of ABC, though, was that CBS, the mighty giant of television, was also starting to fray around the edges. In fact, there were some signs that CBS might be the Goliath of the 1980s.

The TV business was rife with springtime rumors that Columbia Broadcasting was on the verge of some kind of payroll or budget cut. For the first time in many years, there were real financial problems at the network.

In 1982, CBS had refused to guarantee various demographic qualities of the viewing homes it sold to advertisers. CBS would guarantee a certain number of homes, but wouldn't make a money-back offer that these homes would contain the age and income brackets the advertisers wanted. CBS knew that this would cut down on the ads it could sell at the beginning of the year, but it thought that, as the economy improved and ratings soared, the unsold commercial slots would gain in value. It didn't happen. CBS ended up selling many of the time slots at cut-rate prices, and now, by March of 1983, it was feeling the pinch of the lost revenue. A hiring freeze, job cutbacks, and budget chopping were all being discussed as ways to take up the slack.

The network was also hurting from the failure of its CBS Cable venture, a highbrow ballet-and-opera network that had cost tens of millions of dollars. William Paley, in one of his last acts before officially retiring as chairman of CBS, had killed the project. To carry it on, he said, would have "meant another year of agony and another $20 to $30 million down the drain." It was money the network couldn't afford to gamble with. Its 1982 profits were down $37 million from the year before.

And it did not appear as if the programming department was going to pull CBS out of its slump. CBS was embarked upon a very modest pilot-development season, creating fewer than half as many shows as NBC, and five fewer than it had developed the previous year. It was believed that CBS would show movies on three nights, a move widely interpreted as a vote of no confidence in its pilots.

"When I look at the other two networks," Tartikoff said, "I don't see anything that has me shaking in my boots."

Tony Colvin, alone on the *Cheers* set, grabbed an overstuffed chair and wrestled it up a flight of stairs to a room that could be locked. In just a few weeks, the networks would announce their fall schedules, and the furniture from the office and stages of the canceled shows would be pillaged by producer's assistants from all over the Paramount lot. *Cheers* had already been renewed, but the furniture scavengers sometimes got carried away, grabbing whatever was not nailed down. Jim Burrows had told Tony to move the *Cheers* furniture to higher ground. As he puffed up the stairs, Tony's footsteps echoed in the cavernous stage.

The set—a plastic and plywood bar, a few booths and tables, and some one-sided walls on wheels—seemed small and fake and almost inconsequential when it was not highlighted by a balcony of floodlights and was not the center of activity. The bar sat on the cold cement floor of the sound stage, a square shed as big as an airplane hangar, in front of twenty-five rows of box seats. Behind the seats were glassed-in booths where, on show nights, sound and camera directors transmitted messages to their technicians on the floor and on catwalks high above the set. Another windowed cubicle was labeled VIP BOOTH and was the viewing room for network executives, their wives, and their friends.

The ceiling of the stage was a remote, shadowy spider web of cables, cords, and rods; through these channels flowed the electricity that transformed the set. When the electric power surged on, and the actors and extras assumed their places, the set seemed to grow in size and become a real bar—even to the spectators who could see that it was all an illusion.

The *Cheers* company was leasing the entire building, except for a suite in one corner that housed the executive offices of the NBC show *Family Ties*. This suite looked out over a beautiful grassy courtyard and was among the plusher accommodations Paramount offered. It had been Lucille Ball's bungalow in the *I Love Lucy* days. Lucy had her own set of rooms not just because she was a star, but because she and Desi Arnaz were co-owners of the show. CBS had declined complete ownership of the show because they feared WASP America would reject a sitcom with a Cuban husband. To show CBS that America would accept them as a couple, Ball and Arnaz had gone on the road with a nightclub act, playing to grassroots America. The act was a hit. Still, the network was unwilling to bear the show's entire financial risk. So Desi Arnaz and his wife borrowed $5,000 from a bank and became producers. The program, of course, eventually made Desilu into a rich corporation. Desi later sold his part of the show to Lucy, and she resold it a few years later for several times the money she had paid Arnaz.

At the moment, most of the cast members from *Cheers* were on the road, doing grassroots work themselves. *Cheers* had finished seventy-fifth out of ninety-eight programs in the 1982–83 season, just ahead of shows that reeked of cancellation, like *Foot in the Door* and *Small and Frye*. Ted Danson and Nick Colasanto were in Boston, where they were scheduled to appear on *Good Morning America*, and cast members John Ratzenberger and George Wendt, who played the beer-guzzling bar boys, were in Florida to do local TV talk shows and newspaper interviews. *Cheers* had been renewed for the entire 1983 season, but that would be the end of the line unless its ratings picked up. Doing local promotions during the spring hiatus was one of only a few ways that a weak show could strengthen its grip on life. Two years before, many members of the cast of *Hill Street Blues*, then in the ratings cellar, had gone out at their own expense to pump their show. At least NBC was paying for the Florida trip.

Across the on-lot street from where Tony was moving furniture, producer-director Jim Burrows talked about his show. "We're a marginal show at best," he said. "If we get a third year we'll be somewhat over the hump. But if the ratings this year are the same as last, I don't expect there'll be a third year."

Burrows rocked back and forth in a massive, black-leather executive throne, surrounded by the framed announcements of his Emmy victories and a poster from a movie he'd directed, Partners, which had starred Ryan O'Neal. "Both Brandon and Grant love the show," he said. "It's Grant's style. It means a lot to the network. We get lots of mail, lots of public support, which feeds back to the network.

"Cheers," he said, "is the last vestige of sophisticated comedy. If this show can make it, maybe there'll be a resurgence of the comedies of the 1970s, like The Mary Tyler Moore Show and M*A*S*H and Barney Miller. They all started out slowly. We've gone up in the ratings. We started with a fifteen share. Last week we had a twenty-three share. It fluctuates a few points every week. But basically we've had a slow build. Which is nice."

Why?

"Because," he said, "it's better than a slow drop."

In the room across the hall from Burrows's office came the sound of furniture thumping against the walls. It was Tony, now vacating an office that had been occupied by a team of departing story editors, who had revised the scripts that had been assigned to outside writers. The story editors, Ken Levine and David Isaacs, were leaving to become producers of AfterMASH, the M*A*S*H spin-off being developed at Twentieth Century-Fox. Levine and Isaacs's move was dicey, but they'd been offered partial ownership of AfterMASH, which meant they'd cash in if the program beat all the odds and lasted long enough to be sold as a syndicated rerun. Leaving a scheduled show for a pilot was a gamble, but M*A*S*H had made over $250 million in syndication, enough money, obviously, to be worth gambling for.

When Tony did manual labor, he always tried to write in his head. He concocted story ideas and jokes. Tony was a writer now. A professional writer. He had already written a program that had starred Vincent Price. The show was produced by a company that created first-run syndicated shows for cable TV. The subject was cooking. The Vincent Price vehicle was on casseroles. Tony was paid $50.

And he had a partner now, too. He was a member of a writing team. His partner wrote questions for *Family Feud*, and Tony helped him. Therefore, Tony's body of work now included: "When a wife gets mad at her husband, what's the first thing she throws at him?" The question had come naturally to Tony. His wife was not overjoyed with the 600 percent decrease in salary Tony had taken in order to work in TV, or with his being at work sometimes until three in the morning, as he waited to see if any of the writers or producers needed any errands or snacks while they worked late. Tony was now *writing from his own experience*—he was *using his pain*, as a real writer should.

After Tony had moved the last piece of furniture out of the story editors' office, he was called into the office of Richard Villarino, the manager of the *Cheers* production company. "Tony," said Villarino, "you've been here a year, and it's time for a raise." Tony felt a surge of excitement. "How does twenty-five dollars more a week sound?"

Tony felt secure enough in his position to say, "Lousy." Nevertheless, he smiled when he said it.

3
New Blood

"So Corky bombed at the Helm, huh? Ain't comedy a bitch?" Actor Tim Thomerson ripped off his T-shirt, split the fabric right down the middle, and said, "Screw the prop department." He was standing around inside his beat-up trailer at Disney Studios in Burbank, getting ready to eat lunch. Thomerson and Corky Hubbert had known each other for years, ever since they had auditioned for the same part on the ABC sitcom _The Associates_, and Thomerson was one of Hubbert's "heroes," an honor bestowed upon the tall, blond, grizzly, and vaguely maniacal young actor on the grounds that he was (a) a genius, (b) a great guy, and (c) a monied guy, due to his successful transition from stand-up comedy to television.

Corky had done his stand-up routine at a club called the Helm in Santa Monica the night before. This following appearances at the Improv, as well as at the Hollywood Comedy Store, the other top television-audition spot for comics, where Corky had previously impressed owner Mitzi Shore enough to have his name painted on the signature-graffitied wall outside. He'd wowed them at the Comedy Store. It was a college audience and he'd hit them with his special gourmet selection of eighth-grade anal humor. Q: "How do you make a Mexican Jacuzzi?" A: "Eat three enchiladas and sit in your bathtub." But the Helm had been a disaster. Corky had hopes that maybe—just maybe—some network or cable guy might be in the audience, as they often are at the Helm, and that this guy might laugh his ass off, buy Corky a drink, and ask him to sign on the dotted line. But by the time Corky got on, there were only a few people left in the room, including a drunk blond Englishwoman who kept yelling things like "Why do you keep changing? Why don't you be what you used to be?" and "You're wrong! Wrong! Wrong!" The gibberish threw his routine completely off track. Corky began to sweat. He tried to banter the woman to silence—"Lady, you obviously have a big mind, about as big as an empty fucking gymnasium"—but it didn't work. Afterward, he stepped off the stage like a man lost in a maze. His arthritis was acting up again; it felt as if pieces of sand were inside his joints. His shirt was soaking wet and his tie hung around his neck like a plaid hangman's noose. A smile was rigid on his face. He had just enough coins left for a single beer at the bar. "I don't know why, man," he rasped, "but that was just about the worst experience in my whole life."

"Sorry to hear he ate it," Thomerson said, "but that stand-up stuff is lunacy, man." Thomerson should know. Winner of the first Lenny Bruce Comedy Award and labeled "one of the strongest comedy talents I've ever seen" by Improv owner Budd Friedman, Thomerson described his defection to TV as a necessity as he picked his way across the lot on his way to lunch. Like most Hollywood television studios, Disney Burbank is built to look like a combination of Oz, a factory, and a college campus. The *Apple Dumpling* set was adjacent to the set for the NBC series remake of *Casablanca,* and Thomerson strode through a throng of actors dressed in the garb of fifteen decades and several continents and through a litter of cameras, sound gear, and the white flaglike scrims that are used to direct stage lights. He recalled his first crack at TV on *The Merv Griffin Show* at Caesars Palace in Las Vegas. "I fly out. There are tits and diamonds everywhere. And all I got on is a pair of blue jeans and this pimpy shirt." He pulled at the lapel of the shirt he was wearing. "Worse, a lot of my material is pretty blue, so I've got to edit it way down.

So I did my karate beer can number, where I bash a beer can against my forehead at the end. Usually I used one of those simpy aluminum Coors-type cans, but they told me to use a disguised pop can for some reason, so I got stuck with solid steel! I go out on stage—this is like my first time on TV—and all I see is cameras and a sea of faces. Everything's copacetic 'til I get to the head-bashing part. I slam the thing against my forehead and pow! The whole Crab Nebula is right before my very eyes and I'm *this far*"—he raised two fingers—"from hitting the floor. I'm screaming to myself, 'Brace up! Brace up! *You're nationwide!*' So I stagger out and go home feeling like a total zero. But the next thing you know, the Merv guys are back on the phone saying, 'We love the part where you knock yourself out. When can you come back?' "

Now Thomerson was a contented nine-to-fiver on Disney Studios' *Apple Dumpling Gang*. "This *Apple Dumpling* thing is no heaven on earth, but my wife and I just adopted a baby and I want to have a more normal life. Working network TV isn't all that exciting—but sometimes . . . like, there was this line in a *Dumpling* script: 'Whaddaya guys do when you get bored around here?' And the reply was 'Usually we get some bear grease and put it all over the barnyard critters.' That image! It drove me crazy. So TV has, like, you know, its compensations."

Compensations that, right now, Corky Hubbert would have dearly enjoyed. So far he hadn't scored any pilot parts. *The Amazing Adventures of Sparky O'Connor* had bombed at ABC. He and Sacks had gone in to meet with the ABC programming guys and pitched them about what a funny, positive, human show it would be to have Corky play Sparky, the newsboy "from Big Town, U.S.A." And about who his friends would be, like the couple from the ma and pa grocery store down the street. Instead, one of the ABC guys had suggested his friends be Bunny and Bambi from the massage parlor down the street. ABC had scheduled a second meeting, but then kissed it off. Now it turned out to be just a hair too late to pick up anything else and, try as they might, neither his agent nor his manager had been able to turn up anything in TV that paid real American money. So Cork had decided to beat the streets a little himself. Most actors have glossy photos of themselves to hand out to casting agents but—

fresh out—Corky had circumvented the process and saved money by sticking his face into a Xerox machine and having his image recorded New Wave. But the flash had been a killer.

He was seeing nothing but little Hiroshimas—his eyes tearing so badly it looked like he was bawling like a baby every time he walked into a casting office. It hadn't worked out. And money was getting to be a real problem. For the past six months he'd been practically broke. He'd given up his apartment to cut back on expenses, and recently he'd been America's guest, camping out on sofas, apartment sitting, and depending on friends like the chef at Ma Maison who, unfortunately, had just strangled his girlfriend, Dominique Dunne, and was no longer in a very good position to slip Corky leftovers of *haute cuisine*—encamped as he was in the Los Angeles County Jail.

It was time to look up his old buddies. Like John Matuszak. Just before Corky'd left town in the fall he'd gone to see Matuszak in the hospital, where the former Raider lineman was recuperating from a severe back injury. Corky gave him a beer that he'd snuck up in his knapsack. It went down in about three gulps. "What's up, Cork?"

"I don't know, man." Corky sat down on a chair at the end of Matuszak's bed. "Midget consciousness in this town really sucks. The only thing I've got going now is residuals off a garage door ad, where I play one of Santa's elves. It's been regionalized, so I don't get half as much money. I'm beginning to worry about my Q."

"Your what?" Matuszak moved to sit up higher, then winced; he was partly in traction.

"It's a rating system for television performers," Corky explained, shrugging off his backpack. "Kind of like a survey of public awareness. Do people know who you are, and do they love you. What it means to advertisers is, would you rather have Burt Reynolds sell your Ajax foaming cleanser, or some nowhere dwarf? The network guys live and breathe it."

Matuszak smiled. "Hey, it can't be that bad."

Corky nodded. "You shoulda got the Bluto part in *Popeye*. Then it woulda been a smash."

"I was otherwise employed," Matuszak replied, looking up at a TV screen above his head, where the Raiders were manhandling the Atlanta Falcons.

"I gotta get some work. Remember the time Ringo bought all those whips?"

"Those were the golden days of our youth, Cork. Don't worry, though, you're gonna be a star. The portable Belushi, right?"

"Right."

Corky left the hospital grateful for the encouragement and glad to have Matuszak as a pal. He headed back to the house he'd spent the night at, purchasing an uncut roasting hen on the way. With him, he carried his net worth. He traveled light. Almost his entire estate fit into his knapsack. Three or four shirts, a sweater, two pair of pants, a few socks, a tape of *Under the Rainbow*, and a copy of *Playboy*.

He took a shower at his friend's house while the chicken thawed. On his way out of the bathroom he heard a tremendous commotion next door. Screaming. Something slamming against the wall. Corky looked out the window. "A lovers' quarrel," he said, pulling on an Izod golf shirt that fit him like a muumuu. "L.A.'s a poison place. Even the junkies are no fun to watch. Most of them don't even have enough class to shoot real smack." At the sound of water splashing, he looked out the window again. "They've made up and are hosing each other off."

It had been a long day, and it was only noon. He'd spent his last $5 on beer and the chicken. And now, as he rummaged through the kitchen of the house's absent proprietor, comedian Vic Dunlop, it was obvious there was nothing to prepare the chicken with. All Dunlop had in his refrigerator were some radishes and a jar labeled PICKLED MUDFISH. Since the chicken was already half-cooked in its pot, Corky decided to take the whole thing across town and split the bird with his friend Michael in exchange for some vegetables. So he took a crosstown bus, translucent drools of chicken broth coating the side of the stewpot in his lap. But after he got to Michael's his lunch was ruined when Michael dumped half a bottle of Mexi-Pep hot sauce into the pot. Corky's stomach was tender as a wound, so the only thing he could eat was half of Michael's can of string beans.

He headed for a bank, with a roll of pennies sagging in his pocket. The pennies were literally the last of his earnings from *Under the Rainbow*. But when he got there, the teller wouldn't take them.

"Why not?" he asked. "You know me. Remember? When I

was in here last, you said I looked just like Bob Dylan. Remember?"

"What if you're not the same one?"

"How many midgets in Hollywood look like Bob Dylan?"

"We can't take pennies unless we've rolled them ourselves," she said. "I'm sorry, but those are the rules."

Hubbert went back outside. A guy in a purple suit collared him. "You're the little dude from *Fall Guy*, right?" He asked for an autograph.

"You wouldn't happen to have two bits, would you?"

The guy's gaze congealed. "Huh?"

Some time later Corky sat in a pizza parlor in Beverly Hills, sipping dinner. A Coke. Things weren't so bad, really, he ruminated. He'd had a good time that spring up in Anchorage, refereeing ladies' mud wrestling with his pal Vic Dunlop. "Some of the mud ladies were graaa-rate!" His mouth expanded widely, a grin. "I only knew a couple that were outright whores. One of them I liked especially. I wanted to ask her out except that I found out that she was married and that her husband just got his hand ripped off in a construction accident. That kind of took away my enthusiasm. Still, Alaska's a great party town, man." But, things seemed to be passing him by. For example, his pal Dunlop already had a bunch of TV gigs lined up and the LaughCo company had even put out a Vic Dunlop Comedy Kit, complete with nerd glasses, lover lips, dog ears, alien ears, third eye, and heckler handler. Was there ever going to be a Corky Hubbert Comedy Kit? Not at this rate. And the Alaska thing—mud wrestling just wasn't as hip as it used to be. He had to start getting some better gigs. "After I was in *Caveman* and *Rainbow*," he said, "everybody started telling me that I was going to be this big star, and I let it go to my head." The thought plummeted him. Back in Portland, Corky used to write a regular magazine feature: *Corky Hubbert's Conspiracy Digest: Your Worst Fears Confirmed.* He'd broken some hot stories: Howard Hughes kidnapped by Albanian frogmen, Charlie Manson framed, JFK's brain kept alive in saline solution on the isle of Corfu. Corky's column reported all the angles and—through it—he had learned a valuable lesson: In show biz, all things are possible.

But unemployment was weighing on him. He walked out of

the pizza parlor. Porsches, Cadillacs, and Rollses whizzed by on Wilshire. A moment later a bus winged past, ignoring his wave to stop. Corky blew up, chasing after the thing, trying to get its license number. "I'll have your job!" he screamed, a furious midget in the trenches, trying to make it in the post-Belushi world, both fuck-you fingers stabbing the air as he ran like mad. Completely unaware that he was about to score the best job in television.

It was the time of opportunity. Right now, *anything could happen*.

In the television industry, the pilot season is the time of rebirth and renewal, a spring in the perpetual summertime of Los Angeles, when the networks tap proposed pilots for production. These pilots are made over the next few weeks, and are then, during one sudden, dramatic week, purchased or passed over by the networks.

If they are bought, their producers immediately begin to prepare scripts and finalize their casts and crews. In June, the new programs are paraded, like debutantes, before the networks' affiliate stations, and to the nation's press, in orchestrated promotional extravaganzas. In July, the new shows begin production, and in September they appear as parts of the "Most Exciting New Fall Lineup Ever!"

Even before they appear, though, replacements for most of them have been planned, and some have been purchased. The most successful of the new pilots, most of which are now only bought for thirteen original episodes, will be picked up in November or December for the "back nine" shows, allowing them a full season of twenty-two programs in which to make their mark. In December and January, the networks decide which of the new scripts that they have been looking at over the past several months deserve to be produced as pilots. In February and March, the pilot producers prepare to shoot their shows. And then in April, it begins again—the inauguration of a new television season, the time when a television career can suddenly rocket from nowhere to stellar heights.

The 1983 pilot season was less active than some previous seasons had been, but still represented an expenditure over $100 million—enough money to make as many as ten major motion pictures. A total of 76 potential series were being developed by the three networks, down from 96 the previous year, and considerably below the 107 that had been developed during the most active pilot season, in 1979. The networks simply did not have the money to spend on development that they'd had in the years before cable and other nonnetwork threats had begun to erode their audiences.

NBC was developing the most pilots of the three networks; with thirty-three projects in the hopper, it was responsible for almost half of all the season's pilots. The main reason it was so busy was simply because it had the most holes to fill. NBC was putting equal emphasis on comedy and drama; each category accounted for 42 percent of its pilots. And NBC was the only network to actively pursue development of variety shows, a category that had almost evaporated from television.

CBS was working on the fewest pilots, chiefly because the network didn't want to tamper too much with a schedule that was in the process of winning the 1982–83 ratings race, but also because, some critics were saying, the stodgy, conservative CBS had run out of novel programming ideas, and was having too many financial problems to be willing to spend money on a new programming team. Still, CBS was looking for a replacement for the departing M*A*S*H and was trying to find programming that would blunt the World Series and the Olympics, which would be on ABC. Archie Bunker's Place was also showing wear and tear, and if the network decided to cancel it, it might also need to replace Gloria, the spin-off show that followed Archie.

ABC's pilot season was being dominated by typical ABC fare—low-IQ comedies and car chase dramas. ABC was looking to replace Tales of the Gold Monkey, a Raiders of the Lost Ark rip-off, and the comedies Joanie Loves Chachi, The New Odd Couple, and It Takes Two. Even Happy Days, which had been slapped around by The A-Team, was no sure bet to survive for its eleventh season. The only sacrosanct series for ABC were Dynasty, Three's Company, The Love Boat, The Fall Guy, and Fantasy Island.

In all, it looked as if there was room on television for twenty to twenty-five new shows. Room for a little new blood.

*   *   *

A couple of months earlier, in February, a *Cheers* episode written around Harry Anderson's con man character had aired and suddenly, it seemed, the entertainment world wanted to know where Harry had been all its life. So while the recorded message from Danny Robinson that Harry had gotten upon his return from taping *SNL*—the one promising stardom—was exciting, it was not entirely unexpected. Clubs, Vegas, John Davidson, *Saturday Night Live,* and finally *Cheers* had been like pilgrims' shrines on Harry's long climb to the Fujiyama of primetime stardom. Harry was ready. With just a bit more luck, he and his family would soon be enjoying a life like that of most middle-class Americans. The only difference was that, in order to have it, he'd have to become famous.

Climbing with him up the slope was Danny, the good agent, a rare and valuable commodity, easily worth his 10 percent of the gross. The town was full of talented youngsters struggling for even a short interview with an agent of Danny's class. Mostly they were told to come back when they found work.

Danny Robinson worked for APA, which was the third-largest talent agency in the country, though a distant third, behind the mighty William Morris Agency and the mammoth International Creative Management, or ICM. APA, with sixteen agents in its plush Beverly Hills offices, located in a high rise on Sunset Strip, and another seven agents in New York, was respected because it was able to keep a stable roster of clients in a business where clients tend to change agencies as an automatic response to any drift, perceived or real, in their careers. Steve Martin, for example, had been with APA for twelve years, and Liberace had been a steadfast client for twenty years. The agency had enough clout to capture the services of stars as well established as Mary Tyler Moore, Tony Bennett, James Coco, Martin Mull, and Anne Murray.

Though an APA receptionist would quite likely tell a hungry young actor that the agency was "not accepting any new clients at this moment," the standard line that preceded the perfunctory acceptance of an actor's résumé and eight-by-ten glossy, the APA agents did keep their eyes open for unclaimed talent. Danny Robinson, an amicable young man who had worked his way up from the mailroom, first saw Harry when Harry was doing a "middle act" in Las Vegas. The middle act, a seven- to ten-minute time-killing spot between the two name acts, which is needed so that set changes can be made, is a good break-in spot for young comedians, who often do the middle act for musical stars. Harry was middling for Doc Severinsen and Debbie Reynolds, and Danny was in town to watch APA client Steve Martin at a neighboring casino.

When he chanced upon Harry's act, Danny was struck by Harry's hip humor, and the way that the audience instantly warmed to him. Harry, Danny thought, could convey mischief without malice, a trait audiences love. Harry, Danny thought, could pistol-whip your mom and still make you like him.

Danny immediately urged APA to sign Harry on, which to the company meant a major commitment of time and money, particularly in the early stages of a young performer's career. There was some resistance—did the world need another comic magician? But Danny reminded them that when, nine years before, APA agent Bob Klein had proposed Steve Martin, other agents had said, come on, Bob, do we really have room for a guy that does balloon tricks—not to mention the arrow-through-the-head gag?

Danny prevailed and began getting Harry bookings in slightly better clubs, and on as many TV shows as possible.

Harry had to leave ICM to go with APA, but it wasn't much of a sacrifice. He'd been put on the rolls of ICM mostly because ICM represented Kenny Rogers, for whom he frequently opened or middled. But the giant ICM machine had never shown any eagerness to put its immense power behind this incidental name on its list.

Harry, with Danny's help, began to do a string of daytime talk shows, which paid negligible appearance fees, but which began to slowly nudge up Harry's Q, his "likability and recognizability quotient," which is rated by Marketing Evaluations, Inc., and is monitored closely by TV networks, film producers, and nightclub booking agents. The only concrete benefit Harry achieved from the TV appearances, though, was a slight rise in his nightclub fee.

But then came *Saturday Night Live*. Danny helped get Harry

on *SNL* by bringing Dick Ebersol to Budd Friedman's Improv, shortly after Ebersol took over *SNL*. *SNL* was at the make-or-break stage, and Ebersol was famished for specialty acts, like Harry's magic, or Andy Kaufman's impersonation of Mighty Mouse.

At the end of his debut *SNL* show, someone remarked to Harry that the show would be a good stepping-stone in his career. He was shocked. Stepping-stone! To him, *Saturday Night Live* was *it*. The absolute summit of cool show biz, a stratosphere away from the Vegas side lounges and the Pittsburgh toilets he'd been playing.

In 1982, however, *Cheers* called and, once again, Harry found himself in a whole new ball game. Danny sent an *SNL* tape to *Cheers* producers Jim Burrows and Glen and Les Charles. Burrows had already seen Harry on *SNL*, and had also recently gotten a phone call about him from Joel Thurm, the head of casting at NBC, a man Burrows had known and trusted ever since they'd both worked for Broadway producer David Merrick twenty years earlier.

"You guys want your first spin-off show?" Thurm had said.

"Maybe."

"If you do, write a part for a guy named Harry Anderson." Thurm told Burrows that he'd seen Harry at a local club, the Magic Castle, and that Harry was a winner, though whether or not Harry could act was unknown.

Burrows and the Charles brothers met with Harry and offered to write him in as a magician who visits the bar. Harry had a better idea. Why not use him as a con man, since that type of character would be more organic to a bar situation?

Thus was born Harry the Con, a character Harry played three times during *Cheers*'s first season. Harry the Con was always fleecing the customers with card tricks, and being thrown out by bartender Ted Danson, but the people at the bar could not bring themselves to actually dislike the guy. In the final show that Harry did for the series, he ended up saving Sam from losing the bar to an out-of-town hustler, by tricking the other con artist. The sting used to do it was written by Harry. Danny noted, "We suddenly had old friends all over town." ABC and CBS sent scripts to Danny for Harry to read.

In early March, Harry auditioned for the show that seemed

to be best suited to him, a proposed series called *Empire,* a satire of *Dallas*-type shows. The reading had been, Harry was later forced to admit, inept. Painfully inept. He had serious doubts that he was really an actor, doubts he expressed to NBC's Joel Thurm. He was quite capable, he felt, of presenting a persona in his magic act, and of tailoring that persona just enough to get by with something like Harry the Con on *Cheers.* But he was not at all sure he could ever do anything other than a rendition of his stage persona—cool, clever, and mocking.

Joel Thurm did not tell Harry that his fears were groundless. It was quite possible, said Thurm, who had been forced to turn thumbs down on literally thousands of actors, that all Harry would ever be was a nightclub magician. If the disastrous reading for the *Empire* project was any indication . . . well, at least, as a failed actor, he wouldn't be in the minority.

Before Harry even unpacked his bags from the New York trip, he took all the messages off his answering machine and put one of his own on Danny's—communication, L.A. style, 1983. The next morning, the phone rang. "Yo," Harry answered.

"I've got a reading for you." It was Danny. "For a pilot called *Night Court.* It's being done by one of the guys that did *Barney Miller,* and the star role was *made* for you."

Later that day, the script was sent over to Harry's house. Harry went through it once, increasingly thunderstruck as he read. The character, written by a man he'd never met, was flip, hip, and winning. He did card tricks. He answered the phone by saying, "Yo." Some of his friends were con artists. His name was Harry.

Harry, who had always had trouble memorizing, put the script down and found that he could quote entire speeches. He and Leslie stayed up all night, rereading the script in their large, colorful front room dominated by a pool table, a blackjack table, a dart board, a black replica of the Maltese Falcon, and assorted masks, props, games, and collectibles. By the time L.A.'s skyline was visible from his window, Harry had memorized the script.

He drove a few minutes on the freeway to the half-mile-square acreage of the Burbank Studios, the home of Warner Brothers TV, and waited at the studio's drive-in gate, his car idling next to an oversized chalkboard that rates each day's smog level. Harry was cleared to go to the offices of Reinhold Weege, the creator of *Night Court.*

Harry sat in an anteroom, a script in his hand, fidgeting, as actor Jeff Goldblum read for the part Harry wanted. The starring role was that of a judge, appointed on the last hour of the last day in office of a governor who bore an uncanny resemblance to Governor Moonbeam himself, Jerry Brown of California, for the particular reason that he was the only person on the list who was home. As an arbiter, the judge was more enamored of the law's spirit than its letter, enough so, Weege hoped, to make for a half hour of network comedy each week.

Harry cooled his heels until Goldblum, a very respected actor who starred in *Terms of Endearment,* was finished. Other proven talents had also been considered for the part, including Barry Bostwick, who would soon play George Washington in a miniseries, comedian Robert Klein, and William Devane, the actor who had been edged for the lead in *Cheers* by Ted Danson.

Harry shook hands with Weege, a heavyset thirty-eight-year-old, and with Sharon Himes, Weege's chief casting executive for the project. Then Harry exclaimed, "I *am* this character."

Reinhold Weege maintained a polite demeanor, but he felt as if his eyes were rolling back in his head. "Oh, my God," he thought, "another asshole actor telling me he *is* the character. Okay, yeah, fine, let's read and be done with it."

"Whaddaya wanna read?" Weege asked Harry.

"I've memorized the whole script," said Harry. "Pick a scene."

And then Harry began to read, and Reinhold Weege felt a physical sense of amazement, an actual feeling in his body of profound surprise. Because Harry Anderson *was* the character.

Harry read for several minutes, while Weege and Himes stole glances at one another.

Then he stopped, all of a sudden, and stood up, in order to make his way toward the door. He felt slightly sick to his stomach, because he was positive he had failed miserably and had screwed up not just the reading, but maybe his last, best shot at being taken seriously as an actor. He'd blown his chance to get off the road and into a cherished day job.

"Hold on!" Weege exclaimed to the retreating Harry. "Go get something to eat while I see the next guy, and then come back. Okay? Whaddaya say?"

Harry said sure.

Over the next few days, Harry worked on the script with Weege and Jim Burrows, whom Weege had brought over during *Cheers*'s off season to direct the *Night Court* pilot. A number of creative people whose shows were already on the network schedules were freelancing on pilots. The pilot production season offered a good opportunity for these people to make extra money. Pilot work always pays more than regular series work, usually about twice as much, on the general principle that a pilot episode is a series's most crucial episode, and therefore requires the best possible work from the best possible people.

Burrows was especially valuable for the *Night Court* project because of his special ability in working with actors. With his low-key, calm-as-concrete demeanor, Burrows was an ideal coach for a neophyte like Harry. Burrows's ability to make actors feel comfortable was one of the reasons Grant Tinker first entrusted him with the responsibility of directing MTM programs, when Tinker initially sponsored Burrows's sojourn in Hollywood. The ebullient Anderson and the quiet Burrows hit it off immediately, with Burrows convinced from almost the outset that Harry not only could act, but could possibly even become a major TV star.

The NBC and the Warner Brothers executives, however, were not so unabashedly enthusiastic—in part because it was their money that would, if Harry got the part, finance this little experiment in creative casting. Of the approximately $750,000 that it would take to produce the pilot of *Night Court*, the network was in for roughly $640,000. It could make back much of that sum from advertising revenues, if the program was to make it onto the air even as a one-shot. But many pilot programs end up usable only as paperweights.

Reinie Weege, however, thought that *Night Court* was in a great position to become the *Cheers* of 1983, beloved by the critics, a slow-growing hit with a loyal audience of discerning viewers. He thought that one of the show's best possibilities for a scheduling slot would be directly before *Cheers*, at 9 P.M. on Thursdays.

Harry did so well in his rehearsals that eventually the choice for the role boiled down to just him and Barry Bostwick, although a few others were still officially in the running. A read-off was scheduled. The rehearsal hall at NBC was thick with studio and network executives and the reading stretched into the night, with Jim Burrows directing.

When the actors were finished, the executives, plus Burrows, Weege, and Sharon Himes, repaired to a conference room. Most of the talk centered on Harry. There were two major contentions about why he might not be right: (1) He wasn't old enough to be believable as a judge, and (2) he wasn't a seasoned enough actor to carry a show. Weege and Burrows argued in favor of him. So did NBC casting director Thurm. And so, in the end, did Brandon Tartikoff.

"I wanna pitch," said Tony Colvin, pacing his narrow, barren cubicle at *Cheers* like a lifer in a punishment cell.

"We're not ready to pitch," said his partner.

"Speak for youself, man. I've got some primo ideas and I want to run them by Les and Glen. They won't be very busy until Jimmy gets back from *Night Court*."

"Let's just do more *Family Feud* questions," said Scott Gorden, wadding a piece of paper and arcing it into a metal wastebasket. Scott, a producer's assistant and Tony's writing partner for the past few months, liked basketball far more than writing. He'd learned the game on the playgrounds of the black neighborhood that he'd grown up in, and it had paid his way through college. Now he played on the black-dominated playgrounds around Paramount. In many ways, Scott, who was white, felt more at home around black people than Tony did. Scott had grown up poor and now believed in the kind of social welfare and affirmative action programs that Tony loathed. Tony, thought Scott, was such a . . . WASP. Tony was into status, and career advancement, and building investment equity, and all those other uptight, Gerald Ford things that Scott got heartburn just thinking about. Scott's idea of a proper long-term goal was to make a pile of dough and retire while he still had the knees to rebound.

Tony wouldn't even play basketball with Scott, even though, of the two of them, he'd been the major-college jock. Tony hated to be reminded of his athletic prowess, just as he resented being categorized as black. As far as he was concerned, *that* Tony was a case of mistaken identity.

"Screw the *Feud*," said Tony. "We should be out pitching story ideas all over town. Now that the pilot season's starting, everybody's running around like chickens with their heads off. Now's the time for us to *insert* ourselves somewhere." Tony fixed Scott with a gripping stare and wouldn't let go.

Scott sighed. Tony was such a steamroller. "All right, what stories do you want to pitch?"

Tony bent over his desk and slapped open a plastic spiral notebook. "Get this. Carla wants to go out with a high-class dude, so she hires Diane to tell her how to act. *Pygmalion!* Only it's with two women who make each other puke. It's got conflict up the wazoo."

"What else?"

"An old fisherman comes in the bar and tells a ghost story. The bar's haunted. Everybody freaks out."

"Go on."

"The Red Sox need Sam to pitch again. All their relievers are injured, and they know he's off the booze now. He's gotta choose between baseball and the bar." Tony paused and lasered Scott with his gaze. "Those ideas are worth *money*." Scott didn't respond.

Tony opened the office's door and a flood of breeze washed into the cramped space. From the hallway, Tony could hear one of the writers talking. He was saying, "I was trying to do a *Movie of the Week* for ABC. It was just after *Star Wars* came out. Every idea I had, they said, 'Great! But do it like *Star Wars*!' A family gets shipwrecked? 'Great! But do it like *Star Wars*!' An old man wins the lottery? 'Great! But do it like *Star Wars*!' A little girl gets leukemia? 'Great! But do it like *Star Wars*!' "

A secretary walked by with a sheet of ratings for the past week, and Tony intercepted her. Last week, *Cheers* had finished forty-fourth, ahead of only twelve other regular series. The top-rated show for the week had been a brain-numbing ABC special called *Life's Most Embarrassing Moments*. The forty-fourth-place finish made Tony feel a little queasy, because it represented a decline; the previous week Cheers had been thirty-seventh, and thirty-sixth a couple of weeks before that. The one thing Cheers could not afford to do at this stage was to drop. A large part of the reason the show had been renewed for 1983 was because it had not dropped. There had been nothing to convince Brandon Tartikoff that viewers had sampled the show and then left it. Staying near the bottom for most of the year had been safer for the show than climbing and then falling.

But a couple of months ago, the show had begun to get its viewer sampling, and thus could not afford to sink now.

The sampling had been, to an appreciable degree, engineered by Cheers's casting director, whose office sat directly underneath Tony's. Getting a sampling is a trick that TV producers have down pat. It is usually done by offering a "high-concept show," an episode with an obvious, highly promotable "hook," such as (1) a young wife has octuplets, so her mother-in-law moves in, or (2) a stockbroker's mail-order wife turns out to be a chimpanzee (what'll happen when he brings the boss home for dinner?). High-concept episodes are great for TV Guide blurbs, for intranetwork ten-second promos, for word-of-mouth advertising, and, it must be acknowledged, for the less than discerning cultural tastes of many viewers. Even a basically "low-concept" series, such as Cheers, can be manipulated to garner a sampling if it features at least one big name in the cast, or preferably several. Lacking that, it can gain a sampling by clinging to a primal theme, such as sexuality, violence, or greed. It can also gain its initial look-see by being scheduled just after a big hit, like The A-Team, or lastly, by engaging in a plethora of "stunting," such as having the First Lady do a guest shot, filming a segment in China, having a female lead give birth, or, best of all worlds, having the First Lady give birth in China.

Burrows and the Charles brothers were unwilling to compromise their show by injecting audience-grabbing stupidity into it, but they didn't mind doing some stunting with guest stars. So when Cheers's casting director, Steven Kolzak, had a chance to land U.S. Speaker of the House Tip O'Neill as a guest star, the producers jumped at the opportunity.

Kolzak's mother, it seemed, had been the Boston congressman's executive secretary for fifteen years and had mentioned to Kolzak that O'Neill was a fan of the show, partly, of course, because it was set in Boston. "Imagine me down at the end of the bar," O'Neill had once said to Kolzak's mother, "having a beer with that fat guy."

When Kolzak heard that O'Neill was in Palm Springs for the Bob Hope Desert Classic golf tournament, he got on the phone. By 7 P.M. on a Saturday night, he had gotten through to O'Neill. Would the Speaker consider shooting an episode of the show the following Monday?

O'Neill said he wasn't sure. He did know, however, that he wanted to be back in Palm Springs by 2 P.M. Monday to play golf with his wife. Kolzak assured O'Neill that that would be no problem, and O'Neill agreed to do it. In fact, Kolzak hadn't even checked with Burrows yet, and no scene had been written.

But Burrows loved the idea, two scenes were quickly written so that O'Neill would have a choice, the congressman did his work in one take, and he was back on a plane to Palm Springs by lunch time.

The episode aired in early spring, riding a tidal wave of hype organized by NBC publicist Margo Zinberg, which included mentions in *TV Guide, Newsweek, People,* and many local newspapers. The episode was the highest-rated *Cheers* in many weeks, and kicked off the flourish of sampling that the show needed in order to have a chance for third-year renewal.

Now, however, according to the ratings sheet that Tony Colvin was looking at, the sampling audience was ebbing. It was bad news for Tony Colvin, Jim Burrows, and Brandon Tartikoff, to name only three.

Back in the early 1970s, curious if porn star Johnny Wadd was actually the guy who played unctuous Eddie Haskell on *Leave It to Beaver,* and eager to branch from television into journalism, Brandon Tartikoff got a commission from *New Times* magazine, met Wadd in an L.A. hotel room, and Wadd pitched Tartikoff this way:

"I want you to put my address and phone number in the article. I live at 5466 Santa Monica Boulevard, Hollywood, California 90029. My number is area code 213-555-5161. And I want twelve copies of the magazine when it comes out. I produce, act, direct, do script rewrites, stunts, makeup, camera work, sets for television, and ghostwrite magazine articles using a phony middle name. . . . I tried to sit down one time and figure out how many chicks I've done. Start off with an average of a chick and a half per film, that puts me at over three thousand right there. Then I try to hit at least two orgies a week. And, of course, I do tricks. A lot of swing couples. Movie producers and their wives. I figure I'm up to around six or seven thousand. . . . I have several nightclub acts. Modern dance, modern jazz, nude Othello, Picasso, Egyptian, surrealism, everything. . . . The Japanese government invited me to a pornography symposium. They set me up in a booth in this glass-domed auditorium right next to an eighteen-foot-high poster of myself with a hard-on. They were agape at my cock. . . . I don't do drugs and I don't drink alcohol, except maybe once or twice a year I'll have a glass of wine at a Jewish wedding. I like Jewish people. They're gentle. . . . Ozzie Nelson is a transvestite. He really digs gettin' into garter belts and eyeliner. . . . I'm straight, but I like gay people. They're gentle. Like Jews."

Tartikoff said, "There's a rumor going around that you're the same guy who played Eddie Haskell on the *Leave It to Beaver* TV series."

"We won't get into that. *At all!* You're going to let that ride."

"Is there any truth to it?"

"I don't know. . . ."

At the time, Tartikoff was twenty-five and a rising star at WLS-TV, an ABC affiliate in Chicago, where he had just been named director of advertising and promotion.

Tartikoff had grown up in Long Island and graduated from Yale with honors in 1970. At Yale, he was on the fencing team and the basketball team. He fantasized about a career in pro baseball.

Still in college, he wrote a short novel about one Saliva Schwartz, who pursues a quartet of Siamese twins who are spreading a sex disease across America. After Tartikoff suggested to his senior adviser, Robert Penn Warren, that a short story by Tolstoy would have worked better had one of the characters been a car, Warren suggested to his young charge that he might be more comfortable in another medium: TV.

After graduation, Tartikoff moved west, knocked around Oregon for a while, then took a gofer job at an ABC affiliate station in New Haven, Connecticut.

In 1973, Tartikoff landed a job with WLS in Chicago, where he made a name for himself repackaging beach party and ape pictures, Fred Silverman style. "Gorilla My Dreams" week was followed by "Thrilla Gorilla" week. Then he concocted "A Week of Evil" featuring films like *The Evil Garden, Secrets of Evil,* and *Evel Knievel.* He also wrote and produced a local comedy-variety series, *Graffiti,* and involved himself briefly with Chicago's Second City—the ensemble that developed a major portion of the original cast for NBC's *Saturday Night Live.* It was about this time Tartikoff learned he had cancer. Hodgkin's disease.

Tartikoff began a decade of radiation treatments and chemotherapy. Friends remember his throwing up at night from the side effects of the treatment, only to get up the next morning to play basketball, barely willing to acknowledge a disease that would eventually cost him a quarter of his weight and most of his hair.

Professionally, however, he barely missed a stroke. Modeling his moves after those of Silverman and programmer Marty Starger, Brandon came to Silverman's attention by way of WLS's rocketing "overnights"—the station's day-by-day ratings readout. Silverman, then head of ABC Programming, took Tartikoff on as a spear carrier and protégé. "If you want to get out of Chicago," Tartikoff would say later, "go for a role model."

Tartikoff labored first as manager of dramatic development, then as executive for current dramatic programming. Tartikoff moved to NBC a year before Silverman took over there as president of the network, and became director of comedy programming. By the time he was thirty, he had been named NBC's vice-president of programs, West Coast; and on January 15, 1980, Silverman made Tartikoff president of NBC Entertainment, the youngest division chief in NBC history.

By then Tartikoff's mentor was already in deep trouble, however. It was Tartikoff's chore to oversee one of the classic bombs in television history, *Pink Lady,* a show that was so terrible it seriously damaged the reputation of not only Silverman, but almost everyone who was associated with it.

Pink Lady featured two Japanese girls who were variety-show favorites in Japan, but total unknowns in America. Silverman had become so radically enamored of the show's possibilities that he'd hung a poster of the two young women on his office wall—the joke around 30 Rock was that the poster had been the pilot for the series. The two girls, Mie and Kei, had sold twenty-three million records in Japan, but they barely spoke English. Comedian Jeff Altman—a good friend of Harry Anderson's—was hired to serve as interlocutor and translator, but the show went nowhere.

By this time Silverman had accelerated his practice of moving and canceling shows in a wild chase for a greater viewership. Hill Street Blues, one of Silverman's favorite and eventually most successful shows, was moved into five different time slots during its first twelve weeks. Six months after Tartikoff became president of NBC Entertainment, new RCA chief Thornton Bradshaw showed Silverman the door and named Grant Tinker to take his place.

Though the network was on the skids, the mass firings expected by many of NBC's 146 vice-presidents and 9 executive vice-presidents did not occur. If Tinker felt that Tartikoff should be held partly responsible for Silverman's performance, he didn't let on. "I knew Fred had been calling the shots," he explained later. "Brandon hadn't yet been up to bat. What struck me was his excellent sense of this town. He recognized who the best people were and he was willing to let them do their work without meddling. Too many network programmers try to produce the producers."

Under Tinker, Tartikoff's title remained the same, but his responsibilities expanded. Silverman had turned the NBC executive staff into coat holders, and attempted to make most of the important decisions himself. Tinker, on the other hand, was a master at delegating authority. By April of 1983, Tartikoff was directly accountable for the close to $1 billion worth of programming decisions that NBC would make throughout the year.

It was a difficult job, considering the circumstances. Television programming is truly a business where success breeds success, and failure is inclined to snowball. The reasons are relatively simple. A hit show like *Magnum, P.I.* can carry the show that follows it and can even carry an entire evening, because once people tune their TV sets to a particular channel, they're inclined to leave them there. The power of a show's lead-in was demonstrated with the kiss-'em-or-kill-'em CBS action-adventure series *Simon and Simon*, which had debuted and headed straight for the toilet. *Simon and Simon* failed to find an audience until CBS rescheduled it to appear after *Magnum*, where its ratings climbed until it became CBS's fastest-growing hit. Using anchors like *Magnum*, *The Dukes of Hazzard*, and *60 Minutes* in early-evening time slots, CBS had the luxury of placing new or weak-but-promising shows immediately thereafter in order to let them build. ABC could do the same, employing *The Fall Guy* and *Dynasty*.

NBC, though, had only one Top Ten ratings hit, *The A-Team*—a show that had been Brandon Tartikoff's biggest programming success so far. Two years before, Steve Cannell, creator of the James Garner hit *The Rockford Files* and considered the hottest young independent television producer in Hollywood, had come to Tartikoff's office with an idea for a show revolving around a luxury hotel, but Tartikoff had been mulling over something else, having just been introduced to the phenomenon of Mr. T. He'd met T at a reception following the Larry Holmes–Gerry Cooney heavyweight title bout in Las Vegas. In a room containing James Caan, Farrah Fawcett, Ryan O'Neal, and Sylvester Stallone, T was getting all the attention—this after just one film appearance, as Stallone's fierce opponent in *Rocky III*. "Look," Tartikoff told Cannell, "let's give this guy a shot on *Silver Spoons* and see if he can talk, then . . ." He handed Cannell a piece of paper that read: *"The A-Team, Mission Impossible, The Dirty Dozen,* and *The Magnificent Seven,* all rolled into one, and Mr. T drives the car."

Cannell had gone on to invent a character that was mostly just Mr. T himself: a fearsome, outlaw crime fighter who loved justice and kids. Equal parts King Kong, Matt Dillon, and Holden Caulfield, T had created a ratings bonanza, and *The A-Team* had gone a long way toward anchoring Tuesday nights for NBC. It had broken into the Top Ten for the first time in the third week of March 1983.

Tartikoff felt NBC had several potential hits already on the air, including *Knight Rider* (a show about a talking car and the stud who loved it) and *Cheers*. He was confident that if *Cheers* could pull a 30 share the momentum would give *Hill Street* numbers in the 36 range. He believed that while NBC had to do something about its admittedly disastrous weekend numbers, the network needed only "three tough, but achievable" hit shows for its prime-time schedule to be improved dramatically. After all, at ABC, three new hits, *Charlie's Angels, The Love Boat,* and *Three's Company* had turned around the whole network.

Network analysts were also becoming more optimistic about NBC. For one thing, the industry was facing an upswing; Boston Corp.'s Richard McDonald had predicted an overall 1983 profit gain of 13 percent for the combined networks. The U.S. economy was recovering from recession. In similar circumstances in 1976, combined network profits had risen 24 percent, and another 21 percent in 1977. Coming off its lowest-rated season ever in 1981–82, with an average share of 24, NBC was already ascending while the ABC and CBS ratings were down 4 percent each. NBC had also just shown an increase in profits over the previous year for the first time since 1977, up to $107.9 million from $48.1 million in 1981. However, both ABC and CBS profits remained three times as high.

So far, NBC had been most successful in the major urban markets and with affluent viewers in the highly desirable eighteen- to forty-nine-year-old age group. Viewing by women in this group was up 6 percent overall. By contrast, CBS was down in this bracket 9 percent and ABC 12 percent. Tartikoff attributed this largely to shows like *Hill Street Blues, Family Ties,* and *Cheers.* "Once you know what time period you are making the show for, you can design it. It's not totally a science yet, but you know you are going for a certain sophistication in the writing, you want to appeal to the specific *Cheers* demo group, and you want to add an element of attracting kids and teenagers, because you're opposite *Magnum* on Thursday nights and you know *Magnum* goes old."

NBC had begun to attract, too, a significant share of those viewers disenchanted with pay television and cable, a phenomenon industry analysts credited to the relative sophistication of many NBC shows. "The large crime of network television is that we've turned out too much witless, forgettable programming," assayed Grant Tinker. "Against all the network product our shows are just a little more upscale. You have to pay a little more attention to get the jokes or follow the story." NBC's new motto, as announced at the 1982 affiliates' meeting, was Planning, Patience, and Performance. This was reflected in Tinker's refusal to cancel such shows as *Cheers*, despite their mediocre ratings.

"There are some occasions when we have to have the courage to stick by our convictions. In the past, a lot of good, solid, long-running, and successful shows took a long time to find their audience. With *Cheers* we've made an absolute judgment that television these days doesn't get any better. And we are confident that the audience will come along eventually and make it a good business judgment." Consider, he pointed out, that "*The Mary Tyler Moore Show* began poorly; it tested badly. If it had been a pilot it wouldn't have gotten on. I would rather take my chances with the audience."

With that kind of support from on high, Brandon Tartikoff was guardedly confident. He felt the network's development season was shaping up as the best since 1979 and that the momentum was beginning—finally—to shift solidly in NBC's favor. Now the problem seemed to be simply how to keep it going.

4

The Human Cannonball

April 1983

Above the Los Angeles Ambassador Hotel, the sky was becoming blue in parts. Shadows slid across the white sound trucks, the black limousines parked in the drive, and the trim oblongs of grass in front of the Ambassador's regal portico. Sun brightened its marble steps and warmed the mezzanine where Cover Girl, a big, white, precisely barbered French poodle, sat on a pedestal adjacent to chaos.

"News people" were milling everywhere, one chasing the "queen of England" across the room and down the stairs, waving a microphone, and shouting, "Do you and Chuck fool around? Does Lady Di wear a bra?" The queen ran to the Ambassador's lower level; a gray-suited newsman was hot on her heels, knocking nuns and tourists to the rug and finally tackling the queen as she made for the limousine parked out front. Cameramen sprinted after them, their Porta-Paks shouldered like heavy weapons. Extras and hotel guests—the hired agog and the genuinely agog—watched the action on the monitor Allan Katz had set up at the foot of the stairs.

"Okay, cut," Allan Katz said, hands on hips, looking into the monitors. It was the height of the pilot season, and this was

101

the set of *The National Snoop*, an NBC pilot that satirized grocery store tabloids like the *National Enquirer* and the *Star*. Katz, forty-two, had during the last decade written nearly 150 scripts for prime-time TV, programs like *The Mary Tyler Moore Show* and *All in the Family*, and had written and produced *The Cher Show*, *M*A*S*H*, and *Rhoda*. He had a sprinter's build, a tangle of tan hair, and an arrow nose—a kind of linear baby-bird good looks. Jules Feiffer could have drawn him in six lines. "Okay, folks," he said. "Let's try that again." *Snoop*'s director was sick and this was Katz's first time directing a shoot. He gathered his principals around him. "It's my job to tell you what to do. It's your job to tell me when I'm wrong." Then, both arms suddenly flailing, he shadow-wrestled in front of two beefy actors playing the queen's bodyguards, miming the fracas they were about to repeat.

The double for the queen stood nearby. There were two doubles, actually, each wearing the sort of dowager prom dress favored by her majesty, a rope of glass diamonds around each neck, clutch purses in hand. The camera double was regal, accurately barge-butted, and bore a true resemblance to the real Elizabeth. The stunt double, on the other hand, was roller derby stock.

There was a sound problem, and Katz called for a break. Two young extras in nun's habits giggled, lit cigarettes. Then one stabbed hers out and began doing yoga. Other extras, most posing as members of the press, sat in plush chairs waiting to be called. One—lovely, deer-eyed, swan-necked—sat primly, legs crossed, reading a book titled *How to Act for Television*.

One of the sound men drifted over to Katz who, sucking on a yellow lollipop, was conferring with his creative consultant, Bill Richmond. "Jeez, Allan," the sound guy murmured, nodding toward the blonde extra whose lips were moving as she read the acting book, "where did you find that one? I'd like to . . . kiss her brains out."

"I think somebody may have beaten you to it," Katz replied, turning back to Richmond. Allan Katz was a popular man in the industry, a "potential home-run hitter" in Brandon Tartikoff's words, who was known for his unanchored imagination and ability to innovate.

For Katz, time was now an issue. *Snoop* had been licensed by Tartikoff only two weeks ago in a contract through Twentieth Century-Fox, commissioning Katz to create a twenty-four-minute videotape in exchange for $707,000. The fee would cover about 85 percent of *Snoop*'s actual cost, the rest to be picked up by the studio. It was a sizable portion of NBC's huge $30 million pilot push. The network was badly in need of hits. Since the 1975–76 season, NBC had failed to place more than three prime-time shows at any one time in the Top Twenty-five. In fact, of *all* NBC shows combined, only *Sanford and Son, Chico and the Man, Little House on the Prairie, CHiPS, Hill Street, Diff'rent Strokes,* and *A-Team* had made it into the Top Twenty-five network annual ratings in the last eight years. By contrast, under Tinker's administration at MTM, *The Mary Tyler Moore Show* and its spin-offs, *Rhoda* and *Phyllis,* had all placed in the Top Ten for rival CBS. This from a small production company that often employed less than two hundred people.

Tinker's mandate from RCA was to create new shows that would make NBC, if not a dominant force in the ratings war, at least a plausible one. There were myriad costs. Director and cast fees, story and other rights, continuity and treatment, stunts, travel and living expenses, production fees and art direction, set construction and striking, special effects, wardrobe, makeup and hairdressing, electrical rigging and operation, sound men, location costs, insurance, studio overhead, and fringe benefits. Plus editing, music, and last, but hardly least, the producer's cut.

For creating and producing *Snoop,* Allan Katz was to receive approximately $20,000 per episode if it went to series. But that was just the loose change. Should *Snoop* be scheduled by NBC in the fall and become a hit, Katz's profit participation could become colossal. Danny Arnold, the creator of *Barney Miller,* for example, cashed out of his show when it went into syndication for $17 million. Network television offers a venue to cash that is as startling as death. The problem is, the successful-producers' club is a very exclusive one. There are only about sixty series being aired at any one time by all three networks combined. Of those, only about a third are new programming. And of that new programming, no more than half will last a full season. Which means that, during any given year, no more than about ten new productions will get a shot at the veritable cornucopia that is syndication.

Still, there is a fair amount of money to be spread around just while a pilot is taking its shot. The outlays for *Snoop* were average. Creation of main title music—roughly sixty seconds in duration—came in at $12,000, about $200 per second. This included composition fees, musicians, and engineering. Main titles were budgeted at $14,000, wardrobe at $12,000, ground transport at $8,000. Television production costs are dictated by necessity and the unions. Technicians and support personnel must be kept with the production company eight hours a day through most phases of production. If you want truck drivers to take your crew, cast, and equipment on location to the Ambassador Hotel for a day or two, someone has to sign the check for the $35 an hour it costs to keep each trucker behind the wheel to and fro, and then the $35 an hour for sitting in the cab reading *Penthouse* while you make art.

Don Barnhardt, Katz's bearded assistant director, advised the crew, "We're gonna roll."

Katz, hands stuffed in the pockets of his khakis, wondered, "Where's the French poodle with the nice legs? I gave her my ankle bracelet." The poodle was to be part of the chase scene.

Richmond, the creative consultant, stood behind the cameras, talking with the show's publicist. Bill Richmond was a tall, graying, handsome man in his fifties. He had been an Emmy Award-winning writer for *The Carol Burnett Show* and produced *Three's Company* and John Travolta's original vehicle, *Welcome Back, Kotter*, for a year each. "Who knows what works in television? After Travolta hit big in *Saturday Night Fever*, I expected a big jump in *Kotter*'s ratings and we got scarcely a bleep. I produced *Wizards and Warriors* last year and you know what the network did? They made audio tapes of the thing and sent them out to supermarkets, had people listen to them while they were standing in line at the deli department or something, and then asked them, 'Do you think this would make a good show?' "

"What happened?" the publicity man asked. Short, and a tad wizened, he wore an open shirt and a safari jacket.

"We've got another hitch on sound," someone said.

Hotel patrons were watching the film crew. Katz bent forward and gazed again into the video monitor.

"What happened was," Richmond replied, "the network put us opposite *T.J. Hooker* and *Diff'rent Strokes* and we got smeared."

"Allan," Barnhardt said, "you ready to roll?"

"The lunatics are running the asylum," Richmond said. "But in this business people get kicked in the face, get up, and come back for more."

"Hey," Katz said, "do we have sound now or what?"

Somebody said, "Yeah."

Katz said, "Let's do it."

Steve Cannell had two pilots licensed. *The Rousters*, for NBC, starred Chad Everett and was about a descendant of Wyatt Earp who worked as a carnival bounty hunter. The concept was law and order under the big top. His other pilot, for ABC, was *Hardcastle and McCormick*. It was about a retired judge turned bounty hunter played by Brian Keith, and his handsome ex-con side-kick-assistant. Cannell was the creator of NBC's top show, *The A-Team*.

Steve Cannell was also the creator of what was, arguably, the finest television action series ever made, *The Rockford Files*. His specialty is "thugs for good." That is, men who exist beyond the simple realm of rules but who, relentlessly, comically, and violently, seek justice.

At his very own brand-new building on Hollywood Boulevard, Cannell said, "Guys like me are always on one knee. We're begging for money; we're begging for time slots. We're begging for renewals. You're gambling to get an *A-Team*, then to stay in business until it's syndicated. This is an independent company and I'm on the line. I don't have a studio behind me and scarcely a bank. So I have to figure out what I'm doing. I'll sacrifice my writing fees; I'll sacrifice my producing fees. There have been years I've literally lived off my *Rockford* residuals. Last year I put two million dollars of my own money into *A-Team* alone. Right now my stunt costs are in the neighborhood of *one million per month*—just second-unit stuff, and it all comes out of my back pocket."

Cannell was tall, a year or two over forty, frontier tough. Were it not for his pipe and goatee, he would have been the vision of a twenty-first-century cattle rancher and, as it was, he was the most successful young independent television producer in the business. A Cash McCall of the twenty-four-inch screen. He worked hard. Cannell was at his office most days by seven, and regularly wrote a dozen pages of script a day. "Last year we were writing *A-Team* scripts in two days! Three days! I can write twenty pages a day and if I'm really pressed, twenty-five."

Allan Katz stood in the sloping parking lot behind the Old World Restaurant on Sunset Boulevard watching two hippies exchange money in an alley across the street. "It could be a dope deal," he surmised, "but it's not public enough. In this town smart dealers know that advertising pays, so maybe if it was going on in front of Tower Records or the police station it might be a dope deal. But in an alley? Pretty doubtful."

Katz was waiting for the lot attendant to move another car so he could get at his Porsche. *Snoop* was moving toward postproduction and Katz was queasy about its fate. "I got the idea for the show six years ago. First I took it to NBC and Silverman passed, then it went to CBS and they passed, then to ABC—they passed. Then it went back to NBC and they passed, then it went to cable, HBO, and they passed, then it went back to ABC, they passed, then it went back to CBS and they said, 'Okay, do a script.' I did a script and they passed. Finally I was in a meeting with Brandon and he said, 'Hey, we need something completely off the wall. Whatever happened to the *Snoop* script?' I told him, 'CBS is just sitting on it,' and he said, 'Well, see if you can get it back from them. If they release it, we'll give it a go.' "

Katz retrieved his Porsche. It was a Targa, black, road worn. He pulled up onto Sunset and headed back toward his temporary offices, concerned about his show. "The network's always putting in its two cents. We shot a piece with Jerry Lewis in which he played punch-drunk, and the network had one of the executives look at the unedited tapes. He said, 'This whole Jerry Lewis thing doesn't work. We're thinking of pulling the plug on the whole show.' And I said, 'Wait a second. This is only half the tape. Wait'll you see it put together.' So they reluctantly waited until the bit was edited and the executive joked—and I'm not kidding about this—'The bit is great. It just goes to show you can't tell about a comedy piece until it's all edited together.' "

Katz's anxiety was that of an expectant father who realized he had a three out of four chance of witnessing a stillbirth. Though, if *Snoop* failed to sell, he was not at risk. And if the show went above budget, the overrun did not come out of his pocket. But there remained the fact that his future was, at this point, unedited.

"Twentieth Century pays me x amount of dollars to create shows for them, win, lose, or draw. However, they prefer it when you win." One hand on the wheel, he slalomed around a Volkswagen Thing and a chartreuse Dodge Super Bee. "And, believe me, it's a pretty substantial x. But if you want to do your own shows on network prime time, you have to be prepared to fight and get beat up for crazy reasons. I created a show for CBS, *Adam's House*, starring Karen Valentine. She played a social worker in Chicago. It was written as a one-hour dramatic show, much like George C. Scott's *East Side/West Side*, a show that had substance, a show that had real substance. A show about people trying to help other people make it." He braked for a light. "The next thing I know, the network wants to shoot it as a half-hour *comedy*. And I hear myself say, 'Sure.' As soon as I walked out of the network office and onto the sidewalk I knew I was in trouble. Cutting that show to a half hour was like trying to fit the entire Osmond family inside a motorcycle sidecar. We'd had beautiful scenes. Lloyd Nolan played this elderly man who had just lost his wife and was being evicted from the home where they'd lived throughout their marriage. Nolan's real wife had died recently and he cried when he did the scene explaining his predicament to Karen.

"In the hour version the show had room to breathe. You got to see how people's lives were often being screwed up by progress and how they had avenues to control their futures. In the half-hour version it was 'You got a problem, joke, joke, joke, we try to fix the problem, joke, joke, joke, we can't fix the problem, joke, joke, joke, but life goes on, joke, joke, joke, credit crawl.' That was it.

"It was my fault, though. I was the one who said, 'Yeah, I'll make it a half hour! When it failed to make the schedule I got a phone call through Fox. When a show doesn't get picked up the network very rarely calls the creator directly. They notify the studio; the studio tells you. The networks pay you well, but they don't have to be polite. After *Adam's House* was killed I had a meeting with CBS programmer Harvey Shephard. He told me, 'My wife and I watched it the night it was on. She loved it, wanted to know why we didn't put it on.' Shephard never let me in on what he told her."

Over at Universal Studios, Ziggy Steinberg was considerably more positive about *Another Jerk*. He'd just delivered his pilot to NBC and was sitting up in Universal exec Craig Kellem's office watching a tape of his efforts on Kellem's Betamax.

Ziggy was about thirty-five, tall, and a little round. He had an open, optimistic face. Kellem was about the same age, light complected, and wore a plaid suit and saddle shoes. They watched the screen. *Another Jerk* was the TV stepson of the Steve Martin film *The Jerk*. And Martin was serving as executive producer. The story lines were similar and they went like this: Navin is a white boy brought up by a poor black family in the South. He's a rubber-limbed numbskull, expansively naïve and goodhearted, whose single genius is for cardsharping. As the show unfolded on Kellem's Betamax, Navin had just received a letter from his California penpal, Marie, with news that she was going to get married to a European count. Navin decided to head west for the wedding.

Kellem stood by his desk, watching the screen. "I've been trying to get some intelligence on how it tested. We should know something by this afternoon."

"So Perry got it directly," Ziggy said, sitting in a chair, watching, crouched forward, elbows on his knees and chin cradled in his palms. He was referring to NBC exec Perry Lafferty.

"No," Kellem said. "Warren," referring to Warren Littlefield.

On the screen Navin wandered into a camp of hobos led by Ray Walston and, cards fanning in his hands, zapped their grubstakes. "I heard," Ziggy said, "if we go on, it'll be opposite *Magnum*."

Kellem's head tipped to one side. "Terrif."

"It could work, though—we could draw the kids away."

Kellem said, "Could be."

Ziggy's eyebrows lowered. "Though, most likely, we could get killed."

Still, Ziggy was up. He had, after all, come some considerable way from the time just a few years before when he'd made ends meet by writing stories for Magazine Management, cranking out twenty-six pages a week, week after week. Or, even, from the time he'd spent working with his best friend, David Steinberg, cranking out three shows a week for *SCTV* in Toronto.

On the screen Navin had headed west with Walston and they were in the process of simplifying the wallet of Mr. Suicide, a Hell's Angel. Navin was seated at a poker table across from the rock-jawed greaseball, scooping up the pot hand after hand.

"When's the network going to say something?" Ziggy asked Kellem.

"I'm hoping around four o'clock."

On the screen, a luscious, heavy-chested biker mama played by Ziggy's real-life girlfriend had just attempted to rescue Navin from the now apoplectic Suicide. She failed and Navin was heaved off a freeway overpass.

"You think this is too hip for the family hour?" Ziggy asked. "Naw."

Ziggy and Kellem were, it seemed, selling themselves on their own efforts. Not surprising, considering the pilot's major complication. *Another Jerk* had been cut in half.

Originally it had been licensed as a two-hour pilot. Subsequently, NBC had prevailed on Ziggy to ax it to less than fifty minutes. So that now, *Another Jerk* was about 80 percent exposition and 20 percent character development, or, in Ziggy's more euphemistic phrase, "It's pretty heavy on dramatic arc."

On the screen Navin failed to become street pizza because, after being chucked off the overpass, he happened to land in a hay truck whizzing by underneath. After several more rapid-fire adventures he met his warmhearted, beautiful, filthy rich, sexy penpal, Marie, foiled her wedding to the count (who turned out to be bogus and only after her father's money), and then soared away with her in a hot-air balloon, embracing, getting his first French kiss as the screen faded to black.

"Nobody's going to scream about slow pace," Kellem said.

"Yeah." Ziggy nodded. "I heard when the NBC guys saw it several clapped."

"Yeah."

"That's a lot of cutting, though."

Kellem agreed.

Ziggy left Kellem's office in an edgy good mood and talked about his life. His entrée to television had come at UCLA when he bet a friend $1,000 that he could sell MTM a script for *The Mary Tyler Moore Show*, which elevated him from his former position of having to hock stories to pulp magazines under the pseudonym of his sixth-grade girlfriend, Babs Cogan. Ziggy had since sold various properties, including a film script to a major studio. "It's this collegiate sexploitation thing that I did when I was feeling the pressure to get something done. It's a complete embarrassment! But what can I do? They'll sue me if I don't go through with it. Most people would give their first son to get a break like that, but God. It's horrendous! Maybe they'll let me change my name."

Allan Katz sat in his office at Twentieth Century-Fox, writing. His office used to be Marilyn Monroe's dressing room but was now expressive of a different style. Katz's desk, for example, was made from a single piece of glass supported by two huge, stuffed pink pigs. He was bent over his typewriter when his secretary informed him that a friend, Twentieth Century-Fox Vice-President Peter Grad, was on the phone. He'd just gotten a call from NBC. The good news was that many of the NBC execs had liked *National Snoop* quite a bit. The bad news was that Brandon Tartikoff wasn't one of them.

Jennifer Slept Here was another NBC pilot, but one scarcely likely to displease Tartikoff. He'd thought it up. Its co-producer, Larry Rosen, kept a brief journal revealing its genesis. Ten months earlier, Brandon Tartikoff's lieutenant Johnathan Axelrod had told Rosen and his partner, Larry Tucker, that Tartikoff wanted them to consider this idea: A young boy, twelve to fourteen, discovers that the ghost of an actress of the "Marilyn Monroe type" lives in the new house his family bought when they moved to L.A.

Tucker and Rosen had been in television and film for years. Larry Rosen was, among other things, the former producer of *The Mike Douglas Show* and his partner Larry Tucker had been creator—with Paul Mazursky—of two of the most emblematic films of the late sixties and early seventies, *I Love You, Alice B. Toklas* and *Bob and Carol and Ted and Alice*. The basic hook behind the *Jennifer* project was that Jennifer had been run over by a Good Humor truck and, instead of going to heaven or hell, was consigned by the celestials to do some haunting and good deeds before passing through the pearly gates.

A day after talking to Axelrod, Tucker and Rosen had sushi with another Tartikoff lieutenant, Jeff Sagansky, who "suggested the boy be 'hot,' rather than passive and wide-eyed." Tucker suggested Loni Anderson for the Monroe part and NBC offered to purchase a pilot script prior to making a pilot commitment. If the network didn't like the script, it retained the right to give the idea to someone else.

Tucker and Rosen then met with Ann Jillian, who was under contract to their own studio, Columbia. She liked the idea of playing the Monroe role. A script was written and approved, secondary roles were approved, but finding the boy was difficult. Tucker and Rosen auditioned hundreds of kids; NBC ponied up $15,000 to finance the search.

They found the boy, John Navin, and it was all gravy from there. Now Larry Tucker and Larry Rosen were planning to go to New York to be available during the final stages of the NBC pilot selection process.

Tucker and Rosen had been partners for several years and had shepherded several series through development, pilot, and on-the-air commitment. And their position was more secure than most. Their studio had a development slate of eight potential series for prime time and was providing more hours of television to the network than any other studio. Still, they were flying back.

"You never know," Larry Tucker said, "what they want." Tucker was an outwardly relaxed, somewhat overweight man, kindly and competitive. "We've got a good shot. It's always better when the network comes to you, rather than the other way around. They could schedule us for either September or January. There are advantages to both. In January you get special treatment, but have less than five weeks to get into production. In September, you get eaten up by the reviewers."

Rosen carried a four-page synopsis of *Jennifer Slept Here* plot lines in his briefcase. These would go to Manhattan, too. Among the stories:

• Marc (Joey's best friend) and Joey form a punk band. They need a female singer. Jennifer wants to "audition" just for fun. Joey is against it. She's a ghost and, besides, she looks a lot like the late Jennifer Farrell. That might confuse people! Jennifer, in disguise, auditions anyway. She's terrific! Even Joey loves her. They want her to stay with the band. Jennifer is stunned. She only did it for a lark, and the last thing she wants is to be a full-time punk rock star. Jennifer has to start singing badly and causing trouble to disenchant her new admirers and get herself fired.

• Joey sells a photographer the right to take a picture of Jennifer's bathtub. The guy dummies up the picture with a phony nude of Jennifer. She never posed nude and she's mad!

• An Auntie Mame–type character (Bea Arthur does a guest shot) comes to visit. Bea loves parties . . . and hates kids! It's Halloween. Marc is sleeping over. Joey wants Jennifer out of the house, but Jennifer is planning her own party and she wants Joe out of the house. Marc admits he's been having hot dreams about Jennifer Farrell. He wants to hold a seance at Joey's on Halloween night . . . to meet Jennifer Farrell for real. Jennifer and Bea cooperate and give Joey and Marc a Halloween that makes *The Haunting* look like *Sesame Street*.

"I think we have a very good chance of making it," Larry Tucker said. His partner, Larry Rosen, was not as sure. Rosen had black hair, was maybe forty-five, and appeared to be in good shape. He was as voluble as Tucker was calm. "We're taking *Jennifer* back to New York to get the news face to face. . . . If NBC wants to cut my throat, the least I want is that they do it in person."

*   *   *

On the very large TV screen, the newswoman asked Allan Katz, who was dressed up in a bargain-basement Flash Gordon space-traveling costume, how he had broken into the business. "I take after my father," he replied.

"Your father was a human cannonball?"

"No," Katz allowed. "My father was a fool."

Katz climbed into the mouth of a hugh cannon. He was shot high in the air, landing on the far side of a ravine. The newslady rushed to his aid. "Are you all right?"

"Don't worry," he said, head wobbling on his shoulders, "I'll be fine as soon as I'm unconscious."

Snoop tapes continued on the wide screen at the wrap party, and afterward the whole cast and crew clapped for Katz for three minutes straight, yelling, "Speech!" "Get your ass up there and explain yourself, Katz!" So Katz went up dressed in cords, tennis shoes, and sweater, looking like a high school kid. "Look, I don't know what the network is going to do with this show—but I do want to thank you all for your efforts, whether they pick it up or not. All I can do is say I think they're fools if they don't." The lights went up and *Snoop* assistant director Barnhardt asked Katz how it was going. "So-so," Katz replied. "The patient is critical. But I've been told we've still got a shot."

Don Barnhardt was a typical member of an elite working class, employed an average of 120 to 180 days a year to crew series television. Like all lighting technicians, grips, best boys, sound men, and camera operators, he had voluminous dreams that swelled far past the often menial scope of the chores he now performed as Katz's assistant.

He tried to cheer Katz up. "Did I ever tell you about my sex life as a DJ, Allan?"

"No."

"This was when I was working at a radio station up by Santa Barbara. I only got paid a couple hundred dollars a week, but the jocks had groupies and it was possible to get screwed on a cot outside the control room in exchange for record albums. Not that the girls were so hot. Mostly, they'd come in sort of beat up with a couple of moths flying around their heads, but close your eyes"—Barnhardt closed his—"and you were with Brigitte Bardot." It was pretty primordial fornicating, though, and Barn-hardt said he always felt guilty. "They were lousy albums. But what did they know?"

Katz laughed, briefly.

"When are you going to hear something?" Barnhardt asked.

The fate of Snoop was high in the air. Katz looked out the window. The party was being held at the Toluca Lake Tennis Club, a modest facility reflecting a condominium–chain restaurant sensibility. On the courts outside a man in whites was serving. He had a severe case of executive gut. Katz toasted him silently with his beer bottle. "It looks like our chances have gone from lukecold to lukewarm, so I'm semiecstatic." Four nights before he had gone out to dinner with Twentieth Century-Fox VP Peter Grad, and Grad said that Tartikoff didn't like Snoop. The meal had been a wake. Then, Monday morning, Katz got a call from NBC programming VP Warren Littlefield, who said he liked the show and was going to lobby for it. Katz offered to sell it to him for his personal use. At cost.

"Listen, Allan," Barnhardt said, "no matter what happens, I just want you to know I'm giving you my highest accolade ever: I'd work for you for two days for free. That's more than I'd even do for my parents."

Katz smiled.

"Hey, what are friends for?" About forty, halfback stocky, Barnhardt rubbed the back of his hand. His goal was to become a successful double hyphenate. Writer-director-producer. He'd just sold an option for a TV pilot, Mixed Blessings, about a hip minister who moves with his family to a conservative town. "My partner and I got fifteen hundred dollars for a hundred-eighty-day option. We get twenty-five thousand dollars if it's sold as a pilot, with an extra ten thousand if it goes to series. We locked up co-producing positions from day one and 'co-created by' credits, with residuals on those credits. We gave up the right to do the pilot, but are guaranteed the right to write one of the first six episodes, or three out of the first thirteen. Then we'd get five percent of the monies on all spin-offs and merchandising."

Katz took a swallow of Bud. "Now, what's the idea? I mean for the minister."

Barnhardt shoved his hands in his pockets. "He's hip. You know, a little jogging, a little beard."

"Yeah," Katz said.

"The kids," Barnhardt said. "The daughter is the oldest. Straitlaced, wants to be a minister like her dad. The oldest son— not exactly a Hell's Angel, but the kid knows what's coming down. The youngest boy's an entrepreneur, real eighties, owns a fleet of lemonade stands. Barry Krost Productions picked up the options."

Katz grinned. "Hey, it sounds dreadful. So it'll probably get picked up."

"I've got this other property, Allan. A trilogy about my life. Based on a guy that went to this tricolor high school in the Valley—black, white, and Mexican. Primitive: You hang out at the drive-ins, polish your car, play poker, drink beer, be tough, and work in the factory. People do that forever in this country. I grew up in the fifties, hung around with guys that were doing that at age thirty-five or forty. But what changes? Guys are still screwing their girlfriends on the lube racks at the Arco stations. I went back to Encino recently and they're all doing the same thing— polishing the car, drinking beer, hustling chicks, being tough— except some of them are seventy-five or eighty."

Barnhardt was still getting chunky residual fees from directing episodes of *Mork & Mindy,* on which he'd worked for $8,000 a week. Which is bottom scale. Directors like Jim Burrows and Jay Sandrich can command as much as $50,000 to $75,000 just to do a single pilot. Even so, Barnhardt received 75 percent of his flat fee the first time his episodes of *Mork & Mindy* were rerun, 50 percent the second, 25 percent the third, and so on. "It gets down to five percent forever."

Still, Barnhardt was leagues below Katz's position in the industry. A three-time Emmy nominee, for his writing and production work on *M*A*S*H, Rhoda,* and *Laugh-In,* a Golden Globe Award nominee, and a Critic's Choice winner, Katz supplemented his income substantially with a two-year contract at Twentieth, which, like all major television studios in Hollywood, enlists a few successful producers at hefty salaries in exchange for the exclusive privilege of underwriting their work. And to share their profits. There are perhaps no more than a hundred producers in television who enjoy this position. It helped Katz finance, among other things, a beautifully decorated Spanish-style home above Hollywood that spread three or four or five stories over a hillside and was appointed with everything short of cannon ports.

"I think it will be a good year for me, regardless," Katz said, watching the fat man smack the tennis ball with surprising authority across the net to a woman in whites, a gray-haired, matronly little sprite who blooped his sallies back with equally surprising consistency. "But then we'll just have to see."

"I'm thirty-nine; I'm immortal; it's never going to end," said Steve Bochco. He sat slumped in a chair at his office at MTM, in Studio City. His headquarters was on the second floor of a red brick building that—with shade trees around it—gave the MTM facility the feeling more of a midwestern college administration building than of a television production company. "That's the way people of my age tend to think. So the action in St. Elsewhere doesn't bother us. We can watch it and not worry. But the older people? Surveys show St. Elsewhere doesn't have a viewer in the country over fifty-five. It's frightening to people who may be facing a medical trauma themselves. That is a very graphic show. And it's not just Hill Street with stethoscopes. People don't mind seeing graphic or even frighteningly violent police shows, but a medical show is something else. America doesn't go to jail, but America does go to the hospital. And when you get to be a certain age, I think maybe you don't want to be reminded of that. Thanks to technology or whatever, people die in increments in this country. That's the way my father went. A trip to the hospital here, a trip to the hospital there— his end was a collection of stopgap medical moves, not really to save his life, but to prolong it. And to a lot of older people, I think, that makes a show like St. Elsewhere a scary reminder, so they turn the channel to something else."

It was no longer the best of times for MTM, though Mary Tyler Moore Enterprises had been characterized as "television's oasis of quality" in USA Today. The New York Times called it "the most respected production company in Hollywood . . . [where] its knights are called executive producers, who are granted a status and independence unrivaled in Hollywood . . . a kind of Camelot." MTM had, over the course of its fourteen years, created eighteen series and won dozens of Emmys. But the company had money problems. Because it is a privately held company, MTM answers to no public stockholders and is able to spend its money as it pleases. The year before, MTM President Arthur Price had authorized the spending of $1 million to reshoot the premiere episode of *St. Elsewhere*, and the company ended up pumping a full $2.5 million more into the show than it received from the network. After five seasons, *Hill Street* was itself over budget by $6 million, and similar problems plagued other MTM shows, including the ritzy detective series *Remington Steele*. MTM series were the most expensive in television, a fact that had been brought to bear directly on Bochco.

Co-creator of *Hill Street Blues*, winner of a score of Emmy Awards, Bochco had worked in television since being hired as an intern at Universal at age nineteen and had written for several police action-adventure shows, including *The Bold Ones, Columbo*, and *McMillan and Wife*. *Hill Street Blues* had made him nearly famous but the pilot for a baseball show he had created with Jeff Lewis, *Bay City Blues*, was about to be canceled before it had even been shot. "It was the most expensive show NBC had ever commissioned," he said, "so I understand their concern. Part of the premise was that every show would involve an actual game of baseball—it was to be an ensemble production about the lives of a minor-league baseball team in northern California—and the cost for extras just to fill the bleachers was a monster by itself. About eighty dollars per extra per day. Jeff and I felt we could bring much of the quality and human values to *Bay City Blues* we'd brought to *Hill Street*, and Brandon and Grant were both with us. Tinker's thesis is simple: Quality and success can go hand in hand—we felt we could guarantee the former if not the latter. But when it comes down to it, the network business people don't give a fuck how many Emmys *Bay City* could win; what they want is black ink on the bottom line, and I really don't blame them. We tried to negotiate, we cut into muscle, but we still couldn't get an okay on a budget we felt could bring the real grit of baseball to life . . . though, if you want the truth, realism or the absence of realism have never been an important factor in the success or failure of any prime-time television show I can think of."

Bochco looked over to the chalkboard that stood on the other side of the room. On it, a grid outlined the plots and subplots of upcoming *Hill Street* episodes. "I'll take that back a little. Realism probably nearly killed *Hill Street*. People had no frame of reference for it; there were no stars, no formula, and when we went on the air and had a sympathetic character suddenly get blown away, we had letters of outrage all over the place. People couldn't believe that we would actually kill somebody they considered a *good guy*. I think realism, or maybe just the information, we were giving people on *Hill Street* was really a first. We were putting more information on, frame by frame, than any other show in prime time. If I'm proud of anything about *Hill Street*, it's that we redefined for the audience the terms of the agreement under which they turned on the set."

Jeff Lewis had even more to lose from *Bay City*'s stillbirth. He had been named co-executive producer with Bochco. A graduate of both Yale and Harvard Law, Lewis had come to his position at MTM after spending time as an attorney in the district attorney's office in Manhattan. His principal writing credit was *Hill Street Blues*. "This is not a life I would have imagined for myself. I supposed by the time I was thirty-five I'd be living on a Greek island with kids and a European wife, and have had several novels published and some kind of international reputation. The usual horseshit. The Ivy League Dream."

He laughed out loud. "This is personal theory, but I think what TV tends to attract is someone who went into writing on a personal egotistical jag expecting to become the next F. Scott Fitzgerald or whoever, and found themselves in such desperate straits that they're willing to subvert that—at least for a time—into something else."

If *Bay City Blues* was scheduled and made it through even one full season, Lewis stood to make more money than F. Scott Fitzgerald made in a lifetime. "We are forgoing a lot for this show," he said. "You will not, for example, see the kind of violence we have on *Hill Street* . . . that kind of bright palette. NBC didn't want the show to be that much about baseball. I think we sold them on an idea that this was a wonderful magical world . . . small town, a little bit country, a little bit kids—adolescent kids trying to make it in baseball, the flavor of the working class against a colorful athletic background, the problems and drama of soap opera, but funnier." Lewis, sitting in his office at MTM, paused, as if editing his thoughts. A small, tweedy, smooth-skinned man, he had the odd presence of both a professor and a kid. "I have certainly no guarantee that this is going to be any kind of success story either commercially or artistically. The whole thing was presented on the proposition that it would be an ensemble cast like *Hill Street*, and on the fact that Steven had a series commitment to NBC, and on us going in there and say-ing enthusiastically, 'We can do this thing!' and NBC saying, 'I don't know,' and us saying, 'Look, it'll be great! It's America! We can do it!' "

Now it was doubtful they would ever get the chance.

St. Elsewhere was at death's door, too, one of $20 million worth of MTM programs in serious jeopardy as of April 1983. "I'd give it a seventy-five to one shot," producer Mark Tinker said. The son of Grant Tinker, Mark had an office at MTM around the corner from Bochco's. "And that's because I'm an optimist. *Hill Street*, however radical, was still within the timeworn genre of cops and robbers. People still want to see their doctors as mythically infallible. Maybe if we'd made it more uplifting, more optimistic . . . Our only chance will be if NBC has a really great development season—they can carry us for our high demo-graphics. Because we still attract a disproportionately large number of upper-middle-class young people with lots of dis-posable income. On the other hand, if NBC has a rotten devel-opment season, they may have to carry us, just because there's no workable replacement. But if they have a *so-so* season, we'll be bumped, just so they can try other things out. And NBC al-most *always* has a so-so development season. It seems to be a regular tradition over there."

"I'll stay away from New York," Allan Katz said. "I'm a creator, not a salesman. If they like the show and want to buy it we can go from there. If not, it's useless for me to stand up afterwards and say, 'Well, you think it's terrible, but here's why you're wrong.' The studio performs that service, not me." Katz sat behind his desk at Twentieth, worried about *Snoop*'s fate at the hands of NBC. "I'm afraid it may be too different from what they're used to in prime time. Network people are inclined to make decisions on the basis of what they think, rather than the basis of what they've experienced. A lot of them haven't been involved in anything so creative as holding up a service station; they've come out of banking or law or accounting. And many of the most important creative decisions in television—*the decisions of what to air*—are made by people who have never invented a character, never written any dialogue, never plotted a scene in their lives. You'll hear network people tell you, 'Oh, *M*∗*A*∗*S*∗*H* was so innovative, so original, so well written—that's why it was a success.' But they really only care about the one word at the end of the sentence. Success. They have a tremendous problem seeing that it's the *soul* of something like *M*∗*A*∗*S*∗*H* that made it work. They're like somebody who sees a guy run a four-minute mile wearing red shorts and then thinks red shorts are the key to running a four-minute mile."

Katz turned in his chair. "I don't know where some of these things come from. NBC is supposed to stand for quality these days, right? And Ed Weinberger is one of the great writer-producers in television. *Taxi*, and so on. Now, he's doing this thing for Grant Tinker about a genius orangutan called *Mr. Smith*. Maybe it'll be great, I don't know. If I was working at a network and Ed Weinberger came to me and said, 'I want to make a sitcom about Lenny the Talking Table,' I'd say, 'Just wait a sec until I can find my checkbook.' Because Weinberger's that good. Maybe he can make it work. But so many ideas that come out of the networks themselves are simply patched-together retreads of what has already been done before." Katz assumed the stance of a network executive: " 'Okay, we need a hit. So we'll have this big, Mr. T–type black guy, and he lives in a mansion with a little adopted white kid, and there's a maid with a face and body just like Loni Anderson. Except that she's a ghost. Now all we need is a gimmick.'

"So here you are. The network doesn't like your idea and you don't like theirs. You have the choice of doing a bad show or none at all—or fighting with them. The tendency is to rationalize. No matter how crazy the idea is, you're inclined to think, 'It might work.' And if it does, it's the keys to the kingdom. Syndication rights will make you richer than God. So, for a lot of producers, the inclination is to take a shot at it, no matter what. And what usually happens is that the network finds somebody to do the show."

Katz's secretary looked into the room. "Phone. I think it's your attorney."

"Great." Katz leaned forward to pick up the receiver. It was the first of May and *Snoop* was in limbo. But right now, he had other things on his mind. Katz was suing Johnny Carson's company. The conversation with the attorney was brief.

"A year or so ago, NBC thought Carson might go to another network. What to do to keep him? Well, they couldn't make Johnny governor of the state, and they couldn't give him a U.S. Senate seat. And Carson didn't need any more money. So they gave Johnny power—something like a $40 million commitment to on-the-air production at NBC that made Carson overnight one of the biggest independent producers around.

"Through Johnny, Angie Dickinson got an on-air commitment for at least thirteen episodes of a half-hour comedy. Then I got a call from Carson Productions. They said 'We want you to create, write, and produce a show for Angie Dickinson. We guarantee that your show will be on the air for a *minimum* of thirteen episodes.' They were committed to pay me for all thirteen episodes, regardless. Could she do comedy? I hadn't the foggiest. But it was the chance to do my own show. I bit.

"I came up with a concept. Woman in her forties, divorced, no kids, comes to California from Chicago, literally going out into the world for the first time. I liked it. Carson liked it. Angie liked it. NBC liked it. It was a go.

"Because I was under time pressure, the network didn't get a completed script until just before we went to rehearsal. I was told, 'Fred Silverman wants it hotter, wants guys hitting on Angie, wants her boss hitting on her.' I said, 'That's tacky and tasteless. I didn't agree to do that kind of show.' Johnny Carson backed me up. Carson's attorney told the network to cool it on the script demands. Still, I was getting constant calls from the network. Finally, the word was that NBC wanted to have kids in the show. Not her kids. Just kids.

"Fine. I'd have a single male parent with a daughter who attaches herself to Angie as a surrogate mother. NBC said, 'Great.' A few days into rehearsal, it was obvious that the actress playing Angie's girlfriend was not working out. I brought in another actress. She read for a roomful of executives. Nothing. No response. I thought I'd gone deaf.

"Finally, one of the NBC execs, Perry Lafferty, said, 'What about Brenda Vaccaro?' I said, 'She's expensive, and my understanding is she's difficult to work with. I like the actress who just read.' Lafferty said, 'We want Brenda Vaccaro. You'll take Brenda Vaccaro.' I said, 'If you want to do things that way, do the show yourself.'

"Nobody said anything. So I said, 'Either I better leave the room or you leave the room, because somebody has to decide. And I don't think it's up to me.' So they left, came back, and said, 'We'll go with your choice.'

"We shot the show. Angie had never done situation comedy before, and she was tentative. But the audience bought it. Afterwards, I told her, 'Look, don't worry. We can rehearse.' She wasn't convinced. I told the Carson execs, 'I still believe in the show. If Angie doesn't want to continue, maybe we could replace her with somebody like Karen Valentine.' They said, 'We can't do that. We have a commitment to Angie.'

"In private, Angie told Johnny she was uncomfortable. So the Carson people and NBC came up with a new concept: Make her a successful, sexy, high-powered ad exec. A completely different show. The executives at NBC pitched it to me. I hated it. I refused to do the next show. I told them, 'She has no problems, no worries, no struggles. She's successful, she's rich, so who cares? It's gonna sound like *Dallas* with a laugh track.' The Carson company said, 'You're the guy we hired; you have to do it.' I said, 'You hired me to do the show I created. I'm willing to do that show with minor changes, but that's all. Either keep my concept and replace her with a woman with comedy credentials, or put together a show that's more to her form. But that's not my responsibility.' I did offer to help out. I even suggested one alternative: 'Make Angie the wife of a rich private eye who is murdered. She inherits her late husband's rough, boozer partner. Together they solve the murder and then form an agency. That would get her back to the cops-and-robbers stuff where she's been more comfortable, sets up an interesting *Odd Couple* situation that should have some legs to it, and gives her time to ease into the comedy stuff a little at a time.' They wouldn't buy it." Katz tried to explain that you couldn't just declare someone a comedy actress just because she was good at playing George Burns to Johnny's Gracie Allen on *The Tonight Show*. That she could get destroyed. "Considering she'd never done comedy before, I think she did quite well. She could have really improved. But never got the chance. Because the network guys said, 'Freddy wants to have her be a rich ad exec whose boss is trying to fuck her all the time. It'll be hilarious. So you either do it or so long.' And so I said, 'Wait a sec, friends. I have a contract. I have a pay or play deal.' Thus, the attorney."

Katz's life was not always so exciting.

By the time he was twenty-six he had a bleeding ulcer and his parents' hope was that he'd be able to qualify for a job with a future, like as a court reporter, perhaps. By then he'd not only flunked out of college but had been personally barred by the dean from ever returning on the grounds that he was "not a learner, not college material." "I was desperate to get back in," Katz recalled, "if for no other reason so that I could tell people I was a college student instead of a delivery boy."

Meantime, Katz had held every menial job in Chicago. Seventy to eighty different ones. He mowed lawns, drove tractors, worked construction, was a lifeguard and a waiter, bussed tables, became a short-order cook and a cashier, and repossessed television sets. It was a bad time. "Nothing I did worked. I tried to get into advertising. I made these mock-up advertisements and took them to this agency guy. He looked at 'em and said, 'First of all, I'm only seeing you because you're a friend of so-and-so's, and you've already wasted too much of my time. This stuff is the worst creative work I've ever seen. It's unimaginative, boring, and shows no talent whatsover.' I was floored. I said, 'What about something basic, like writing catalog material?' And he said, 'Forget about it. I don't even want to talk to you anymore. I saw you because of a friend and that's it. Goodbye.' I walked out of there and was totally destroyed."

Katz was born on Chicago's North Side. His father was in the garment industry; he worked for a company that made costume jewelry for dress manufacturers. Before that he'd been in vaudeville, a tap dancer. His act was Mills, Malcolm, and Mills.

Jerry Katz was a Mills. "My father kept a lot of the vaudeville, show business attitude toward life. He wasn't very materialistic. Whatever extra he had he'd blow at the racetrack, buying drinks for friends, or on presents at Christmastime. He was generous when he had it and generous when he didn't."

Katz's elder brother was a musical prodigy who enrolled in the University of Chicago Lab School at age eleven. Allan's experience was more mundane. He went to a middle-class public high school where, at the conclusion of twelfth grade, he was named "most humorous" graduating senior by the school newspaper. After graduation, he enrolled at the University of Illinois and flunked out. "The only genius I had was for screwing around. I even flunked writing. They told me I couldn't write."

At age twenty he took off for Europe with a backpack and $200. He slept in youth hostels and on the streets. In France he met an illustrator and got a job coloring Christmas cards and sold them on the rue Rivoli next to the Louvre, posing as a French artist. Then he got a job driving cars from Rome to Copenhagen and ended up ricocheting all over the Continent. England, France, Belgium, Luxembourg, Germany, Sweden, Austria, Spain, living on scraps. He got caught taking coins out of the Trevi Fountain. Trying to get into England on a ferry, he was turned back at the dock. "I didn't have any money. I was a derelict. I was dirty, had a beard that looked like spaghettini. They sent me back to France." Where he was thrown in jail.

Shortly thereafter, he was classified a "destitute American" and sent home on a troop ship. On board, he got staked to a poker game and made enough money to buy a bus ticket from New York to Chicago.

There he enrolled in Roosevelt University, where he caused minor havoc by dumping 120 tablets of Salvo detergent into the Buckingham Fountain. Shortly thereafter, he was not only bounced but banned. So, more jobs. Working graveyard at the Sara Lee kitchens cleaning stoves, checking records in a law office, working at MCA in the publicity department mailing out pictures of José Melis and photos of people who were joining Tex Beneke's band.

Finally, he was hired as a subject for hypnosis. Only to discover, "I wasn't even together enough to be able to go into a trance." Shortly thereafter he got a steady job measuring pupil dilation at the University of Chicago. "I wondered, 'Is this what God made me for?' I got really depressed. I became a loner. I was single before single was hip, living that lonely life ten years ahead of my time. I would have jobs where people didn't understand my humor or have any idea who I was. I never fit in. I was a person who needed to do eccentric, funny kinds of things and those were the places where nobody was looking for that. All my friends were becoming doctors, lawyers, and Indian chiefs, while I was going nowhere."

He ended up burning a hole in his stomach. When he got out of the hospital, Katz went into the poster business. Taking photos at frat parties and weddings and then blowing them up to make personalized posters. "Mostly, these were the kind of guys that would hear the pitch and say, 'Well, frankly, I don't need any posters. I'd rather just drink too many beers and throw up.' "

Shortly thereafter, Katz took that first stab at advertising and was told he was useless. "The guy was the most rude and devastating man I'd ever met. To this day that experience burns. I mean, if some poor guy comes into my office and gives me a script that's so bad that it looks like he had to hold his pencil in both hands in order to write the thing, I'll try to give him some kind of useful advice that doesn't obliterate him in the process. There is just no human excuse for that kind of behavior."

But Katz's life was about to take a turn for the much better. He walked into the offices of Hurvis, Binzer, and Churchill, then the hottest small ad agency in Chicago, showed the creative director his homemade mock-ups, and was hired immediately. Within months he had become co-creator of Screaming Yellow Zonkers—a caramel corn "box of hype" that became a wall-to-wall national fad—and his ideas were being interpreted by America's preeminent art directors and illustrators, Charlie White, Milton Glaser, and Seymour Chwast among them. "It happened so fast that I didn't really even know who these people were. All I knew was that I had a real job where people that I liked, liked my ideas." Within six months Katz had won a slew of national and regional awards, including several from Communication Arts, and the Chicago Art Directors' and New York Art Directors' awards.

"And you won't believe this: The same guy that had stomped on me, the same guy less than six months later was on the panel of judges at the Chicago Advertising Show that gave me ten awards, best for this, best for that. I couldn't believe it. I never talked to him, never confronted him. Why? I don't know. I wanted to ask him, 'How come?' But I never got the chance. Because it was right about then that I got the call from George Schlatter, the executive producer of *Laugh-In*. He called me at the office and I didn't even know who he was. He had me flown out to California and hired me to write for the show. By then, I knew who *he* was. I just wasn't quite sure who *I* was."

5

Maalox Time

April 1983

Harry Anderson, who was so close to the day job of being a TV star that he could almost taste the residuals, sat in his black, sweeping judge's robes, as Jim Burrows, Reinie Weege, and an NBC representative conferred at the perimeter of the stage. There was a poisonous atmosphere in the room.

NBC, in the person of compact, red-haired young Warren Littlefield, was skittish about this pilot, more so than some of the others in production. For starters, the network was biting its corporate nails over Mr. Harry Anderson, and whether or not the American public would buy this overage hippie as a judge. Weege had even had to write in a scene where the newly appointed judge's own secretarial staff won't believe he's the new judge, until he sprays them all with spring-loaded snakes out of a gun, insisting, "If I wasn't really the judge, would I have had the nerve to do that?"

The network was also a bit leery of Weege, who was known as a guy who could be a bear to work with—stubborn, to put it mildly.

And lastly, of course, the network had the very concept itself to worry about—that of a weird, whimsical judge. Wasn't it like trying to sell a sitcom about an eye surgeon, where the funny part is that he has cerebral palsy?

Therefore, what Warren Littlefield, huddled with Burrows and Weege, wanted to inject into the character of the judge was a touch of erudition—a bit of book learning. Couldn't the judge just spout a few stanzas of memorized law text? Just an odd paragraph or three—word for word?

But that's not realistic, replied Burrows and Weege. Judges don't do that. It would have the paradoxical effect of undermining Harry's believability.

Nevertheless . . . it might help the show. The underlying meaning here, of course, was that it might help the show crawl onto the NBC schedule.

Jim Burrows considered himself adept at compromise, and was scornful of producers who wouldn't bend a little. But he could tell that Weege felt strongly about the matter, and it seemed important to him, too, and so he put down his foot. Littlefield didn't make a big deal out of it—that wasn't the reigning style at NBC—and the rehearsing resumed. But Harry Anderson wished that the whole episode had never happened, because he could feel his one big shot at fame, fortune, and bankers' hours slipping inexorably through his fingers. He figured he was just one more bad vibe away from being back on the road, spending night after night in dingy clubs, having to tie Leslie up and kick her all over the stage.

Rumor had it that NBC was only serious about picking up two or three new comedies, or "half hours" in the argot, and competition among the thirteen comedy-pilot makers was intense. And Harry was aware that several productions were already in trouble. One, called There Goes the Neighborhood, a takeoff on The Beverly Hillbillies, starring Buddy Hackett, was walking death. Although officially still a contender, the show had been written off as a $500,000 bath after its first viewing by Tinker, Tartikoff, Littlefield, and the less than half-dozen others who constituted the voice of NBC programming. The flop of the pilot had its genesis in problems similar to Harry's own. Throughout the rehearsals the show had brimmed with possibility; bystanders watching the actors go through their paces giggled uncontrollably. Hackett's casual, throwaway style translated into the type of adventurous, wing-it humor that works so well on the small screen. But when the sound stage door's red warning light went on, and the sound stage buzzer blared, and the cameras rolled, Buddy Hackett froze up. Suddenly, his screw-it attitude congealed into a stilted, painful tightness that brought the entire production down around him. An older NBC executive had warned that this type of thing occasionally happened with Hackett, in spite of all his years in the business, but the disaster was a shock to most of the people involved with the pilot.

Another falling star was Don Rickles, who had been the main course in a dubious video feast that never went beyond the stage of being known as the Don Rickles spin-off. Rickles had appeared in 1982 on the "warm comedy," or "warmedy," *Gimme a Break.* The character he played was then used as the tent pole of a pilot, but the pilot was of such negligible substance that it collapsed of its own weightlessness.

By 1983, the networks had, in most cases, given up on the idea that a star could carry a poor show. The recent failures of Rock Hudson, Doris Day, Henry Fonda, Glenn Ford, and a clutch of other Big Faces had confirmed the adage that "TV makes stars, but stars don't make TV." The theory was that the twenty-one-inch tube humbled major entertainment deities, just as the large movie screen exploited their glory to its fullest. The small screen thrust actors into a format that rewarded performers the audience felt intimate with, and punished those the audience felt in awe of.

As the actors in the Warner Brothers studio started exchanging their lines again, something felt out of tune. Harry, the others in the cast, and even Burrows could feel it. The timing was off; the jokes weren't landing. The script seemed suddenly flat.

Warren Littlefield watched from a stage-side chair, making occasional notes.

On the opposite side of Hollywood from the Warners studio is a conglomeration of sound stages, offices, and broadcast studios that is generally referred to as KTLA, after the local television station that is housed there. As Harry rehearsed his pilot,

Fred Silverman was working at KTLA on *We Got It Made*, the program that was shaping up as *Night Court*'s primary rival for a 9:30 slot.

Like *Night Court*, *We Got It Made* had no stars. But it did have that one element that is bulldog powerful in breaking through viewer alienation. Sex. Fred Silverman, the father of Jiggle TV, had a hot little T-and-A farce up his sleeve that he believed would not only be the hit of the season, but would also prove a point he believed was sorely in need of establishing: that Freddy Silverman, banished in disgrace from his most recent network position, still possessed the gut that glitters.

In order to prove that, though, he would have to sell his show to Tartikoff, who now sought distance from the memory of his Silverman servitude, and also to Tinker, the man who had once called him "a greedy kid"—and also the man who had taken his job. He would also have to slide ahead of a gaggle of arty young writer-producers who made the kind of cerebral character comedies that Tinker and Tartikoff found so appealing. Fred Silverman, therefore, had his work cut out for him.

But Silverman had some aces of his own. Chief among them was his deal. Silverman had a commitment from NBC that allowed him to place one show on the air for thirteen episodes; it had been part of his severance pay. His riding out of NBC with a production commitment in his pocket was fitting, because Silverman had pioneered the networks' practice of offering production deals to people they owed something to, in lieu of cash. It had seemed like a great idea in the 1970s; it saved direct dollar outlays, and since the networks had to make production deals anyway, why not make them with people they were already familiar with? Johnny Carson had gone from performer to power producer with the production deals Silverman had given him. By the 1980s, though, the practice had mostly evaporated. The networks had realized that not a single popular show had come from Carson Productions, and there were many even more glaring examples of performers being miscast as producers. For example, child star Gary Coleman had been named the executive producer of the video version of Broadway's *Evita*.

But Fred Silverman was not Gary Coleman. He'd had the good sense, right off the bat, to hire one of the most proven writer-producer teams in TV. Gordon Farr and Lynn Farr Brao had written for dozens of TV shows, even though they were just into their forties, and had produced *The Bob Newhart Show* and *The Love Boat*. Working with them was Alan Rafkin, who was second only to Jim Burrows among working TV directors. Rafkin, in fact, had a sweat shirt that celebrated his one Emmy; it said I BEAT JIMMY BURROWS.

After the script conferences earlier in the spring at NBC, and the approval of the third story line—the one without big John Matuszak—the major chore had been casting. Casting the pivotal role of the maid was tough, because the girl not only had to look like Miss Universe, but also had to have a personality compatible with that of the character. Which meant finding a woman who was smart enough to achieve fame and fortune as a TV actress, but who could also genuinely merge with a character who perceived things like walking around naked on window ledges as perfectly reasonable acts.

But if this actress came off as just plain stupid, she would alienate female viewers, a catastrophe, considering that women tend to dominate viewing choices. Women watch four times as much daytime TV as men; they also predominate in the evenings by an almost three-to-two margin, and are particularly avid watchers of comedy. So a dingbat beauty queen was not exactly what the situation called for.

Unfortunately, that was how most of the actresses they auditioned were playing the part. And some of them were doing a bang-up job of it.

Gordon, Lynn, Silverman, and the MGM and NBC casting directors had come up with several hundred names for the part, and dutifully talked to each girl. At the same time, they were seeing actors for the other four parts. The work was exhausting, but absolutely vital to their chances of making it onto the schedule. The wrong mix of actors, or even one poor choice, could easily doom the pilot.

When they met Teri Copley, they knew their problems were over. For starters, she was beautiful, and had a body that men would commit crimes for. More importantly, though, she leaked almost palpable puddles of naïveté and vulnerability, traits that would fit with the character without offending the female audience. And, unlike other gorgeous young actresses, Teri was not an impossible prima donna; she'd been born poor, had worked hard, and wasn't about to let this chance slip away.

Even though the logistics of casting and producing the pilot entailed an endless string of late nights and early mornings, the process seemed relatively easy to Gordon and Lynn, because their last production had been three years of *The Love Boat*, a monumental exercise in controlled chaos, in which almost fifteen tons of equipment, 130 crew members, numerous high-voltage guest stars, three interwoven plot lines, and the governments of numerous foreign countries had to regularly be controlled, organized, appeased, and made sense of week after week after week. Customs agents had to be paid off, lovers' quarrels, both straight and gay, patched up, guest stars flown in and out, and new locations constantly scouted. On top of that, each show had three story lines: the "heart story," the "tears story," and the "laugh story," all of which had to be somehow melded into a cohesive plot. And every show had to have an appeal to each of the three major age groups; a guest star like June Allyson would be hauled in for the older group, a Tom Selleck type would be brought in for the young adults, and somebody from a kids' show, like *Eight Is Enough*, would be shoehorned into the plot to attract the youngsters. To Gordon Farr, doing *Love Boat* had seemed like waging a weekly World War III. Puffy-eyed and achy, he'd ask himself, "Why do I live like this? Why do I feel this way? I'm paid an exorbitant sum of money, but who cares? What's it going to get me? The best intensive-care room money can buy?" So he and Lynn had quit.

But now, as the weeks of preparing *We Got It Made* dragged on, Lynn and Gordon found themselves chain-drinking coffee and reaching reflexively for the nicotine energy of cigarettes. Silverman appeared to be feeling the strain, too. When Gordon had first started working with Silverman, he'd thought that Fred, who was the same age, looked old enough, on some days, to be his father. But now, when Gordon looked in the mirror, he wasn't so sure.

Only Alan Rafkin, at fifty-four the old man of the outfit, seemed impervious to the pace. An ex-marine, small and stringy-tough, he was still standing at the end of a day at KTLA, as the rest of them, their eyelids swollen with fatigue, went over the script changes one more time. Rafkin had just quit *One Day at a Time* after five years and was looking for one more hit series before he called it quits. He had more money than he knew what to do with, but, for some weird reason—probably the poverty of his childhood—he never felt like he was worth more than whatever few bucks he had in his back pocket.

Current fortune notwithstanding, Alan Rafkin was determined not to ease up on the production schedule. He was a pro. All four of them were. They had the scars to prove it. And none of them intended to be beaten up by a bunch of thirty-year-old kids.

Gary Nardino, president of Paramount Television, had made it his job for the past several years to try to corner the market on the hot young producers whose services Brandon Tartikoff coveted. And now, as the pilot season reached its peak, he was grooming his final presentations to the network of several of these producers' works. Within two weeks, he would be in New York, where the network would make its final decisions.

In a business where "Two bad years and you're out" was an axiom, Nardino had never been fired, had run Paramount TV for six years, and was one of the most successful TV execs in the business. Only a veritable four-star general of the industry would have had the balls to call Tartikoff in 1982 with the notion that NBC should pick up the low-rated *Taxi* when it was axed by ABC.

"If you pick up *Taxi*," Nardino had promised, "there will be a flow of creative people to NBC that you won't believe. You think you are seeing everyone, getting every idea, now. You are not. There are people holding themselves back because they don't think NBC is the place to be. But if you schedule *Taxi*, there will be a march to you that will astonish you."

Tartikoff had scheduled *Taxi*, which was a very unusual move; networks simply do not adopt their rivals' castoffs. And, sure enough, he had caught the collective attention of the Young Turk producers. Now Tartikoff was the first person most of them thought of when they contemplated marketing a sophisticated project.

The selling of *Taxi* to NBC had been a major coup for Nardino, because it made him the undisputed leading middleman between the quality producers and the networks. He had at Paramount men like Burrows, the Charles brothers, Gary Goldberg, Michael Weithorn, and Lloyd Garver (of *Family Ties*), Ed Weinberger, Stan Daniels, and David Lloyd *(Taxi)*, Jim Brooks *(The Mary Tyler Moore Show)*, Dan Curtis *(The Winds of War)*, and more.

Nardino had, over the past couple of years, shifted most of Paramount's business away from ABC, which carried the studio's Garry Marshall shows, like *Happy Days* and *Laverne and Shirley,* and toward NBC, which aired *Cheers* and *Family Ties.* The shift had been engineered as the Marshall shows worked their way toward a natural death due to old age. These shows, from their syndication sales, had brought enormous wealth to the studio, but they were running out of steam, and Garry Marshall showed no real interest in toiling for another decade on replacements for them. As principal owner in several syndication sales, each of which hovered around the $100 million mark, Marshall had no financial need to work. Half hours are more desirable than hour-long shows as reruns for local stations, since they allow greater flexibility and less viewer involvement, and Marshall's half hours had made him one of the richest men in show business.

Almost as rich was his sister, Penny, star of *Laverne and Shirley.* In fact, Nardino had lately been able to lure her into reenlisting only by going for her soft spot: major appliances. When cash would no longer motivate Penny Marshall, Nardino had to hold under her nose a fifty-cycle washing machine, or a microwave so complicated it could do your taxes. And thus far the ploy had worked. But how long could her appetite for porcelain and chrome be expected to last?

So Nardino was trying to make the future as lucrative as the past had been for Paramount TV by cultivating a new crop of shows, and it was a wearying job. ABC had just made him an offer for a show that he had great hopes for, but he'd turned the offer down. The show was the brainchild of a pair of his white-hot quality producers, Ed Weinberger and Stan Daniels, who were working with David Lloyd, the man Nardino considered the best comedy writer in television. The show, called Mr. *Smith,* was based on the notion that a monkey could talk, but Nardino had seen enough of the show's proposed story lines and its pilot script to know that the program would be good in spite of its premise. ABC had offered to buy but wouldn't make a commitment for thirteen episodes, which Nardino considered the minimum needed to get the viewers past the belief that this was just another inane sitcom. Turning the network down had been difficult; most sellers of TV programming jump at any chance to get their shows on the air. Furthermore, the pass on thirteen shows by ABC would, Nardino knew, stigmatize the show as a loser, and make its sale elsewhere more difficult.

Nardino had hopes, though, that Brandon Tartikoff would welcome the services of Weinberger, Daniels, and Lloyd. After all, when they were with the marginally rated *Taxi,* Tartikoff had picked up the show not just to impress the production community, but also to grab the contracts of those particular individuals.

Another project Nardino was working on for NBC that was a last-minute hopeful for scheduling was a prime-time version of *SCTV.* The avant-garde comedy show, which resembled *Saturday Night Live,* was acclaimed by the critics but was on the brink of extermination in a very late Friday-night slot. Tartikoff wanted to inject some children's appeal into the show, and then counterprogram it against *60 Minutes,* a show that had been murdering everything that had been thrown up against it for years. If Tartikoff didn't go with *SCTV,* he'd probably have to take the opposite tack, which was to try to blunt *60 Minutes* with his own newsmagazine show, *Monitor.* But he had only the thinnest faith that the rather scholarly *Monitor* could go head to head with the black-hats-versus-white-hats "news cowboy show" that was *60 Minutes.*

But the *SCTV* people were not acting ready for prime time. In fact, they were, Nardino felt, almost perversely obstreperous about not wanting to tailor their show to attract kids. They were fueled in their vision of artistic integrity, it seemed, by an offer from Home Box Office. It was one of the first times that a cable outlet was seriously competing with a broadcast network. HBO had tried to buy *Taxi* the previous year, but at that time the cable giant just didn't have enough money. A year had made a significant difference in HBO's appetite for original programming, though, and in its ability to finance its own shows. The HBO competition for *SCTV* was having a chilling effect on not just NBC, but also the other networks. The monopoly on original programming enjoyed by the Big Three broadcasters, which were, in effect, "the OPEC of Information," was being threatened.

On top of all that, two graphic designers in New York who liked to watch *SCTV* had kicked off a petition and phone-in campaign to save the show. They were introducing to the media, which had been very kind to NBC, the notion that perhaps NBC was not the ultimate purveyor of high-quality programming. Worst of all, these two TV watchers were planning an actual protest, a demonstration—a March on Burbank!—to bitch about the show's threatened cancellation.

Nardino found the whole episode trying, and it dragged him away from preparing his sales pitches for his new pilots. Now ready for inspection by NBC were *Little Shots,* a kids' show produced by actor-director Ron Howard; *The President of Love,* about a wealthy eccentric; and a variety show called *The Sunday Funnies.* He was also trying to sell a show called *Webster,* about a little black kid, to ABC.

Nardino also had to convince the networks to retain old shows, including *Joanie Loves Chachi,* a spin-off that was going nowhere at ABC, and *Taxi,* which had not grabbed ratings after its transplant to NBC. To convince the networks to keep these shows on, and to buy the new pilots, he would have to show them audience research, letters from viewers, attractive new story ideas for the upcoming year, and any other information that would spotlight the assets of his shows, in particular relation to the weaknesses of each network's competition.

In the past, this crucial part of the year had invigorated Nardino. But this year, for some reason, it just made him feel . . . tired.

At Twentieth Century-Fox Television, Peter Grad sat at his desk in his corner office, an abiding Hollywood power symbol, and reflected uneasily on NBC. At times that network seemed to have a bureaucracy that made the federal government look

streamlined by comparison. Though Tartikoff was, of all network programming chiefs, the one most intimately involved in
actual program development, Grad was having a hard time trying
to tell what NBC actually wanted. "Quality" had been the
watchword at the beginning of the Tinker regime; now it seemed
more and more to be replaced by "quality of execution." There
could be a significant difference between the two. "Quality"
meant substance. "Quality of execution" meant only form. An
Army training film could exhibit "quality of execution" if the
camera work was performed with a little ˏflair. Another recent
watchword was "balanced menu"—which seemed to mean NBC
was commissioning every kind of program imaginable. So what
did NBC want? Not only was it hard to get a fix on the network's wants and needs, but it was also becoming more and more
difficult to get to the man who dictated most of them: Tartikoff.
More and more of late, he seemed to be insulated by lieuten-
ants. It was getting harder and harder to pitch an idea to him
directly, and by the time a proposal got to him, who knew what
Frankenstein it might have turned into.

Nonetheless, trying to sell a show to NBC was now better,
in Grad's opinion, than it had been in the Silverman years. The
bad old days. In Grad's opinion, Fred Silverman is "the most
difficult man I've ever met in the business." Worse, Silverman
was inefficient. And inefficiency was the sin Grad found hardest to forgive. He had come to Hollywood from Wall Street, where
he was a broker and securities analyst, and he harbored dreams
that he could one day help see that the trains ran on time here
in fantasyland.

At forty-three Peter Grad was, in many ways, a grown-up
boy who had run away to join the circus, leaving the button-
down universe of Wall Street for the zoo of show business. Married to Jim Burrows's sister Laurie, a tall blonde who had been
a successful model and was now the resident cooking expert on
the nationally syndicated Hour Magazine, Grad had already attained most of the rewards America can provide. Except job security.

If things didn't turn around at Twentieth Century quickly,
he knew he could find himself back on Wall Street speculating
about pork bellies and soybean futures in practically no time at

all. Right now he was worried that NBC had pigeonholed him as a developer who could provide only action-adventure, when Grad's first love was comedy. A slightly chubby man who looked ten years younger than he was, Grad was dying to promote something to the networks that really had some laughs in it. But NBC wasn't biting. For half hours the network looked to Nardino at Paramount, and a few independents. As a result, Grad had only one comedy pilot in the works at NBC, and he was starting to get negative feedback on it.

The pilot he had given Tartikoff was *The National Snoop*, done by his colleague and close friend Allan Katz. Grad thought the pilot was great, possibly the breakout hit of the season, but now all he was hearing from NBC was that they hadn't gotten what they'd ordered. At first, when he'd given the pilot to Warren Littlefield, he'd been told the network was delighted. And just a few days ago, there had been a screening in Burbank for a number of NBC execs that had everybody rolling with laughter. Now, though, he was hearing, secondhand, that Tartikoff didn't like the show.

Grad was a little worried about Allan Katz. Katz was as independent as anybody in TV—one of the most intelligently creative people in the business—but sometimes independence and intellect were not conducive to success in television. Part of Grad's job was to run interference for guys like Katz. About fifteen writer-producers were under exclusive contract to Twentieth, and sometimes—what with their egos, ideas, and concerns—it was enough to make Grad feel like a sergeant at arms at a solipsists' convention.

The shoeshine guy came in to pick up Grad's shoes. Office-to-office service was one of the minor perks of the trade. Grad set aside the script he was reading and fixed his eyes on the black goop in the little cask on his desk. He really didn't want to eat it; it tasted awful but was one of the things his wife, Laurie, was always concocting to help him keep halfway trim. Space-age diet food. Maybe if you froze it. Yeah, he could get his secretary to pop it in a freezer somewhere and then he could swallow it like ice cubes.

Grad got up and looked out the window. Twentieth Century-Fox had been the brainchild and fiefdom of Darryl F.

Zanuck, whose phosphorescent wheelings and dealings had created a movie empire like the world had never seen. Darryl was long gone, but his signature remained in the permanent buildings and sets that made the lot at Twentieth an odd and garish conglomeration even in a town that was perhaps the oddest and most garish in the world.

Grad looked down the street toward the offices of Glen Larson. Grad liked to think of all the producers in his charge as equals, but Glen Larson had to be the equalest of all. And especially valuable now that Grad and Harris Katleman, Grad's boss as head of the TV division, were both in the biggest put-up-or-shut-up position in their careers. Larson was an action-adventure producer so successful that he had his own personal building—not office, actually, but edifice—on the Twentieth lot. Larson, who had become Grad's only consistent calling card at NBC, was now putting finishing touches on an action pilot called *Manimal,* a program based on the idea that a man had the power to change himself into various beasts. The hokey concept, which NBC had brought to Grad on the condition that Larson produce it, was a welcome one to the studio, since it had good possibilities for foreign sales. To a large extent, America's television is the world's television, but the programs that the studios can sell abroad most readily are those involving high, melodramatic action. Comedies generally lose their yucks in translation. Shortly after the networks would choose their new shows, the foreign buyers would descend on Hollywood, and Grad was confident that *Manimal,* if sold to NBC, would further enrich Twentieth's coffers from foreign sales.

Another Glen Larson production that Grad had hopes for was called *Automan.* It was an "ABC-type" show, starring a sexy-looking hologram who fights crime. Utterly devoid of ideas, it was a sure bet to keep ten or twenty million kids off the streets for an hour each week.

The producers of *Dynasty* had cooked up another steamy nighttime soap for Grad, which they called *Navy,* and ABC was enthusiastic about a family comedy, *It's Not Easy,* as well as a mystery called *Masquerade.*

All in all, the Twentieth pilots, which Peter Grad had been nursing along for almost a year, looked fit. And it was a good thing they did, because the previous year had been a disaster. A zero. Twentieth hadn't sold a single pilot. When the season was over, Harris Katleman, a self-described street fighter, took Grad and a few other Twentieth executives to Palm Springs, where he sat them down amid 109-degree heat and said, "This is the closest to hell you're ever gonna get. We did a shitty job, myself included. It was lousy. We stunk. The whole operation will have to be turned around, and immediately." It was then that Grad had begun to fully appreciate the axiom "Two bad years and you're out."

This year, though, would be different. Grad was sure of it. But whenever Grad gazed out his window, in the direction of the office of his pal Allan Katz, he had a sinking feeling.

As the deadline for pilots approached, production companies were working deep into golden time—triple wages for union members now asked to work fourteen hours a day—in order to meet due dates set by networks.

At Fort MacArthur, dark and pretty actress Rachel Ticotin rubbed her brown eyes as she reached for the energy to do another take. Above, the sky was a hazy bruise blue, and below, San Pedro jutted against the ocean—a regular tropical Pittsburgh. San Pedro was a vast smear of oil rigs, oil tanks, a grid of pipe, warehouses, container trucks, fast-food stands, and gas stations stretched to the gray spine of mountains on the horizon.

Cannon guarded either side of a red sign that read:

HEADQUARTERS
1ST BATTALION (AIRBORNE)
316TH FIELD ARTILLERY
EXCELLENCE IN PEACE AND WAR

Ticotin was working on a pilot about the military called *The Whole Nine Yards*. Rachel had had a large role in a movie with Paul Newman and Ed Asner, *Fort Apache, the Bronx*, several years before. She'd gotten good notices and had lain back and waited for the next big offer. But nothing had happened. Now, still in her early twenties, she was trying to get things rolling again. But she felt dead on her feet. It wasn't the work that was beating her into the ground, though. It was the traffic—the god-awful noisy snakes of metal that ate up hours of every day as she fought her way down to San Pedro, where a military base was letting them film.

Jim Brown, the producer of the pilot, was living in Marina del Rey, an expensive oceanside condo-burg an hour away from greasy little San Pedro. Brown also hated the commute, but he had a thing about being at home now, even though his kids and their mother were long gone. The fourteen-hour workdays that are necessary to produce an on-location, hour-long drama had totaled Brown's marriage, and his chance to watch his kids grow up. So now he liked to be home even though, in some ways, it was too late to matter.

Brown was almost certain that they would have to go into the New York selection process with just a work-print of the show—without the complete postproduction process of adding in music and looping out mistakes. Even so, he had high hopes for it. He had been getting good feedback from the network, mostly through David Gerber, who was executive producer of the show and a good friend of Grant Tinker's—they had once sailed down the Nile together. The word from Tinker was that Tartikoff was impressed with the clips he'd seen from the daily shooting.

Gerber was a big name in action shows, so Brown hoped the network would realize that the rough cut they would be getting was not the very best that he and Gerber could do. But you never really knew what the networks thought.

From the dark insides of a boxy, rolling sound studio, not much bigger than a bread truck, Brown stepped out into the biting shine of the postlunch sun. Before he could stop squinting, a big, baby-faced guy with a walrus moustache flipped Brown a football and then went down and out. Brown cocked his arm and threw a high, arching spiral that thunked into the kid's sweaty, T-shirted chest.

"Jesus," said Brown, who was about fifty and bearish and straw-blond, "I better quit while I've still got an arm." The kid who had caught the pass kept going with it, like a deer, hurdling equipment boxes and castoff costumes. He was exuberant.

"Morale is high," said Brown. "You can feel it when you've got a good one going."

From the studio truck came the soundtrack to a porno film. Primordial groans, gasps, and sighs of female pleasure wafted across the Army's crewcut greensward as Jim Brown waded into a crowd of actors and began to set up the next shot.

Rachel Ticotin, listening, ground the ball of her hand into first one eye and then the other.

Under the same hot yellow sun, on the same afternoon, producer Chuck Bowman idled away the day at one of his favorite places, an outdoor showbiz café called Patty's, not far from the Burbank Studios, NBC, and Universal. Bowman was the producer of V, an NBC miniseries that was a candidate for a full thirteen-show order. V was, to date, the most expensive set of hours the network had ever bought and had—prior to airing—been identified as "Tartikoff's *Supertrain*" in the press. In any case, NBC had invested hundreds of thousands of dollars building a bogus fleet of flying saucers in order to create an opus about Nazi, rat-eating lizard aliens from outer space who conquer earth and impregnate comely teenagers, etc., etc. Thanks to promo savant Steve Sohmer, the miniseries had been the beneficiary of tremendous amounts of hype—there were "vandalized" V billboards all over L.A. showing the alien leader's face emblazoned with the "spray-painted" V that symbolized earthly resistance to intergalactic fascism. America, apparently, was ready for a bit of E.T. cold war. V, finishing first among all shows with a monster 40 share, was the highest-rated NBC miniseries since Shōgun, and had powered NBC to a lead in the May sweeps. So Chuck Bowman, "Mr. Sweeps," at least for the moment, was content to kick back on this clear, hot afternoon, knowing that the network people wanted him, for once, just as badly as he wanted them. He was certain his show would be scheduled in 1983–84. Brown's only concerns were about what format V would be shown in and how much money NBC would pay to show it. Those are the kinds of worries producers like to have.

Brown's friend Jonathan Winters, who came to Patty's almost every day, strolled by. "I saw your series," said Winters. "Great! At least the first twenty minutes. After that"—he made a face—"history. You do well with it?" Brown nodded. "Big points?" Brown nodded. "Lotta money?" Brown was starting to get embarrassed by the stares of people at nearby tables. Even in a showbiz café, Jonathan Winters is a magnet for attention.

Suddenly, Winters seemed to become aware that he was wearing a baseball cap, and he went into an impromptu impersonation of local Dodger pitching star Fernando Valenzuela; he faced an imaginary batter and went through all manner of pitching maneuvers. "Santo Domingo, *muy bueno*, treeple-A ball, Gaylord Perry ees a queer." Then he became two snooty patrons in the stands. "Tell me, is that the Mexican? Hey! I don't need that kind of talk! I have box seats!" Then he was the manager in the dugout. "Tell me about it, queer. Go up to Frisco and get yourself a pink little cable car. Cost you five thousand. I got that much on me. Here, asshole. Take it!" Then he was the batter. "Hey! Ya'll show me somethin'. Less see what you hay-yuv." He was Fernando. "*Aquí. Cómo dagos.*" He was in the stands. "Jesus Christ, eight hundred thousand dollars and he can't speak the goddamn language!" Then the manager. "Good work, kid. Now go put away your uniform. You get the white chick with blue eyes. Show her the dragon! Make the iguana talk!"

Bowman jumped in with the only thing he could think of that would quell this performance, which was starting to draw a crowd. "You hear what Jack Paar said about you this morning on CBS?"

"No," said Winters, "what'd he say?"

"He said you were absolutely the best and funniest man of this time."

That seemed to strike Winters's fancy, and for at least ten seconds he was reflective and calm. Then he spotted some painters touching up the front of the restaurant and went over to them and did a painting routine.

Chuck Bowman watched Winters, happy for this rare chance to sit like a big cat in the sun, not even remotely concerned about the daily hassles of shooting schedules, locations, and contract negotiations. Chuck Bowman had it made, temporarily.

Two and half days later, in the cool evening quiet of Beverly Hills, Earl Hamner, the fifty-nine-year-old writer-producer who had created *The Waltons,* fed his two dogs, and felt uneasy. All day long he had felt uneasy.

In the morning, he had planted his yearly vegetable garden. It was one of the few vegetable plots in this enormously expensive city's confines; every square yard of bean and corn was worth about $1,000. Near his garden was a small shed where he had until recently kept chickens. He was certain he was the only person in Beverly Hills who raised chickens. When the last chicken had died, about two months before, Hamner and his wife had cried. The chicken, seventeen years old, had been hatched by Hamner's son, and its death had seemed to the Hamners to terminate a link with their children.

At the moment, Hamner's wife was out of town, and the house sounded empty to him, which made him even more uneasy.

Then, as nighttime shadows started to blacken, Earl Hamner sat down at his typewriter. He began to write story ideas for *Boone,* a pilot that he would present to NBC within a few days. *Boone* was not, as *The Waltons* had been, an exercise in autobiography. The John-Boy Walton character had been closely modeled on Hamner. Tom Byrd, who played the lead in *Boone,* thought that Hamner's new show was an expression of young Earl Hamner's dreams. "John-Boy was Earl," Byrd had said, "but *Boone* is what Earl fantasized about being."

As Hamner started to write, the uneasiness lifted off him.

The following Sunday, a line of people shuffled their feet outside a Sunset Boulevard building known as the Preview House, awaiting an afternoon of free entertainment and door prizes. Some prostitutes in the bank parking lot across the street provided visual diversion as the minutes of waiting added up and the afternoon got hotter. Every few minutes the line would inch ahead, but the doors of the Preview House were not open; the line was just compressing itself. Almost all the people in the line looked like less-than-middle-income wage earners. Their clothes were shiny and imitative of style, and their handbags and hats and shoes looked cheap and synthetic. They looked like people who were used to standing in line and were here because it was free, so when they finally started to trudge slowly ahead, actually moving into the open doors, there was a feeling of having achieved a victory.

Inside, they filled out forms and were assigned seats. The managers of the Preview House then knew that seat 11-B, for example, was that of a woman in the eighteen-to-twenty-five age group, with no children and an income of between $11,000 and $14,000.

The motion picture and TV industries use the Preview House to test audience response to their offerings and decide the fate of films, programs, actors and actresses, directors, producers, and, indirectly, select real estate agents and car dealers. Audience members hold a dial that relays their sentiment of "I like this" or "I don't like this" to a control room where a computer records, for example, how many women who make $11,000 to $14,000 a year like a particular actor. The computer also records the responses of the "overall laugh meter," which grades each joke in the story on a one-to-one-hundred scale. This information is correlated with questionnaires that are filled out by the audience at the end of the program, and with the responses of discussion groups that are held after the showings.

Bud, a twenty-three-year-old unemployed janitor, was one of the representatives of American cultural tastes this last Sunday of the pilot production season. Bud was a friendly person, a very decent man who offered to change places with a stranger so that she could see better. But Bud was not very smart. He was not retarded, but he probably could have carried on a satisfying conversation with someone who was.

When the show started, Bud chuckled enigmatically at the opening credits. Bud was easily satisfied. Then a young, attractive woman came on the screen and Bud grabbed his dial and twisted it as far as it would go on the "I like this" scale. "She's pretty!" he said.

Throughout the program, which was attractively filmed but was slow and shallow and pointless, Bud cranked his dial in ecstasy.

After this preview, the ABC network, which was considering the show Bud liked for scheduling in its 1983–84 season, was convinced it had a hit on its hands. It began to hype the show as its can't-miss smash, ladling a fortune in promotional money into it and scheduling it in a golden time slot after the hit show *Dynasty*.

The show that Bud liked so much was *Hotel*. Its executive producer was Aaron Spelling, a man who had seven similarly shallow and successful programs on the air, most on ABC. Aaron Spelling's salary from those shows—not his ownership interest in them, but his salary, on which he had to file a 1040 like any other working stiff—was somewhere around $280,000 per week. At one time in his life, Aaron Spelling had been poorer than Bud was now. But Spelling was a clever man with a story line, and that was all it had taken.

After the show was over, Bud had to stand in line again. He lived in Canoga Park, a good twenty-five miles away, and since he couldn't afford a car, he had to stand in line for the city bus. But he was happy with his afternoon's entertainment. It had, after all, been free.

Another person exiting the Preview House, stepping toward a big new Oldsmobile with bucket seats that would look right at home in the first-class section of a Boeing 747, was pharmacist Herbert Goldman. Goldman's drugstore in Tarzana catered to singer Michael Jackson, actor Walter Matthau, and a clutch of other celebrities; in true L.A. style, Goldman should have hung autographed eight-by-tens of these Big Faces on the wall, as did the Baskin-Robbins around the corner from him. But Herbert was a published poet and one of approximately ninety-five living Americans who still read more hours per day than they watched television, and he had no intention of billing himself as Pharmacist to the Stars. So it was not proximity to glamour nor a free afternoon's entertainment that had brought him to the Preview House this Sunday afternoon. Goldman was there on the off chance—admittedly quixotic—that he might somehow influence the viewing choices of two hundred ten million Americans in 1983 and 1984. He'd voted adamantly against the scheduling of *Hotel*. But he was realistic about his voice being heard, mostly because of a previous experience at the Preview House.

At that time, several years earlier, he had been invited to fill out a questionnaire and participate in a discussion group after the screening. The show he'd watched had loser written all over it; it was so hopeless, in fact, that the questionnaire was totally inadequate. It allowed him to express the opinion that the pilot was (1) excellent, (2) very good, (3) good, (4) fair, (5) poor, or (6) very poor. But there was no place to record an opinion of (7) offensive, (8) appalling, or (9) god-awful.

Moreover, Goldman noticed that a great many of the two hundred people who had also just watched the pilot did not seem overcome with unalloyed joy, either. This garden variety detective-show-with-a-sexy-twist had provoked an unusual amount of coughing and squirming throughout the screening, and afterward most of the video guinea pigs looked eager to participate in the drawing for the door prize and vamoose into the early-summer evening cool of June 1975.

The viewers in the discussion group seemed genuinely relieved when informed that no one in the room was connected with the show, and therefore would not be hurt by negative opinions. Actually, the discussion group's director revealed, this was just one of several evaluative screenings. What he was soliciting were absolutely frank, gut-level opinions—so don't anybody be shy; speak right up.

All around Goldman, people were sticking up their hands and denouncing the show. "If I come down hard enough on this dreck," he told himself, "maybe it could actually be the straw that breaks the camel's back; maybe somebody at the network will get wind of my opinion and voilà! they'll say, 'That's it! Let's can this dog.' " Not likely, but possible. After all, that was what he was here for, right? So . . . up went his hand.

After a moment, the director pointed to him and said, "Yes, sir."

"Well," he replied, suddenly nervous as his voice echoed around the room, "I think it would be a really bad thing to put this show on the air. I wouldn't want my children to see it, because it's violent and cynical and extremely sexist. I mean, I have three daughters and, well, the scene where the woman is posing as a hooker and kicks the pimp in the groin, I mean, I don't think I'd let my girls watch it. And if kids don't watch it, I think it's got a real *problem*, because it's not done intelligently enough to appeal to many adults." Goldman had coughed and cleared his throat. His critique wasn't coming out quite as succinctly and

eloquently as he might have hoped. "The central problem with
. . . I'm sorry, I've forgotten the show's title. . . ."

"Charlie's Angels."

". . . with Charlie's Angels is that it just isn't interesting
enough to make adults want to watch it. I don't know anything
about television programming, but I do know that if this show
gets scheduled, it's bound to flop."

Norman Goldman left the Preview House in 1975 feeling
good. For one thing, he felt as though he might possibly have
prevented a real cultural travesty from occurring—and saved
some poor, multimillion-dollar Hollywood production company a few bucks in the process. And, for another, he'd won the
door prize. It was $25. Cash.

At the beginning of the last week of the pilot season, they
shot We Got It Made.

Teri Copley hadn't been completely invented yet, though she
already had most of her moves down, and was sashaying around
the set at KTLA with her shirt pulled down around her back,
both shoulders bare, her cumulus mop of yellow hair silvered
by the klieg lights. It was the break between the first and second
tapings of the show and the audience was a mix of old folks from
an Episcopal church, St. Thomas of Canterbury, Long Beach, and
a busload of marines from Camp Pendleton. Because it's against
the law to charge admission to the filming of network shows,
everybody gets in free. As the sound system boomed Elvis's
"Suspicious Eyes," the guy who had been hired to do the audience warm-up surveyed the crowd skeptically. "Episcopalians!" he snorted. "Give me a black church group any day! Yeah!
Holy Rollers! Get down! But not," he sniffed, "Long Beach
Episcopalians."

The warm-up guy was Mark Maxwell-Smith, or simply Max,
a bouncy, excited game-show developer whose one true love was
doing audience warm-ups. He was considered the best in the
business. "There's Silverman," Max whispered hoarsely, his voice
a mess from doing two Tic Tac Doughs earlier.

Silverman paced the wings, looking tan and imploded,
smoking, drinking Pepsi, casually dressed in a golf sweater among
a militia of network men in dark suits. Wider at the stomach

than at the shoulder, he had the appearance of a fat man who had just lost weight.

"The first time I ever saw Silverman," said Max, "was about four years ago during pilot selection week in New York. I was at '21' with the vice-president of NBC daytime, at about the time NBC daytime was starting to go down the toilet. When the veep I was with saw Fred, who was president then, he got white and dived under the table. I mean, he went under like the *Nautilus*, ready for a long-term stay down there. I mean, what am I supposed to do, order him a drink with a three-foot straw, or what?"

Silverman and the men in suits marched up a stairway to a control booth. This show would be shot with three videotape cameras, and would be edited almost as it was shot on the sound stage. It was a much easier, much cheaper process than shooting a show on location with just one camera. Three-camera in-studio shows offer great profits to their producers.

Max took a gulp of air and stepped in front of the audience. "I'd like to give you our old New York–style welcome," Max said. Then he paused and shrieked, "Stick 'em up, give me your money," and was off and rolling.

Teri Copley, in a denim miniskirt, her blond cloud of hair now raining down onto her clingy, robin's-egg-blue top, bounced up and down, as if she were cold. "It gets me up," she said.

Teri looked at the audience. Her huge blue eyes were the very definition of innocence and insecurity. "What if they don't laugh?" she said.

But they did laugh. The biggest laugh of the night was when a toilet seat, under the impact of pressing thighs, played the first four notes of "Here Comes the Bride." The second-biggest laugh was when it did it again.

After the show the principals wandered around the stage, posing together briefly for pictures, as the crowd filed out. Teri Copley stepped into the wings, where Fred Silverman stood, cigarette in hand, smoke curling and twisting around his head. Copley seemed now both princess and puppy dog, wriggling up to her lord and master. "What did you think?"

Fred Silverman took a quick pull on his cigarette. "They loved you. They loved you."

"I was nervous," she said, one hand grasping the other.

"The worst is over," Silverman advised.

A young man walked up to Fred and offered, "Congratulations, it looks like a TV show."

"A successful TV show," Silverman amended.

Brandon Tartikoff had one more corporate duty before he could leave for New York to make NBC's final pilot selection for 1983. He had to make an appearance at an industry luncheon where Grant Tinker would tell about five hundred people in Century City's Century Plaza Hotel ballroom why he had flip-flopped on the question of who should own syndicated reruns. The matter was the largest single issue of the year in the TV industry. In 1970, the Federal Communications Commission had banned the networks from owning the prime-time shows that they ran. The networks' purchase of them from the studios became nothing more than a rental fee. The studios inherited the right to sell the shows as syndicated reruns, after the networks had shown them twice. Syndication sales were a windfall to the studios. Now Peter Grad's Twentieth Century-Fox could make $250 million on *M*A*S*H*. Gary Nardino's Paramount could make almost as much on *Laverne and Shirley* and *Happy Days*.

When he was head of MTM, Grant Tinker had railed against the network's owning the shows. Now that he was head of NBC, though, he had changed his mind. On this day in late April, he was going to tell the industry why. The reason he would offer his peers was that now he "knew more."

Brandon was expected to be there for the unveiling of this wisdom.

On the way to Century City, a good forty-five minutes from Burbank in the noontime traffic, Tartikoff, riding with Gene Walsh, NBC's head of press relations, talked about the pilot season. He looked tired. "I now believe in the Dave Kingman School of Broadcasting." Dave Kingman was a baseball player who hit many home runs, struck out many times—and didn't do much in between. "After doing this for the past six years, I've finally realized that when I come to bat, it doesn't count if I get close. It only counts if I hit it out. To get a twenty-three share is worth nothing—it's better to risk a fifteen share going for a forty share.

"We're not going to get to the promised land by doing a well-made domestic comedy or a well-made detective show or cop show. Look at years past, when everybody used up about half their budget covering their ass, making sure they had a doctor show, a cop show, a standard domestic show. The idea was, if everything falls apart, we'll have this stuff to fall back on. But now it's time to throw the long bomb."

Brandon Tartikoff had spies. All the networks did. He had people who got as close to the insides of the other networks as possible, and then called him up. And his spies were telling him that CBS and ABC were playing it safe this year. According to his informants, and also according to the legion of agents, producers, writers, and others who called him with information—hoping to curry favor—the pilot seasons of ABC and CBS were soft. And, it seemed to Tartikoff, their entire schedules were as vulnerable as toothless dogs. The paradox of having a schedule of veteran winners is that proven shows are often approaching the end of their natural shelf life. *Happy Days,* long the strongman of ABC's Tuesday night lineup, had folded when Tartikoff threw the vital, young *A-Team* against it. In one strike Tartikoff had won the whole of Tuesday night, with the previously weak *Remington Steele* now flashing swiftly along in the *A-Team's* powerful wake. *Fantasy Island, The Love Boat,* and *The Dukes of Hazzard* could fall, Tartikoff thought, if only he could come up with the home run again.

"I think we've got some shows with real breakout potential," Tartikoff said. "*Mr. Smith* could be the breakout show of 1983. So could *Manimal.* Harry Anderson could be the breakout star of the eighties. *Boone* has Earl Hamner behind it and *The Whole Nine Yards* has David Gerber. *We Got It Made* has the most attractive cast I've seen in a show in five years. Teri Copley tested through the roof at the Preview House."

But what does Grant Tinker—television's knight on the white horse—think about *We Got It Made?*

"What would be more interesting than watching *We Got It Made* would be to watch a movie of Grant watching that show. And that has nothing to do with Fred Silverman; it's just totally not the kind of comedy I know Grant to like.

"There are projects you get in the business to do," he said, "such as *Hill Street Blues.* You go to a party and your friends and family say, 'Gee, you had something to do with *Hill Street!*' But that's not to say that *The A-Team* doesn't pay the bills.''

So far, Tartikoff had seen about half the new pilots. "I'm just coming out of that initial feeling of euphoria where you think, 'My God, what if everything comes out great?' Now I'm sort of in the phase where the blue sky is still there but I'm starting to get a little nervous. What if I don't have enough? It reminds me of my days when I played football, and we'd be prepping for a game with the second string, who'd be wearing the jerseys of the team we'd be playing on Saturday. We'd always run great against their defense!

"Actually, I do feel quite solid. *The Whole Nine Yards*, for example—which we're now calling *Ride with the Wind*—I think it has the potential to be extraordinary. I saw a documentary last January called *Soldier Girls*, about six girls in basic training, that won a lot of prizes. Springboarding off of that, I fell in love with doing what would amount to a working-class drama, set against the modern Army, with men and women. We've got an extraordinary cast, and the final script, as Grant would say, is publishable.''

Color flushed into his face, and the look of tiredness lifted out of his eyes. "There's another one!" he said. "It's called *Boone.* When I first got involved in trying to work up a series with Earl Hamner, I had just spent two years working with the Mandrell sisters, and had gotten to know the world of country music a little better. So I asked Earl if there was a way to loosely fabricate a sort of Conway Twitty–Elvis Presley kind of story. Instead of having John-Boy, who you knew would grow up and be a big writer someday, you start with somebody that you just knew, three or four years down the line in the series, would turn into a singing idol.''

Tartikoff was interrupted by the soft honk of a stretch limo passing on the left. Inside were the producers of *V*, gliding by on their way to the Century Plaza. The whole carload of them flashed the two-finger V-for-victory sign, which was the signature of their miniseries—or was possibly their acknowledgment of just having kicked the competition's butts in the ratings. Whatever the sign meant, Tartikoff returned it, grinning.

Then Tartikoff explained why things looked bad for *The*

National Snoop. "Allan Katz had good credentials coming into the project—*M*A*S*H, Rhoda, Laugh-In*—when we first started working on the script three years ago. Fred Silverman was my boss at the time. When we got the script, I liked it, my head of comedy liked it, but Fred thought it was too late night. It was a phase Fred was going through. Fred went through different phases. His first phase was quality—*Lifeline, United States.* Then, let's get back to basics, which was *Sheriff Lobo* and *96.* Then, after that, it sort of became anything. *Real People* to *Diff'rent Strokes* to *Hill Street Blues*—anything. Let's just get circulation up. *Snoop* fell into the crack separating back to basics and anything goes.

"But this development season, with me using the hit mentality—the Dave Kingman School of Broadcasting—I thought maybe *Snoop* would be a home run. But it was just too zany— there was never any reality to work off of.

"If it were going to be a monster hit, you would have known. There's a *smell* to them. The first piece of film that I saw on *The A-Team* was a twelve-minute cut put together for the affiliates in early January. And when Mr. T went into the fight scene with the seven-foot Mexican, I knew there was something going on there. And when Howling Mad Murdock started singing Mick Jagger, I was going—" His face contorted in triumph. "Those two moments . . ." Tartikoff's voice trailed off, conceding the inexpressible.

When Walsh and Tartikoff pulled up at the Century Plaza, dozens of Tartikoff's colleagues descended upon them and began glad-handing him. Many of them were producers—some known as men who really cared about what their peers thought of their artistic abilities. Others had the reputations of being guys who just wanted to pay the bills. Tartikoff greeted them all with equal enthusiasm.

6
Manhattan Heat

May 1983

The NBC executives, sprawled in comfortable chairs, smoking cigarettes and fidgeting, were not laughing at Teri Copley in *We Got It Made*. The pilot, which was being screened in a conference room off the main office of NBC's head censor in Los Angeles, provoked the occasional mirthless chuckles, but, all in all, it just wasn't working for this very important audience—the first network pilot selection group to see it.

Just a few years ago, during the reign of Fred Silverman—the executive producer of *We Got It Made*—a pilot tape getting a reception like this might have been jerked, half played, out of the machine. If it had been played all the way through, it would have been the target of a barrage of barbs in a noisy hubbub of scattered attention. But the current NBC executives were mindful of a memo that Brandon Tartikoff had issued a couple of days before the pilot selection process had started. The memo began, "Welcome to the gladiator wars!" and then went on to remind the executives that they owed each of the pilots "an open mind, and the requisite attention span."

"Naturally," Tartikoff had written, "all of these projects will not be active contenders to the end. It is, however, worthwhile

154

to screen the products we have spent good money on." The memo said that the pilots were "geared 90% to the hit mentality. Naturally, when you go for home runs, you get a lot of strike-outs (Dave Kingman comes readily to mind)." The memo represented one more move by Tartikoff to distance himself from the memory of the golden gut.

Therefore, the execs were keeping their traps shut about the frothy sexual farce that they were watching—but there was no memo from the boss saying they had to laugh.

The testing on *We Got It Made*—done at the Preview House and also over a cable system on which NBC rented time—indicated that audiences liked Teri Copley, but didn't like her goofy co-star, Tom Villard, and were lukewarm about the show as a whole. But the testing results were not as damning to this program as the current frigid silence. NBC programmers were essentially of the gut-instinct school, a *modus operandi* that was a carry-over from the Silverman days. Fred was fond of saying to Brandon, "Fuck the testing. This goes on the air." Now Tartikoff himself had the reputation for being a gut-instinct or showbiz programmer, rather than a research programmer, and this attitude sifted down to his vice-presidents.

A number of those present were lower-ranked executives whose opinions would carry little weight, if any. When it came time for the absolute yes or no, Tinker and Tartikoff would have the final say, with the advice and consent of Burbank programmers Jeff Sagansky, Warren Littlefield, and Michelle Brustin, and New York business executives Robert Mulholland (president of NBC), Irwin Segelstein (NBC's vice-chairman of the board), Robert Blackmore (head of network ad sales), and William Reubens (head of research). The men from sales, research, and business affairs would point out which shows would be the easiest to sell to advertisers and promote to the public. They were indispensable conduits to the opinions of the major New York ad agencies and sponsors. The network would be soliciting millions of dollars from the ad agencies in just a matter of days, and it was important to know which of the new pilots Madison Avenue liked.

Not all the pilots, though, would make the plane to New York. After the first few days of screening in Burbank, it appeared that Allan Katz's *National Snoop* was an early casualty.

There Goes the Neighborhood, the Buddy Hackett fiasco, was left for dead and so were *Me and Mrs. C.*, a yawner sitcom about a white woman and a black woman; *President of Love* (a Gary Nardino pilot), about a wealthy eccentric; and *Yazoo*, a show that used puppets.

But most of the pilots survived the first cut: *Night Court*, *Jennifer Slept Here*, Ron Howard's kids' show *Little Shots*, Ziggy Steinberg's *Another Jerk*, Jim Brown's *The Whole Nine Yards*, Earl Hamner's *Boone*, Peter Grad's *Manimal*, and about twenty other hopefuls. Also up for consideration were a series based on *V*, and the pilot scripts for Paramount's *Mr. Smith* and Steve Bochco's *Bay City Blues*.

The mood at NBC, just before the mass exodus to New York began, was buoyant. NBC, Tartikoff thought, had the best pilots of any of the networks. Certainly it had the most; of the seventy-six total pilots, thirty-three were NBC's. NBC had the quality, Tartikoff thought, in potential shows like *Boone*, *The Whole Nine Yards*, *Bay City Blues*, and *Mr. Smith*. And it had possible breakout hits in shows like *Night Court* and *Manimal*. By the time he left for New York, Brandon Tartikoff had begun to think that this pilot season would mark the emergence of NBC from last place.

But he remembered an old saying in the TV business, one that applied to L.A. programmers winging their way to New York with suitcases full of can't-miss product. The saying was "Everything turns to shit over Denver."

Gordon Farr took one more call before leaving his office. He was trying to get away to Alan Rafkin's condo in Palm Springs. The call was from the agent of one of his cast members. A friend of the agent had flown back to New York with an NBC executive who said he had "loved" *We Got It Made*.

"Loved it, huh?" said Farr, from his black leather chair in his KTLA office. "Well, that's great. We heard last night that the West Coast head of research loved it, too. Hey, thanks for calling, and I'll let you know first thing when I hear something." Farr cradled the receiver and traced his finger in a circle around the top of his coffee mug. His face, its skin slack from fatigue, was as droopy as a bloodhound's. The phone had been ringing three, four, five times a day with people, even from other net-works, saying, "I hear you're on on!" But years of experience had taught him that that and $2 would buy you a cup of coffee in Beverly Hills.

The next day, Farr slept late and then dragged himself out onto the Palm Springs condo's patio, where he intended to spend the afternoon, dozing in the sun. The phone rang; it was Warren Littlefield. The pilot had by now been screened for everyone, Littlefield said, and the overall reception had been "good." But changes were needed. Could Gordon make them?

Sure, Farr said—what kind of changes?

Warren Littlefield didn't know what kind of changes.

But he promised to get back with them as soon as he could. Farr was encouraged. They wouldn't bother with changes if the program was already a ghost. So he lurked inside by the phone, waiting to find out what they wanted. The phone didn't ring until 4 P.M. The network wanted the first act tightened. That was all.

Farr got into his car and fought the traffic back into L.A.

The next morning he met with a couple of NBC program-mers, then edited his show for seven hours. Then, until 11 P.M., he worked on the program's audio track. Finally, he drove back to Palm Springs, the expensive golf town that sits two hours out of L.A. on a green square of irrigated desert.

On the last day of the Burbank portion of the pilot selection fortnight, Farr got a call from Tartikoff. It was usually a good sign when Brandon himself called—a signal of interest. More audience testing had been done, Brandon said, and now both Teri Copley and Tom Villard were testing through the roof. But there was concern about the program's "legs," its long-term po-tential. Could Gordon call him back with some future story ideas?

Within four hours, Farr had three stories. One was a char-acter piece; the other two were mistaken-identity pieces. Tarti-koff seemed satisfied.

But then Farr got a call from Fred Silverman, who had bet-ter spies at the network than Farr did. It looked bad, said Sil-verman. The network was only going to pick up three new half hours. And the reaction to the screenings had not been nearly as good as Farr had been led to believe. It was fifty-fifty, said Silverman. At best.

Farr went out for a round of golf. But it was extremely difficult to play. His mind was elsewhere.

On the first day of the final New York meetings, Harry Anderson and his ventriloquist buddy Jay Johnson addressed an Artists' Lecture Series audience at the University of Southern California. Both of them had pilots that were up for decision—Jay's was called *Sutter's Bay*—but Jay Johnson had already done four years of series work on *Soap*, and was holding his nerves a little better than Harry was. When one student asked a question, Harry screamed, "What are you, a cop!" And he shrieked at students who tried to edge out the back doors of the hall unobtrusively. Jay patiently answered questions. "We make a good team," observed Harry.

When he got home, Harry began to pace his living room, hugging a cup of night-black coffee, chain-smoking unfiltered Camels, and pouring out his nerves to Leslie, who had just dropped Eva off at day school.

"I can't get this image out of my head. There's all these NBC guys running around the Park Lane Hotel wearing red fezzes and falling down in the hallways and *deciding on my future*. My fate is going to be determined by a guy whose chest says, 'Hi! I'm Larry.' "

Harry vaulted up into his dentist's chair, which sat between the pool table and the blackjack table in his living room. "They do this scientific research, right? Okay, then, if it's so scientific, tell me why they have a seventy-five percent failure rate. They've got all these people wired up in chairs, with buttons that go, 'Do you like Harry Anderson, yes or no?' Then it goes into this huge demographic computer and condenses down into three-by-five cards. Then they put the cards on a board and go 'Eenie, meanie, minie, mo. Oh, my kid likes this one; let's put it on! Where's that *Night Court* card? Is it under your chair? I had it a minute ago. Well, fuck it, what cards *do* we have?'

"Damn!" Harry said, suddenly pensive. "I'm really getting caught up in this. At first, it just seemed like a way to get off the road, and to have some fun. Jeez—having the best guys in the business write your lines for you. Being the center of attention. That's fun! But the longer this drags on the more I start feeling like I've got to have it—no matter why I wanted it in the first place." Harry pulled three giant playing cards off the pool table, cards as big as the walking cards in *Alice in Wonderland*. "This is what's good for my nervousness!" he said. "Look! The biggest goddamned three-card monte set in the world." He began to slide the two black cards and one red card, face down, on the green felt. The idea of the game is to keep your eye on the red card. "A little game of hanky-poo, one for me and one for you. Where's the red card?"

"There." Leslie pointed to a card.

"Wrong! Here it is." He flipped the card to her left. "You can beat the cards, but you can never beat the game. Play me again."

Leslie had better things to do, but Harry, stuck with the cards, kept them moving in his hands, relaxed by the familiar motions of shuffling and dealing, shuffling and dealing. For a time, a deck like this had been his meal ticket. Though Harry, with his clipped blond hair, pressed shirts, and fine features looked like an Ivy League graduate student, his background was blue collar. Dark blue collar. He was fond of telling people that he had spent the last years of his adolescence and the first years of his adulthood fleecing honest people out of their hard-earned money. He had played three-card monte, the shell game, and other games of quasi-chance on the streets of San Francisco, Austin, L.A., and points in between. He only abandoned the illegal aspects of the games when he found out he could make more money by turning his street hustle into comedy. He added magic tricks, upgraded his patter, and moved from city sidewalks to college campus quadrangles. Only after several years of street performing was he able to get his first club bookings.

"I want to stay self-employed!" Harry said, putting down the cards and picking up an attaché case, as Leslie speared a needle's eye with thread. "The beauty of self-employment is that you retain the right to say fuck you. Now Brandon's got me on the dangle, and I resent it. The network is going, 'Either expect a major change in your life, or not.' " He flipped open the attaché case and it became two place settings, complete with wine and matching goblets. Harry had just built it, along with a two-hundred-piece paper model of the Chrysler Building. "If this is any indication of how I'm holding up," he said, "then God help me." Harry toured his living room, which was specifically designed to soak up nervous energy—in it were an old-fashioned slot machine, an assortment of carnival games, a dice cage, a nut dispenser, an octagonal card table, a home computer, an overhead string of scoring beads for the pool table, and a gallery of posters: Jean Harlow on bearskin, a shackled man upside down in a water-filled cage, and magician Harry Masters.

A newspaper thumped into the porch and Harry said to his basset hound, "Fetch!" The dog didn't even blink. Harry went after the paper himself, muttering, "How can you respect an animal that eats its own vomit?"

He unfolded the paper to the entertainment section, said, "Shit," then "Great," grabbed his phone, and pressed buttons frantically. "Jay! Yo! Was *Sutter's Bay* for ABC? No? Good! Because I just saw the new ABC schedule and *Sutter's Bay* isn't on it. CBS, huh? Okay. Later!" Then he looked up a number in the phone book and punched it into the phone. While it rang he pointed to an article in the paper headlined NEW SHOWS TO LOOK FOR, which included a list of the new pilots most likely to get scheduled. On the list was *Night Court*. He introduced himself to the writer of the story and pumped him for information. "Tartikoff says *Night Court* could be another *Cheers?*" Harry's face lit up. "Did he say where in the schedule it might go? But he did say it would probably get on? Oh." Harry's upturned mouth slowly slid down. "Okay, well, thanks for the info.

"He didn't know," Harry said to Leslie. "He's just trying to create some news."

Harry turned on the afternoon showing of *The Twilight Zone*, taping it for his collection. "I don't even know why I want this thing at all. I sure as hell don't want to get hassled by everybody that owns a TV set every time I go down to the store to buy some smokes. I don't want to be on the front page of the paper if I ever get caught with a roach in my ashtray. We're not even going to move.

"If I get it, I'll probably go down to the magic store and buy a new piece of equipment. You can fix the Volvo. Eva will get a toy. The dog will get a Milk-Bone."

He was silent for a while. "On the other hand," he said. "The first year I make ten thousand dollars a show. Right? The next year I make twenty thousand. And so on and so forth. You know, guys like Alda make a hundred fifty thousand a show. That's enough money," he said, his eyes on fire, "to really get in trouble with.

"I've been thinking," he said thoughtfully. "Ronald Reagan may be just the man for our times, now that I'm going to be a corporation. Know what I am?" he concluded happily. "I'm the kind of guy we used to spit at."

Earl Hamner, on the second day of the New York week of the pilot selection process, got on a plane at LAX. He was on his way to New York to try to help sell *Boone*. He had a meeting scheduled the next day with Tartikoff, Jeff Sagansky, and Lee Rich, his partner at Lorimar Productions on *The Waltons, Falcon Crest,* and *Boone.*

Hamner had been getting good feedback from the network. The market research on *Boone* had been positive. The tape, directed by Paul Wendkos, who did only high-paying movie-of-the-week and pilot work, had the visual gloss and aural richness of a feature film.

Hamner thought that he had found a market neglected by almost all other TV producers—the market segment of people who were hungry for family-oriented, emotional entertainment. His one venture into flashier programming, *Falcon Crest*, had been a fluke. The show was originally intended to be "the Waltons make wine," but the network needed a nine-o'clock adult show, rather than a wholesome eight-o'clock family show. Hamner told himself that he was doing "Gothic drama" on *Falcon Crest*, but joked that "my wife won't let me watch it."

"If *Boone* isn't scheduled," Wendkos had said before Hamner left, "it will be because of some weird perception that it's too emotional for our comic strip society—that it needs more razzle-dazzle."

As executive producer, Hamner could assign himself a salary in the $30,000- to $40,000-per-show range if *Boone* made the cut. And even those sums would be dwarfed by his ownership percentage if the show was successful enough to go into syndication. But "it's a matter of career," he said, "not of life or death."

Earl Hamner, however, was *nervous*. He had a deathly fear of flying. His brother-in-law, though, who was a parapsychologist, had created a meditation for Hamner to repeat to himself on the way to New York. As Hamner said it, over and over, he also made up subplots for *Boone*. The meditation worked remarkably well. By the time he reached New York he had twenty subplots.

Alan Rafkin, producer-director of *We Got It Made*, sat in his $1 million Malibu house on the third day of the New York meetings, letting his answering machine screen his calls. At the moment, he was taking calls only from Gordon Farr and Fred Silverman, either of whom might phone with the fate of his show. Rafkin, who put in many hours on the courts at the Malibu Racquet Club, was fit and brown in jeans and an expensive yellow sweater. "It won't sell," he announced. "I've got a gut feeling. It's exactly what the network wanted. I did a very good job on it; there's nothing left in that script. That script is left for dead. But I don't know why anyone *wants* that kind of material. *I* don't watch it.

"But what I watch and what I do for a living are two different things. My father was a clothing manufacturer, who made what's called a number six suit, the least expensive you can buy. But he didn't wear a number six suit. He did, however, make the best number six suit he could.

"*We Got It Made* is cute, but not very substantive. I really don't think it will sell."

But he certainly wished that it would. He wanted one more hit show, and then he wanted out.

"There hasn't been a new hit sitcom for two years, and the last one that made it big, *Too Close for Comfort*, made it because it followed *Laverne and Shirley*. People are watching TV less, because it isn't very good. There are no new *Maudes* or *All in the Familys*. Now it's all *Kinderspiel*. I looked at *Goodnight, Beantown*; it's TV 1961. You could get diabetes watching it, it's so precious. Crapola!"

Nevertheless, Alan Rafkin was on pins and needles over his own personal *Kinderspiel* crap. "You work on a show for a while, like I did on *One Day at a Time*, and suddenly you start getting residuals from something you did five years ago. I mean, you go out there to the mailbox and pull out a check that's obscene. For all the people in the business, I wish them three to five years on a hit. You really get well, financially."

The phone rang and Rafkin's voice, on tape, announcing his momentary absence, carried through the living room, the seaside half of which was all glass. Waves crashed into foam below his deck. One of the breakers brought up a freshly painted board. Malibu Beach had just been savaged by shattering waves, and the meadows across the highway had been decimated by a devastating brush fire. Millions of dollars' worth of mansions still lay bare, exposed and charred. Malibu was at once the most financially secure of communities and the most physically threatened.

"About this time of year," Rafkin said, "I start to get calls from kids I know. Asking for money. Everything is shut down now, waiting for the new production season. Guys have been out of work for a few weeks and are starting to miss payments. I'm talking about guys and girls in their fifties.

"Some of these kids make maybe five thousand dollars a year. You just can't get by in the business without a series." His phone rang, his machine clicked on, and Rafkin strained to hear if the caller was Farr or Silverman. It wasn't. "Where was I? Oh, yeah. You just can't make it."

While Alan Rafkin listened to the roar of the surf, Richard Sakai, a producer on *Taxi*, rummaged through the story files in an office that was quiet as a mausoleum. Sakai was afraid that the *Taxi* offices would never be noisy again, because the show's ratings were bad.

Sakai was in the office to do some editing of episodes that hadn't been aired yet. After years of working on *Taxi* as little more than an errand boy, Sakai had at last worked his way up to the job of line producer, the person who was in charge of the day-to-day activities of the production company. And now all his work seemed to be going down the drain, because the network feedback on *Taxi*'s renewal didn't sound promising. The studio air conditioning shot gales of arctic wind into the room, but Sakai's forehead was shiny with sweat. "There are a lot of reasons why they would keep us," he reasoned. "First, I have enormous faith in NBC's striving for quality programming. I also know that advertisers are lining up to get a piece of the show. And I don't think NBC will be able to replace us with a better show. Also, NBC gained a lot of their image as the quality network with the TV reviewers by picking up *Taxi*, and keeping us would preserve the support of their programming."

All of Sakai's reasons were logical and strong. "I'd say we have a sixty percent chance of getting picked up." He smiled confidently. His reasoning was good and his smile was brave, but the perspiration on Richard Sakai's face said goodbye, farewell, amen.

The black guy's switchblade, gleaming against the night sky, was just a tad shorter than a broadsword but the two white guys he'd just punched in the face were still plenty game.

"Motherfucking nigger-loving whore!" one screamed, backing up on the sidewalk. He was addressing the black guy's girlfriend, a pretty and elegantly dressed white woman of about twenty-five.

A helpful suggestion issued from a network programmer, observing the scene as if it were just one more clip of action-adventure: "Whaddaya say we settle this over a toddy at Elaine's?" But he was ignored, because this was Manhattan, where business is business and if you're not part of the deal you're just in the way.

So there was little for the trio of network employees to do but step back as the black guy went chasing after the white guys. Talk about heat. Talk about hooks. Talk about instant backstory. Why, this dynamic young racist had blood running down his cheek from where the black guy had popped him a hard left— yet he was still ready, willing, and able to stand up for his lead-headed nostrums, even in the face of a monster switchblade. Really willing to put his gizzard on the line for what he believed. Which was "You dirty spook! If I had a gun I'd blow your black ass to kingdom come!" This he was screaming as he ran down First Avenue, the black guy—slashing the foul air with his dagger—in truly hot pursuit.

It was hard for the network men not to chuckle at all this

and wonder at the miracle that was New York City. Here it was, three o'clock in the morning, and the town appeared to be just warming up. It had been an exhausting day. So much to see. Such as: a shopkeeper exploding from behind his cash register in pursuit of a rat the size of a ghetto blaster, scurrying down the sidewalk for its very life.

Which it got to keep for about ten more seconds before the shop guy poked it artfully to death with a broom handle—a skill you can go for days without seeing in downtown Beverly Hills. The contrasts between New York and Los Angeles are intriguing. L.A., for example, is an exciting feudal society where the middle class is invisible and society is run by the rich and the poor. The moguls and the Mexicans. Slam your $113,000 Lamborghini CML convertible into a guardrail on the way home from the studio and Pedro or José or Manuel will be buzzing at your servants' gate in a trice offering to fix the whole thing for $160. Everything is either laughably cheap or unbelievably expensive in L.A. Maids are a dime a dozen, so long as you pay cash and don't ask to see anybody's green card. L.A. is a great place to be a fatcat. Everybody knows his place. So long as he can afford it. A Malibu condo not much nicer than the honeymoon suite at a Motel Six could go for $750,000.

In New York life is not so well defined. In fact, one of the really irritating things about Manhattan, compared to Los Angeles, is the lousy job they've done at separating the haves from the have-nots. In L.A. the poor live in their barrios and the rich nestle in the great ghettos of tract mansions in Rolling Hills and Brentwood. But in Manhattan, my God. You can work your manicured fingernails to the bone clawing your way to the summit of NBC or CBS, purchase a penthouse on Park Avenue, and still have to endure the weird strains of Puerto Rican bebop echoing from the magazine stand down the street when you hop into your limo in the morning.

Anyway, the night was still young and the crowd spilled away from the sight of the fight. The black man came huffing back up the street, frustrated and victorious, having been able to chase the white men away but not fillet them. The three guys stood with their drinks in hand there on the sidewalk.

"You hear NBC is going for an all-white night on Fridays?"

"No way. Freddie tried that at ABC," another said, "and it went over like . . . like . . ." Comparison escaped him and he yanked his tie open at the collar.

"That joker was serious," a third guy said.

"Who?" the man pulling at his tie asked.

"The spade, who do you think?"

Rockefeller Center in midtown Manhattan is a gem of Gotham Deco architecture, and its centerpiece, the RCA Building, is a true 1930s futurismo building among boring sixties monoliths. The doors to NBC headquarters are abutted with statues of nymphs, naked and pneumatic, girlfriends to a huge golden nude statue of Prometheus dominating the plaza. Few creative decisions—except the most important ones—are made here. Chiefly, NBC New York is responsible for news and sports and administration. But once a year the top brass of business and programming powwow for four days in the network's central conference rooms to decide the shape of the network's fall schedule. "Your bat, your ball, your ass, your call," as the New York street kids chant.

For four days the executives manipulated the "cards," magnetic rectangles representing about sixty possible shows. Some, like *Cheers* and *Hill Street*, were maneuvered only in terms of changing time slot, and were never in jeopardy. Others, like the sitcom *Gimme a Break*, were discarded, then picked up again. *Night Court* was scheduled for Sunday, then taken off the board entirely, while the heads of the network decided what to do with it. If anything. There was plenty of grumbling about Harry Anderson's not being able to carry the show.

The hard calls were made last, and the hardest was *Fame*. NBC had received over fifty thousand letters from viewers begging that it be retained, despite poor ratings. But Bill Reubens, NBC's head of research, somewhat sadly revealed that the only people who seemed to be watching it were brilliant high school girls. Finally, this and other decisions devolved upon two people, Brandon Tartikoff and Grant Tinker.

The Parker Meridien is lovely. Ceilings in the lobby, though several stories high, are hand painted; a mirrored hallway is twenty feet wide. Marble floors. Trees growing indoors. The trill of a piano in the dining room accompanies waiters in black and white floating among the tables. Fresh flowers everywhere.

Above all this, on the forty-first floor, Larry Rosen and Larry Tucker sat pensively on the edges of their beds waiting to hear news about their NBC pilot, *Jennifer Slept Here.*

Tucker was big in the stomach, bearded, fifty. He wore a blue caftan. "We came to New York for one thing," he said, "to get an order; to hear, 'You're on the air.' Four, four and a half words. And we can't leave until we get them. Though that doesn't always happen."

Rosen propped his feet on an overstuffed blue chair. "You come back here so that if at the last moment, somebody says, 'We like you, but have you really got legs?' you can hand them a stack of story lines." He held a stack of them in his hand. "You run into a network guy at 21," he said, "and have a drink and put an idea, a lie, a hope, or a dream in his pocket, so that he can go back and tell his people, 'I just had a black Russian with the Larrys down at 21 and it turns out they forgot to tell us they got Barbra Streisand to guest on *Jennifer* for eleven straight episodes.' " Rosen wore jeans and a work shirt. He was trim, dark. "You come back here, you pray, you go to a show, you go out to eat, you pray, and if you make it, you make it, and if you don't, you pray you've got the guts to go out and congratulate the ones who did."

Larry Tucker glanced out the window. Below was the boxed sprawl of Central Park. "We're going to make it."

Rosen: "Yeah."

Tucker: "We're on the board. I know it. We've been told."

Rosen: "I know."

Tucker: "So."

Rosen: "Warren Littlefield can love the show. Perry Lafferty can love it, Joel Thurm can love it, Steve Sohmer can love it, but if Brandon Tartikoff doesn't love it, there's going to be no *Jennifer*." Rosen stood up. "Gary Nardino is probably cutting his wrists today because out of seven pilots, I think he got one on the air. Things can change overnight. One day they love you, the next day . . . You can be sure Gary Nardino is sweating over his studio."

Tucker: "What I want to know, what interests me, is, how do you be successful in a business that fails ninety-eight percent of the time?"

Rosen: "We're pretty well paid for our tears."

Tucker: "Just for the record, last year we produced *Mr. Merlin*. It was canceled by CBS with a twenty-seven share and a seventeen rating, and nothing they've put on this year at eight o'clock has come anywhere near those numbers. That hurts, because we know that some fucker down there made a mistake. And it wasn't us."

Rosen: "It's a year out of your life."

Tucker: "I've been doing this kind of thing for quite a while now. I also play poker. *Jennifer's* in."

Credit for which, Tucker and Rosen give largely to Ann Jillian. Rosen said, "Ann knows how to package Ann. We were in a limo the other day and the driver was saying what a big Ann Jillian fan he was, and he didn't even realize she was sitting right there in the backseat. Because she had her regular hair, her regular makeup, her regular clothes. Ann Jillian with the platinum hair and the lips and the flashy clothes, that's Ann Jillian for television, and while we are creating that TV for her, we live and breathe those characters—the ghost, whatever—twenty-four hours a day, seven days a week. We live, eat, sleep, and dream Jennifer the ghost, the kid, and the parents. Then we come here to New York and hope for a payoff."

Peter Grad and his boss at Twentieth Century-Fox, Harris Katleman, were encamped with five other Twentieth executives at the Ritz Carlton. Grad's job was to answer questions Katleman might have about the creative aspects of the shows they were trying to sell. Time was running out—the network would decide on their shows in less than a full day—but things were looking good for Grad.

Grad had developed twelve pilots, and it seemed as if at least half of them, maybe even two thirds, would sell. If that happened, Twentieth could be the year's top supplier to the networks, a prestigious coup for Peter Grad. CBS had already bought *AfterMASH* and *Navy*, and it looked as if ABC might buy *It's Not Easy*, *Trauma Center*, *Masquerade*, and *Automan*. If they took *Trauma Center*, they wanted to change its name to *Medstar*, a move that Grad thought was stupid. *Medstar* meant nothing to the viewing public, and a bad name could kill a show.

Grad liked working with Katleman because of the latter's efficiency. Katleman could present a wealth of information from the Twentieth research department on who would watch a show, what other programs the show would be compatible with, whom they would be able to line up to write it, and dozens of other selling points—a tidal wave of information—but in a way that was so clear and low key that it didn't even sound like a pitch.

The best news of the week, though, had not come from New York, but from Allan Katz in L.A. NBC's Warren Littlefield had told Katz that *Snoop* had experienced a rebirth in its chances of being scheduled.

Peter Grad was having fun.

"Rumors on the street, running into people—I'm waiting on pins and needles, I'm up 'til four or five in the morning because I can't sleep." Reinhold Weege, producer of *Night Court*, sat slumped in a chair in the lobby of the Park Lane Hotel. It was midday, but Weege's eyes sagged, and he appeared to be settling into the sofa as if in the clutch of supergravity. Olive-skinned and a little chubby, wearing a leather flight jacket, his eyes diagonal slits, his black hair uncombed, Weege looked like an Eskimo fighter pilot recovering from a hard night on leave.

His head tipped one way, then the other. "Telephone calls, rumors, waiting, waiting, waiting; this is incredible, nonsensical, totally unbelievable, fantastic. It's gambling. Exciting. Come on, let's find a better place to talk."

He stood up. People were sitting and standing all around the lobby. Luggage was piled everywhere. A collage of doormen and cab drivers. Several in military-style uniforms. One dressed up like the popinjay on the Beefeater gin bottle. "I think *Night Court*'s the best shot I've got in terms of salability. Jimmy Burrows directed, but the big factor is going to be Harry Anderson. The guy has got star potential."

Weege walked out the Park Lane's front doors to the panorama of Central Park. Young men stood in livery beside horse and carriage. It was sunny. "*Night Court* is the same show I always do. Basically, I look for an idea I can do over and over, a situation that lends itself to different things happening under the same basic circumstance. In that way, *Night Court* will be much like *Barney Miller*." On which Weege had been a producer. "I left after writing fifty to sixty episodes," he explained as he walked down the street. On the corner, a black man in worn jungle fatigues was preaching a gospel—"You think I don't know war! Bullshit! I spent three years in a prison camp in Laos."

Weege ignored him, oblivious. "I absolutely loved that show; it was the most creative period in my life. I got a Writers Guild Award, a Peabody one year . . . but seven days a week for three years . . . I had what I call a nervous breakdown." He headed back into the lobby of the Park Lane. "As fat as I am now, I was ninety pounds heavier. The script commitment was heavy, forty to fifty pages every two weeks. . . . The stories were confined to a single set; it was an ensemble cast; stories had to be meaningful and—I think it was generally agreed on *Barney Miller*— potent. They also had to be hysterically funny, had to bring lots of guest people in during the course of twenty-four minutes, had to get them on stage and get them off." He walked through a clutch of old women as thick as sparrows in front of the registration desk. Jewelry. Furs. They seemed to be speaking in every language but English. "Sometimes we'd start scripts on a Friday for a Monday reading, knowing that you had a quarter-million-dollar show waiting for a script in two days. To do as we did caused physical and emotional problems."

He fidgeted out through the glass back doors of the Park Lane. Three black limousines with smoked black windows sat parked in the hotel turnaround. "*Nothing* scares me anymore in terms of time and task." In front of the limos, a skinny man of indeterminate race and age stood gripping an empty bottle of wine by the neck, upside down, as if it were a green glass baseball bat. He was an almost stylish ensemble of filth. His clothes were uniformly covered with mud. Dirt appeared to be actually woven into his hair in corn rows, Rasta style. His eyes were blazing tomatoes. A designer wino. But Weege was not seeing this part of the world. "People ask me, 'Aren't you in awe of your ability to do this, at your age?' And I have to say no, because it seems as natural as being the class clown in high school." Weege turned around, a slow pirouette. "For a writer, there's no way to work up to TV, no evolutionary process to go through. If any other business was done this way, it couldn't exist."

As for his own chances, "We did a good job. Harry was great. We're going to win."

Eight A.M., May 10.

The mezzanine of the Waldorf-Astoria, where $32 million in television programming was on the line that morning, is large and palatial, seeming built for a race bigger and more elegant than our own. Towering ceilings, glittering chandeliers. The Waldorf is everything permanent and rich. Waiters served eggs, sausage, and champagne to dozens of yellow-clothed tables jammed with advertisers and the press. Mr. T stared from a screen above the stage—a bejeweled Big Brother—about $500,000 worth of gold hanging around his neck in a sheet of chains that created a sort of bullion bib over his monster chest. There was an announcement that NBC had won twenty Emmy Awards, five People's Choice Awards, and four Golden Globe Awards the previous year. Then a tan, neatly suited Brandon Tartikoff was introduced as the "Mr. T of programming."

Behind him T. vanished and was replaced by this list:

1. Get the best people to do what they do best.
2. Most comedy ever on NBC schedule.
3. "Youngest" schedule of all the networks.
4. Continued emphasis on key demographics.

Tartikoff began explaining NBC's new schedule and the network's Rx for the future, beginning with Monday night. "*Little House* was pulled because it was trending downward, but *The Monday Night Movie* is averaging a thirty share. . . . We feel Monday night will be very strong for NBC. . . . We have a chance to win every week." Earl Hamner's *Boone* was announced to replace *Little House*. "As for Tuesday. First, a word about *The A-Team*." Tartikoff put a hand to his chin. "A lot has been written in the press about what we're proud about and what we're not. . . . We're proud of *The A-Team*. . . . It got the highest Q score of any show on TV and has driven *Happy Days* below a twenty share. . . . The synergism between *A-Team* and *Remington Steele* is finally working."

Then Steve Bochco's face appeared on the screen, two stories high. Tartikoff said, "If the face seems familiar it's because you've seen him on two Emmy presentations collecting twenty-one statues. He heads the finest staff on a dramatic series on NBC." Tartikoff announced that *Bay City Blues* would go on the air after all. "We believe Tuesday nights will be permanently in the win column for NBC."

On to Wednesday. There was an update on *Real People* and various *Real People* events, followed by the news that *St. Elsewhere* had been renewed. Tartikoff said, "It achieved its highest-rated evening last week . . . a twenty-eight share. . . . We looked at the data and it matched *Hill Street's*, during *Hill Street's* first season, exactly. . . . I had twenty thousand pieces of mail saying, 'Remember *Hill Street*.' . . . I just hope *St. Elsewhere* is remembered at Emmy time." A doctor from *St. Elsewhere* came up on stage and said, "The chance to work in quality is all we want to do."

Now Thursday. *Mama's Family*. Tartikoff said, "This show got the highest Q score of any comedy show last season." Next, *We've Got It Made*. They did. Tartikoff offered, "Not since *Three's Company* have we had these kinds of breakout characters." Teri Copley minced onstage, cooing and squeaking, "I wanna say thanks to Grant and Brandon and . . . ah . . . ah . . . that the guys are gorgeous and the girls are really great." More giggling. Exit. Tartikoff predicted, "Thursday should work like *Three's Company* and *Taxi* did for ABC." The strategy for Thursday? "Good family comedy."

Friday was revealed as a complete overhaul from the previous year. "Well," Tartikoff shrugged, "we roll the dice again . . . but the shows opposite us are weak. . . . *Mr. Smith* is the first comedy since *Mork and Mindy* that could really explode." Tartikoff reminded his audience that Ed Weinberger and other *Taxi* people would be involved in the ape's debut. "*The Dukes* are down nine share points since last year and *Mr. Smith* is smarter than both the Duke boys put together." As for plot, "Mr.

Smith lives with a normal orangutan brother based loosely on Billy Carter."

Tartikoff then announced the scheduling of *Jennifer Slept Here*. "*Jennifer*," he said, "was the highest-testing comedy this year. . . . We could score a *Sanford and Son*, *Chico and the Man* hit block of comedy." Ann Jillian came onstage, all bright and glistening, her blond bell of hair reflecting the house lights. She had the presence of a star from the fifties, a creature from a voluptuous, more chromium time. She said, "Larry Rosen and Larry Tucker are the finest writer-producers I've ever worked with."

Tartikoff again. Announcing that *Manimal* would replace *Knight Rider*, which was to be moved to Sunday nights. A *Manimal* promo flashed on the screen behind Tartikoff's head, explaining in graphic, exciting terms *Manimal*'s backstory. In everyday life, Manimal is a dynamic, handsome, lady-killing fruitcake college professor, but his dad was "sole heir to the secret link that binds man and animal." On the screen there were images of a human hand metamorphosing into the claw of a beast. "He's creeping up on crime and when he pounces the fur will fly."

Tartikoff moved on. The selection of *For Love and Honor*—formerly *The Whole Nine Yards*—was announced. Again, backstory on the screen focusing on Rachel Ticotin's character being hassled by male recruits. The voiceover: "She's got the guts to take it, but does she have the courage to find the love she needs?"

It would be an "all-white Friday night," concluded Tartikoff, a reference to the color code on the screen behind his head, white letters indicating a new show. A tactic that many network programming analysts consider suicidal. "Friday," Tartikoff said, "will be as successful this year as it was in the fall of '82."

Saturday. The blond boy from *Silver Spoons*, Ricky Schroder, got up on stage and said, "I just want to thank all the advertisers." Tartikoff announced selection of Steven Cannell's *The Rousters*. He predicted that this action show about a relative of Wyatt Earp who is a "bounty hunter–carnival cop" could come in against *Love Boat* the way *A-Team* had come in against *Laverne and Shirley*. A promo clip was shown on the screen: Chad Everett playing Wyatt's twentieth-century relative. An old lady blowing away things with a shotgun. A weirdo playing dumb, but really being crazy like a fox. A chase scene. Chad Everett took the stage and pitched to the advertisers.

Yellow Rose was announced, followed by a *Yellow Rose* clip whose theme was (1) Cybill Shepherd is sexy, and (2) don't fuck with Sam Elliott. Finally, Sunday. Tartikoff: "We toyed with a new name for *Monitor* but the best thing we could come up with was *NBC's Merrimack.*"

The presentation wound up with previews of upcoming NBC movies, including a clip from one titled *A Haunting Passion*, featuring a housewife apparently in the throes of sexual intercourse with a ghost, her husband sitting beside her in bed looking somewhat astonished. Tartikoff concluded, "I think we have the horses this year. I think our ratings are just going to get bigger and bigger and bigger."

Fred Silverman, at his home in Connecticut, nursed the telephone, waiting for former protégé Tartikoff to call. If the phone rang and it was someone from NBC other than Tartikoff, such as Jeff Sagansky or Perry Lafferty, it would be a bad sign. As a rule, the boss called to say you were on the air, and his underlings called to say, "Sorry." On the other hand, however, Brandon might call either way, out of respect for the man who had brought him to network television. Silverman's nerves began to fray as the phone stayed silent, so he walked down to the lake near his house.

He made himself stay down there a few minutes.

As soon as he opened the front door of his house, he was handed a message from his maid: "Brandon Tartikoff just called."

Silverman grabbed the phone and punched in Tartikoff's number. It was busy. For the next hour and half, he tried the number every few minutes.

Gordon Farr, in his Hollywood Hills home, which overlooks the flat stratum of smog that is generally clamped over the city, sat holding his newborn baby. Today was Mother's Day, and Farr's wife, Hillary, a mother for eight weeks now, had just left the house. Farr had been jumpy and irritable all week, as the NBC programmers decided his fate, but today he'd pushed *We Got It Made* out of his mind. He'd given Hillary a present, and they'd had a leisurely breakfast together.

The telephone rang and the baby started to cry. Farr squashed the receiver into his ear to drown out the baby's wails. It was Fred Silverman.

"We're on!" said Silverman.

The baby drooled down Gordon Farr's shoulder.

Gary Nardino, in his hotel room, took the last series of calls. It was from Brandon Tartikoff. *Mr. Smith* had been scheduled. Nardino's mood lifted.

But after the initial excitement of selling *Mr. Smith*, the dominant feeling he'd been experiencing all week returned—it was fatigue. Failing made him tired, and most of his efforts this week had failed. *SCTV* would not stay with NBC—the network was losing the show to cable, which would hurt NBC in particular and the networks in general. ABC had passed on *Laverne and Shirley*, a rejection that caused a particularly sharp sting, since all episodes of the show were assured of being salable for syndication. Another year of *Laverne and Shirley* would have made the studio several million dollars. ABC had renewed *Happy Days* for another year, but with the clear caveat that this was the show's last chance to show some strength against *The A-Team*. Likewise, NBC had made it clear that *Cheers* and *Family Ties* were in their last year, unless their ratings pepped up considerably. The comedy called *Webster* was the only other Paramount pilot that had sold.

Gary Nardino, as he packed his bags and prepared to fly home, began to wonder seriously if he wanted to stay in this job for another year.

Earl Hamner was not surprised when he was told that *Boone* would be on the air. He had had a meeting the first day he'd been in New York with Tartikoff, Sagansky, and his partner Lee Rich, at which they had discussed not why *Boone* should be on, but when it should be on. Tartikoff asked if nine o'clock would be satisfactory, but Hamner said that at eight he'd be able to pick up children looking for a family show, as well as the adolescents who would be interested in the birth of rock music. Tartikoff said that if the show was scheduled for eight o'clock, the roles of the children in the show should be increased. Hamner had backup scripts in his briefcase that provided exactly that.

So when Hamner phoned the good news to Claylene Jones, who would produce the show, he felt, he would later say, "almost arrogantly optimistic."

Reinhold Weege was in the bathtub when the phone rang. He was looking forward to a call this Sunday—hoping it would be Tartikoff—but also dreading the call. He had been hearing a wealth of conflicting information about the network's reaction to *Night Court*. He had heard that the NBC sales department loved the show, and thought that Harry Anderson would haul in just the type of upscale viewer that sponsors love. He had also heard, though, that NBC's testing of the show over its cable outlets indicated that the judge Harry played lacked credibility, because of his age. To counter that appraisal, Warners had done testing of its own at the Preview House; its testing had said that no such credibility gap existed. There had been inconclusive talk of putting someone else in the judge's role.

Weege had heard that *Night Court* had been on the board most of the week. But lately, it seemed, *We Got It Made* had been edging it for the slot just before *Cheers*. Some late testing had been done on *We Got It Made*, and that made Weege nervous, because it was his belief that "sex always tests well." He'd also heard that promo visionary Steve Sohmer had put together a trailer for *We Got It Made*—a snippet of scenes from it—and that the trailer had reawakened Tartikoff's interest in the show.

The phone call was from Larry Little of Warner Brothers. *We Got It Made* had beaten them. They were not on the schedule.

NBC, however, did want them to make six episodes of the show, which would probably be sandwiched into the schedule at some vague point in the future. If that brief appetizer of episodes met with overwhelming audience approval, perhaps more episodes would be ordered. Weege's first thought was that a low-concept character comedy like *Night Court* usually took twenty-five to thirty episodes to catch on.

There was, furthermore, a condition to the purchase of the six shows. Some recasting, which they would discuss in detail at a later date, would have to be done.

Weege got dressed, made a gloomy call to Harry Anderson, and immediately left for the airport. Before he got on the plane, he called Harry again and apologized for being so depressed.

The rudeness. The arrogance of those bastards. No one had bothered to call Allan Katz when *Snoop* failed to make the cut and the thought of it was enough to spoil Peter Grad's flight home from New York. Almost.

Actually, Twentieth Century-Fox had sold five shows outright and two more had been chosen as backups, which put Fox ahead of all the other studios and imparted such a ruddy glow to the mood of Grad, more successful now than he'd ever been in his life, as to blunt his analysis of this latest behavior on the part of Tartikoff and his minions.

Which was: that it was almost as if Fred Silverman had slithered back on board at NBC.

On the day NBC announced its new schedule, Tony Colvin's writing partner, Scott, returned from a brief vacation to resume his job as a *Cheers* gofer. He was in a sunshiny mood. This would be the year, he figured, when he and Tony would sell *Cheers* a script. It would also be the year when he would buy a new house—negotiations were near culmination. Scott's brother and his wife had just arrived in L.A., and they would be living with Scott in the new $109,000 place Scott had found in the Valley. With Scott's salaries from *Cheers* and *Family Feud*, plus his wife's earnings, they were just able to afford it.

Early in the morning, Scott went down to the *Cheers* office to find out what his schedule would be. He took a seat in the office of his supervisor, Richard Villarino.

"So when do I come back?"

"Well, as you know, Scott, the show's down now—hardly anything's going on. It's just the producers and secretaries up here."

Of course Scott knew that. But the question was still, when did he start, and what were his hours?

Villarino hedged a little longer and then blurted it out. Scott was finished. His services were not vital at the moment, and they didn't have the money to pay him.

Didn't have the money? Scott almost laughed. With guys like

Burrows pulling thirty or forty grand a *week*, compared to Scott's three hundred bucks, how could Villarino sit there and *say* he didn't have the money? But he felt too sick to laugh.

"Can I at least split the time with Tony? So that he and I can keep writing together?"

"No," said Villarino. "Tony's a freebie for us. Paramount pays him as part of the PA program."

So that was it. Scott was out of a job. Scott was out of a new house.

And Tony was out of a partner.

PART TWO Staying In

7
Affiliates

June 1983

Poor Sohmer. There it was, his first big chance to shine for the NBC affiliates—sober men of heft and clout, many Caesars in their own principalities—and Steve Sohmer, the rising promotions star at NBC, had to stand there with a stiff smile stapled to his face as an orangutan gripped him firmly by the privates. What to do? Strike the beast? Have a minion throttle it?

Not likely. Because no small part of the new NBC schedule rested upon the hairy shoulders of this overgrown monkey, and others like it in the newly scheduled *Mr. Smith*, a program that would show the TV world how lovable, controllable, and *promotable* these animals really were. This simian was valuable property—a star! Besides, Sohmer had ambitions that would not have been well served by wrestling in public with an ape. So he cooed some words into the black, waggly ear, and the primate slowly relinquished its prize. Bringing the ape onstage to jazz up his presentation had been Sohmer's own idea in the first place, so when it freed him, he let the animal continue to share space on the stage, though he kept a close eye on it as he plunged back into his speech.

Sohmer was boasting to the affiliates, the group of station owners who assembled every year in L.A. to hear that happy days were here again, that NBC was cleaning up on the competition every Saturday morning. Cartoon time was under Sohmer's direction, as was the rest of the daytime schedule, as well as all of NBC's promotional activities and specials. "We've been number one, twenty-five out of the last twenty-eight weeks," he announced, and then waded into the grislier territories of his job, which consisted of every day of the week *except* Saturday morning. The ballroom of the Century Plaza Hotel grew still, because the day slots were the most lucrative parts of the affiliates' schedules. But lately, their day parts had been grim. In the most recent ratings week, NBC had owned six of the bottom-ten shows; only three of its daytime shows were *not* in the last ten. Its top show was *Days of Our Lives*, which did only about half as well as ABC's and CBS's leading soap operas. For the season, NBC had a 17 percent share of the available audience; CBS and ABC had 25 percent and 26 percent.

The NBC daytime schedule had been a rambling wreck ever since the mid-1970s, when the network had let Monty Hall's *Let's Make a Deal* slip away to ABC rather than cough up the extra money that the show's producers were demanding. The schedule had been further abused by Fred Silverman's insertion of *The David Letterman Show* into a morning slot. But the sorrow and the pity of NBC daytime was history, Sohmer boomed evangelically. "Any show not pulling its weight will be thrown out of the boat." Sohmer, in contrast to the ever gentlemanly Tinker and Tartikoff, was sounding like a real butt kicker. "We've told daytime directors and producers that we've got to get it together, and we've told them in words *cats and dogs* can understand!" The affiliates murmured appreciatively, and Sohmer beamed.

Specifically, what he had in mind was merely upping the ante on the two staples that eternally dominate daytime programming—sex and greed. The soaps, he said, would feature more "heat," which is to say, more people in dramatic situations with their shirts off. And the game shows would hereinafter become gushers of cash. Sohmer proudly announced that the producers of the *Queen for a Day*–style afternoon show called

Fantasy had talked a hamburger chain into helping raise 8 million bucks of giveaway money, and that *Dream House* would soon tickle recession-weary housewives' fancies with a free $97,000 home. Times being what they were, though, the ninety-seven-grand house, Sohmer had to admit, would be a mobile. Sohmer leaned forward over the podium, his face enlarged behind him on a screen almost as big as a theater's, his eyes, each no smaller than a man's head on the screen, sharp, dark, kinetic. "The other networks," he thundered, "are going to run into a whole medicine show and flying circus of promotion this summer!" The affiliates, tired of seeing their local ABC and CBS counterparts outearn them, scared to death by the menacing specter of cable, thundered back with applause. The commotion loosened several balloons hung from the ceiling to symbolize festivity, and they floated down in slow motion.

Sohmer quit while he was ahead and led the troops to lunch. The station owners, their wives, and the accompanying local executives streamed out of the ballroom, up an escalator, and onto a beach-umbrellaed veranda, where a salty June breeze filtered in from the shoreline a couple of miles away, and the bar was open and free. The affiliates dismembered toasty, roasted game hens, asparagus in Bordelaise sauce, and fat cream puffs. For the most part, their moods were good, in some cases positively giddy. The newspapers were carrying stories every day about CBS's budget problems and hiring freeze. And, at a courthouse across town, CBS anchorman Dan Rather was being grilled like a common criminal in a libel case. Furthermore, nothing in the just announced ABC and CBS schedules seemed to them as good as the new NBC shows, which the affiliates had been watching in their hotel rooms over closed-circuit TV.

"We've been in third place for a long time," said Fred Paxton, head of the NBC affiliates' association, "but the affiliates seem to be pleased that NBC is making progress." Much of their pleasure came from NBC's recent close-second-place finish in the May sweeps, which would allow most of the affilliates to jack up their ad rates. Sweeps success, even more than game hens and cream puffs, is what keeps the affiliates happy.

Dixie Whatley, the glamorous blond reporter for *Entertainment Tonight*, table-hopped with her cameraman, collecting material for the evening report. A few days before, at the CBS affiliates' meeting, an *Entertainment Tonight* reporter had cornered a CBS executive and said, "The headline in *Weekly Variety* says, 'Network Overestimates TV Sales.' Some critics are calling this a disaster."

"I wouldn't use the word 'disaster,' " the man from CBS answered. "I think what we're doing is taking prudent action."

"But in the broadcast group, a freeze on all new hiring . . . ?"

"Yes. Well. Again—prudent action."

A station manager from central Missouri waved an undersized drumstick at Whatley. "Hey, Dixie," he yelled, between gulps of Chablis from a plastic glass, "I got somethin' to tell you." She stood up, smiled amiably, and began to trek over to the table of Missourians. But as she approached, the station manager choked a little on his wine and crowed, "God Almighty. Dixie has jugs that just won't quit!" Whatley veered off course and didn't stop to talk to another table until she was far away. "Hell's bells," said the manager, "I wanted to tell her somethin'."

Inside, in the near dark of the deserted ballroom, Brandon Tartikoff practiced the speech he would give the next day, trying to coordinate with slides flashed onto the mammoth screen.

"Now it's time to present," he said, reading from a script, "with a great deal of pride, the 1983–84 NBC prime-time schedule. You want me to run through the whole thing?"

"Yeah," boomed the voice of a director in an unseen booth.

"It's the most balanced schedule for NBC ever," said Tartikoff, "in terms of . . ."

"Just a second," thundered the voice.

Tartikoff paused a moment, then said, ". . . in terms of . . ."

"Just a second!" Tartikoff, for now, was just one more dancer in the chorus line. His face, magnified on the screen, looked tired and blurred. Above his head, burst balloons hung from the ceiling like shriveled rubber rags.

The next day, just before the meeting's big finale, RCA head Thornton Bradshaw lounged against a chalky concrete pillar, some of which rubbed off onto his black suit, near an exhibition demonstrating Teletext, a video newspaper. Each of the three networks had showcased its own version of this invention at its affiliates' meeting. Having helped spawn the electronics revolution, the networks were now being forced to run for their lives through the realms of cable, home computers, video games, and electronic print. None of the networks had sponsored any glowing successes, though, and there had been some stinging failures, such as the CBS culture-cable experiment.

The affiliates, on their way back to the ballroom after filing past countless yards of breakfast buffet, paid no attention to this old man in a black suit. They knew Thornton Bradshaw's name, but not his face.

Bradshaw, who had been expected by many Wall Street analysts to sell off NBC when he'd taken over in 1981, said he was committed to the network's success. "We have a long way to go," he said, "but when NBC turns around, RCA wants to be there."

Nevertheless, NBC was just one part of the RCA group, and Bradshaw's primary interest, he said, was in exploring new areas of electronic communications. "We are going back to our roots," he said, "to what made this company famous for so long, to what made it the pioneer in the early days of electronics. We're going back to dependence upon the laboratories as the source of our future.

"During the last two-year period of time," he said, "when we were suffering from a lack of funds, we still paid out record sums each year for electronic communications and entertainment.

"We also kept research and development at a record level of expenditure. So we think we know where we're going."

When Brandon Tartikoff took the podium in the ballroom, he had on a tie, a rare piece of apparel for an NBC programmer. As slides flashed behind him and outtakes of the new shows lit up the room, Tartikoff, the master programming technician, explained the strategies that he'd engineered the previous month in New York. He walked the audience through the schedule, night by night.

On Mondays, he said, *Boone* would work because it was a family show and eight o'clock on Monday was the ultimate family slot. Strengthening its *Waltons*-flavored family appeal would be

Elvismania. The audience would flow perfectly into the movie that followed it, and wipe out *After MASH*, which, he said, "is simply not *M*A*S*H*."

Tuesday night would still be *A-Team* night, and the well-plotted *Remington Steele* would become less plotted, in order to capitalize on the presumably nonthinking *A-Team* watchers. The male-female *Remington* audience, he said, would flow into *Bay City Blues*, a pansexual show with a great writing staff.

Wednesday night would be ladies' night. *Real People* would pull in the women and kids, while the men watched Aaron Spelling's *The Fall Guy* on ABC. Then would come the frilly comedy *The Facts of Life*, followed by the intelligent, domestic character comedy *Family Ties*, both of which had strong female appeal. Last, *St. Elsewhere* would try to show why twenty thousand viewers—many of them women—liked it enough to send in letters clamoring for its renewal.

Thursday night would be "demographic night." *Cheers* and *Hill Street* had the most affluent young audiences of any shows on television, and NBC was inordinately proud of its programs' demographics, its consolation prize in a low-rated schedule. *We Got It Made* would be *Cheers*'s new lead-in, and Tartikoff told the affiliates that for this addition to American culture they could thank Fred Silverman, "whose singular vision put *Three's Company* on the air"—a compliment that could be taken at face value *only* at a TV affiliates' convention.

Friday night would be all white—it would feature three new shows, a strategy that had worked rather well the previous year. *Mr. Smith* could become "an explosion," said Tartikoff, pulling in the kids who liked monkey shows and the adults who actually listened to a program's words. *Mr. Smith* came from Paramount, Tartikoff reminded the audience, which had given the network "our best comedies—*Family Ties*, *Cheers*, and *Taxi*—all quality shows." The audience flow to *Jennifer Slept Here* could be enough to make that show a hit, he said, particularly since *Jennifer* was the "highest-testing comedy concept we had this year." Friday would also include *Love and Honor*, which tested with high teen appeal, and *Manimal* from Glen Larson, who theoretically could do no wrong with an action show.

Saturday would be highlighted by *The Yellow Rose*, which

had "the best cast ever assembled for this kind of television show," and *The Rousters,* from Steve Cannell, "the man who gave us *The A-Team.*" Invoking the name *A-Team* was explanation enough for the affiliates.

On Sunday, *Monitor* would be rechristened *First Camera,* would hire some hotshot investigative journalists, and would go head to head against *60 Minutes,* hoping to capitalize on viewer disaffection from that smugly successful show, which was by now as much giant as giant killer, an image reinforced by the ongoing libel trial.

By the end of the presentation, the consensus of the room, judging from applause, was that young Brandon had cooked up a hell of a battle plan. Program compatibility. Audience flow. Quality. High concepts. Hooks. Heat. Sex. Violence. "We've got a shot next year," he concluded. "Look at our shows. They can be distinctive, competitive, and commercial. I looked across the street and I don't see anything that's innovative or intimidating.

"I want to thank you, our very supportive partners in the fallow years, and I'm pleased that I don't have to tell you it's going to be different next year. I've already done that for three years. This time, I get to tell you that we're sticking with what we all started last year. Because it works. It's working now, as we can see from the May sweeps, and it's only going to work better in the future, as these shows get stronger and stronger and stronger. Exciting, isn't it?

"And it only gets more exciting when you think who we've got promoting these shows like no one else can. Mr. Be There himself—Steve Sohmer!"

Sohmer jumped up to the podium and he, on his arrival, and Tartikoff, on his departure, shared the waves of applause that rolled up from the back of the ballroom, crested, then rolled up again.

"Be There," blasted Sohmer, who had originated the epigram himself, "is not just another slogan. It is a fundamental change in the way we promote television at NBC." Sohmer paused as, on screen, the camera created a dramatic close-up of his dark, intense face. "It is no longer sufficient for us to compete with ABC and CBS for our share of the three-network audience. We have got to promote our product to bring people back to our network, away from *cable*." He'd said the magic word: "cable"—and the affiliates roared their approval. "That is the lesson of *The Winds of War*—when we have the event, the public will come to us. So Be There ushers in a whole new phase in network promotion—*event orientation* in everything we do! Be There is a call to viewer action!"

Sohmer told them they would get Be There T-shirts and balloons and satin jackets for a theme party. They would get *Cheers* swizzle sticks and napkins. They would get a complete package of on-air promos, with "windows" for their own messages, and daily feeds from the network with teasers for the evening news.

"News promotion," Sohmer told them, "will be our number one priority." That suited the affiliates fine, because their local news shows' ratings were crucial to their stations' finances.

And, last but not least, each station would get a commercial featuring a family flipping disconsolately through an endless stream of pay channels, only to find repetition, boring programs, and amateurs. Then the family dial in NBC, a nirvana of stars and slickness, and melt into their seats with ecstasy. The tag is "The stars belong to everyone; the best things in life are free."

The affiliates were pounding each other on the back. They were in heaven. For years they had had, as a Seattle executive put it, "a license to print money," and now these . . . *wires* were threatening to turn them into ordinary struggling businessmen. But not if Steve Sohmer had anything to say about it.

Then the music started. It was a single guy at a piano, the composer of the Be There theme song, crooning slowly, with his whole heart, "You got to Be There, yah-ah, you just got to Be There, be what you wanta be, see whatcha wanta see, you can NBC there, Be There." Then he went into the reprise, stretching each phrase soulfully, his carotid standing out on his neck as he sang a paean to television watching, his face twisting in agony or ecstasy as he belted, 'You gotta Be There, Be There!'" Conga drums cut into his bluesy riff, punching rhythm into it, then more percussion came in, and suddenly there was a whole band celebrating Be There in a reggae beat. As the music crashed around the room a troupe of dancers bounded down the aisles from the back of the ballroom, dressed in wild Reno-calypso costumes, each of them a cross between Carmen Miranda and a Sunset Boulevard whore. Women with a drizzle of glitter over their chests cavorted around a pink man on stilts wearing a white cowboy hat and plumage, as they all boogalooed past befuddled NBC executives in suits. The Be There composer's face came onto the big screen, twisted in ecstasy, sweating, as he wailed his anthem.

A voice overrode the deafening music and began crying out the names of NBC stars, who danced out from backstage and paraded one by one through the ballroom. Mr. T. Michael Landon. Cybill Shepherd. Shelley Long. Veronica Hamel. Ted Danson. Dabney Coleman. George Peppard. Ann Jillian. Daniel J. Travanti. Until soon the corridor outside the ballroom was aflame with face recognition. The affiliates streamed out after the stars, to rub elbows with them. Later that evening, there would be a dinner for the affiliates that the stars would be required to attend.

After the extravaganza had blown over, Grant Tinker said, "I didn't know whether to enlist or go have lunch." Tinker would make no grand predictions for his network—only that it "might surprise some people." Tinker was not impressed with the development seasons of ABC and CBS. "I get the feeling they're really trying to help us; they're trying to come back and meet us. Which is okay with me."

Would NBC stick with its dedication to quality programming? "We want quantity of audience as well as quality," he said. "Let me say it straight out: We are satisfied with nothing. Programs, people, policies, plans—all are subject to review and change. We are not a bunch of laid-back losers."

As the convention wound down, affiliates filling the lobby with their luggage, USA Today television writer Ben Brown sipped a glass of beer and talked about the networks. "TV is no longer a three-network monopoly," he said. "NBC is getting hit from every side. Mostly by cable. HBO is hot; it's got both entrepreneurial expertise and top management. The steal of SCTV from NBC was very important, because it brought network fans to cable. The independent stations are hurting the networks as well. They're sharper, leaner, and they know the game.

"The affiliates," said Brown, "are getting scared."

Portland, Oregon, is, in the eyes of the network, flyover country. It is an affluent middle-sized city full of joggers and Volvos, anglophiles and liberals who hate niggers, logging execs who wear Beatles haircuts and charge their cocaine habits on their Visa cards; it's a white-bread bohemia fueled by the timber and garment industries, shipping, and computer software.

It is also the twenty-third-largest television market in the country, in the top quarter nationwide.

Key institutions of this clean and fairly pleasant place tend to be owned elsewhere, however. The local pro basketball franchise belongs to a Mr. Larry Weinberg of Los Angeles. Likewise, the NBC outpost in Portland—KGW—is a fiefdom of Seattle-based communications giant King Broadcasting. Like many other NBC affiliates nationwide, KGW had for several years been going down fast. But a top-notch manager named Irwin Starr had arrived from Spokane recently to take command, and his task was to save this little corner of the world for Grant Tinker. So, back home after the affiliates meeting, Starr allowed himself a mood best termed modified rapture.

"Morning is a problem. NBC's poor ratings really hurt us. We've got about a fifty-fifty commercial split with them at that time, and if their shows can't pull the numbers for us, well . . . Network prime-time news is also very important, because it really reflects over the whole network schedule. It's a matter of reputation and prestige. Right now, NBC News is in trouble. And when NBC gets cut, its affiliates bleed. News is right in the middle of a station's most profitable time, the hours between four and eight P.M. We sell the most expensive spots in prime time, but we sell the most spots between four and eight. Prime time is second—because the only local ad availabilities are at midbreak or terminal break in the shows. Daytime is third.

"Anyway . . ." Starr turned in his chair to glance at the three television monitors mounted flush with the east wall of his Portland office—one for each network station. "The affiliate meetings give the network a once-a-year chance to rally the troops. NBC has lost a number of affiliates in the last several years to the other networks, and if that ever snowballed, the results could be ruinous to the whole NBC network system. So you bring everybody together. Not just for the sake of influence—it's better to be heard one on one on the phone—but for morale. You can't have affiliates out in the boonies going, 'NBC stinks.'

"So they try to get us to take their merchandise, to stop preemptions of network shows. They don't mind if you do it on a one-time-only basis; it's when you preempt them weekly. Billy Graham drives them crazy—he preempts for hours."

Starr was not convinced that NBC's prime responsibility to its affiliates was raw numbers, however. "Raw numbers are bullshit. ABC proved that years ago. Demographics are the key. Agencies say, 'We're selling baby food; what can you deliver us in men aged one to three?' " And NBC was still, as the affiliates had been told repeatedly at the convention, delivering demos.

Starr put both elbows on his desk. "I can find out to three hundred how many people in Portland make forty thousand dollars a year or more. Then I match that information to what's available from the networks, the syndicators, and from what we can produce ourselves. Age definitely plays a part. Game shows skew old; soaps skew younger; news skews middle; *People's Court* skews young; *M*A*S*H* hits eighteen to forty-nine; *PM Magazine* is in the middle. So we try to match our viewership with what there is to view. Then we try to sell the result of that combination. Because TV is like any other kind of business: It all comes down to what you bring in versus what you take out. My key to solving that equation is always being based on just three things: Is it technically feasible, is it legal, and can we find somebody to buy it?"

KGW and its parent station KING-TV, are big-money operations—the average affiliate station makes a profit of $1 million per year. KING was created when the widow of a wealthy realtor paid a reported $325,000 for KPSC-TV in 1947. This at a time when there were only six thousand TV sets in all of Seattle. Nevertheless, the Federal Communications Commission placed a temporary freeze on station-license applications shortly thereafter, a freeze that lasted five years and allowed the newly named KING to get a vital jump on its competition.

For many years KING was a cash gusher. Most of the revenues came out of Seattle, a heavy, handsome city, resembling a cleaner, cooler, more compact Los Angeles. Its ganglia of freeways are margined by clouds of vivid green vegetation. King Broadcasting headquarters are large, new, and plushly space age, though King President Ancil Payne's own office looks more like a den than an executive suite.

"Our success depends on cooperation," said Payne. "Network affiliation is like two scorpions locked up in a bottle: They either learn to live together or sting each other to death. And news is what holds the system together. Nationally and locally. But now the system is changing."

Payne resembles Lyndon Johnson, and is considered by many to be that rarest of television executives, a man who knows what he's doing more than 80 percent of the time. The creator of the hugely successful, nationally syndicated *PM Magazine* and past president of the NBC affiliates' association, Payne's endorsement of Grant Tinker is said to have been pivotal to Tinker's being named chairman of NBC. And, in the early summer of 1983, Payne was of the opinion that television in general and network television in particular faced a radical, uncertain future.

"If you can peddle a product direct from a satellite into the home, you don't have to pay anybody as an affiliate. . . . The networks pay us. Procter and Gamble can bounce its soap right off a Telstar. Direct-broadcast satellites could rob fifty percent of local advertising, and that could ruin local news, local programming, because there'd be no longer the monies to pay for it. So where does the FCC draw the line? They're going to have to control this; somebody has to keep a fence up to keep the elephants from crushing the mice.

"And what if the networks themselves turned on us? What if NBC turned on us? What if they said, 'We're tired of catering to the whims of two hundred, however many affiliates; we're going to sell directly to satellite'?"

Payne gave a historical analysis of the network scorpion in his bottle. "It goes back to when Julian Goodman was president, which was only two eons ago. NBC was just not attentive to what they were doing demographically. NBC programs were aging and nothing was coming out of them. Bobby Sarnoff decided he was going to be number one on NBC's fiftieth anniversary, so he fired

Julian Goodman and brought in Herb Schlosser from the coast. A very bright man, but he had never been a manager. He was a lawyer and fell into the grip of Marvin Antonowsky. Who was a statistician who used to do a magical act with figures." Payne's hands bob up and down, as if juggling unseen balls. "Talk about a case of the emperor's new clothes, Antonowsky just snowed everyone. Everyone believed everything that Antonowsky said. After that, nothing was stable at the parent company; the longest anybody held the chairmanship of RCA was about fifteen minutes.

"Finally, a few years back, Griffiths takes the helm. He hires Silverman to handle the network. Fred's a genius, but not a manager. A guy that always did well if he was under a strong thumb from above, but if you let Fred loose he's like a mad dog in a meat house. Then they brought in Jane Pfeiffer. A colossal error. She'd been an assistant to Thomas Watson at IBM, but that's much different than being a manager to Fred Silverman. The upshot was that they destroyed the central part of the system."

So, Grant Tinker. "He's the first person that's been an actual producer in almost a double decade. He's never run a huge corporation, but at least he's had the experience choosing talent, buying time, and so on. Now, he's considered the greatest thing that ever happened at NBC. Faith is very important."

Payne believed that vital sections of the NBC infrastructure remained intact.

"The key executives Ray Timothy and Bob Mulholland— the survivors—are there; the team is there. Steve Sohmer is really putting the energy into daytime. That's where the money is, not prime time. It's been a mess. NBC made two or three enormous errors. Putting David Letterman on during the daytime. He was way too sophisticated; there's no hip audience at nine o'clock in the morning. That was ninety minutes gone at a crack when what people want then is the excitement of games. But that wasn't the big screw-up. The big screw-up was Lynn Bolin. NBC was solid in daytime, not number one but sure not number three, and was making lots of money. But she took care of that. Started switching everything around. Swept the schedule, went to all games and it didn't work. The girl in *Network* was Lynn Bolin. It wasn't even a parody, except the actress had a better figure.

"Lynn forced a lot of affiliates to go to syndicators. And by doing that she forced a lot of affiliates to program through syndicators." A group that Payne generally has little sympathy with. "I think the affiliates overwhelmingly support the network position. Networks are going to have to remain healthy enough to stimulate programming, and the network function—just in news alone!—is invaluable to holding the whole system together. And if I have to choose between networks and syndicators—most of whom spend their nights sniffin' coke 'cause they can't think of anything else to do . . ."

Payne shook his big gray head. "They take their money and run. They don't do a single goddamned thing for anybody except make money. That's the only thing they produce. They're a bunch of pirates on top of that. The syndicators are always talking about the poor stations and the poor producers. Nonsense. When the programming dried up at the turnback time, they escalated their prices to the point of piracy. Programs we'd take, build up, promote, they'd come back and increase rates on as much as a thousand percent. And the producers. The producers make their money off the networks. And if the network is excluded from participation, you're going to see more and more sales to HBO and Showtime. Syndicated people are going to take their product to pay TV. And I, for one, don't want to see free television come apart."

"NBC's battle to pull itself out of its hole is irrelevant, because it's all over." Paul Klein, Brandon Tartikoff's predecessor at NBC, current president of the Playboy Channel, sat on a white lawn chair on the patio of a suite at the Beverly Hilton and had a drink. It was high noon. He wore a charcoal suit and a blue shirt, the skin of his neck puddling a little over his collar. Black hair thinning on top, Klein has a friendly pear-shaped face and nothing to suggest outwardly his reputation as the most creative manager in TV history. "I predict that the guy who wins the ratings battle for the next season will have a lower rating than the guy who finished third last season. How do I know this is true? Because the year before it was true. And the year before that it was true.

"Competition will kill NBC." Klein took another sip. "This

is all a repeat of what happened in radio and magazines in the fifties. I know, because I was there. At one point there was a guy who advertised himself as the largest advertising medium in the history of humanity. Do you know who that guy was? *Life* magazine, that's who. *Life, Look, Saturday Evening Post.* They all folded at one time or another.

"Not that I have anything against network TV." Klein put his hands in the air. "I mean, I started half of them." He wasn't joking. "I love networks. That's my business. To start networks. Who do you think started daytime TV? Who do you think started color TV? Who do you think came up with TVQs? So I like networks just fine. But times are changing. If RCA had a brain, it would simply tell NBC, 'Hey, folks, the party's over. So long.' But they haven't got the guts to lay off ten thousand workers. People think that God created network TV. Well, it's all gonna change within three years.

"I think it'll come down to a combination of dish antennas and fiber optics. How strong a force can AT and T be in this market? That's the only question." Klein sighed. "It's a crazy business. Cable's going off in so many directions. Having to pay off all these local city guys. It's a very corrupt business, but telephone was like that in the beginning." He sighed and shook his head, signed again. "Playboy is a program service that will seek to reach its audience through any means it can. And I don't want to deal with a thousand direct-broadcast franchises, a thousand shmuck middlemen—I want to deal with the *phone company.* That'll take Playboy direct to the consumer."

Organization man, anarchist, oracle—no figure in television programming has cut a wider swath than Paul Lester Klein. Graduating as a "mathematician and philosopher" from Brooklyn College in 1953, Klein came to television by way of the Doyle Dane Bernbach ad agency. He made his mark in the industry as a connoisseur of demographics. Klein was among the first to point out the necessity to advertisers of ignoring raw TV rating points and, instead, discovering exactly who the audience is for any particular show before sponsoring it. But he is most remembered for two theories. The first: Least Objectionable Programming. Which means that people don't watch shows; they watch television, and will stay with a channel all night if it doesn't offend them. The theory in part created network television's hunger for video Pablum. But there is a corollary to the theory: While the *most* people may turn to the least objectionable program, the *best* people, in terms of advertising market, will opt for shows that are ambitious and even controversial.

Klein became such an influence as NBC's vice-president for audience measurement that his notions about "quality demographics" inspired rival CBS to unload such programs as *The Jackie Gleason Show, Petticoat Junction, The Beverly Hillbillies, Green Acres,* and *The Red Skelton Hour* even though these shows in 1969 were giving CBS the largest television audience in the world.

Equally as important, however, was Klein's invention of Q scores, the system by which shows and actors are given recognizability and likability quotients. The Qs don't always correspond to a show's ratings, but are often predictive of future popularity. "I was able to show that just because a show was being hurt by positioning didn't mean it couldn't be a big success. *The Man from U.N.C.L.E., Laugh-In, The Monkees* were all high-Q shows."

Klein's current enthusiasm was for cable and he discounted the argument of Steve Bochco, of *Hill Street Blues,* that there is not enough talent to fill multiple cable channels. "I want you to go to the graduation ceremonies at USC if you don't think there are the talented people available to create top-quality cable TV. There's talent all over the place. It doesn't matter what Steve Bochco says; he's protecting a franchise. If he says he's screaming for writers, that he can't find them, then he's not really looking. But if he says he can't get writers to give him a thirty share . . . well, it just doesn't matter anymore. Soon, a ten share will be enough.

"There are a hundred thousand writers out there"—Klein's hand swept toward the city of Los Angeles—"who couldn't write that stuff. My own daughter, she's a fine writer, but she couldn't write it. So what? Bochco doesn't even know who's watching his own show, because he's not facile with the numbers. He used to work for me. Who cares what goes on in a precinct house? I'll tell you: The largest audience for *Hill Street* is *blacks* and *Puerto Ricans,* who couldn't care *less* about intellectual writing. And it wouldn't matter a whit if you lost all the intellectuals watching TV, because they only tune in randomly."

And, as for NBC's current programming, "Grant Tinker gave up on quality with *The A-Team*. Now what's he doing? Renewing this hospital show. This *St. Elsewhere*. It's in the basement and they commit to thirteen episodes. Thirteen million dollars." According to Klein, what Grant Tinker was doing was "building MTM into the strongest production company in the business." A production company that presumably, as soon as NBC closed its doors, would be selling its shows to the new moguls of television.

Paul Klein believed that he had seen the future of television—and that it was cable.

"The network executives rely on bravado and wishful thinking in confronting cable," suggested Thomas E. Wheeler, the young president of the National Cable Television Association. "But cable is unstoppable.

"A while ago one network vice-president exclaimed, 'If anyone tells you they know where the viewers are going, they're wrong.'

"Well, sir, *you* are wrong. There is *no* mystery as to where the viewers are going. *We* don't need your *Magnum, P.I.*, your *Quincy*, your *Hart to Hart*, or your men and women of *Hill Street Blues* to solve this mystery. Your *lost* viewers have been *found*, and believe me, they are alive and well as contented subscribers to cable television."

As of the summer of 1983, the medium was no longer a weak sister to network TV. Over thirty million American homes were now wired to some kind of cable service and cable companies were adding subscribers at the rate of four hundred homes per month. Cable's staple products were movies and sports, with news, sex, and Jesus not far behind. Numerous large cable companies, such as Time Inc.'s Home Box Office and Viacom's Showtime, were offering not only recent theatrical releases but first-run films financed by the companies themselves. Other cable concerns were contracting with professional football, baseball, and basketball teams to show hometown games in large urban markets, a service the networks had so far been unable to provide.

But the true attraction of cable TV is its relatively limitless world. Unlike network television, which is fenced by censorship and the confines of three channels, cable TV has hundreds of channels at its disposal and almost no censorship. What's your pleasure? Hourly stock market returns, sex with furry forest creatures, jai alai lessons, the latest technique in open heart surgery? Cable television has them all. Cable's sheer breadth allows even the most narrow of viewer interests to be satisfied. A fact much in evidence at the Thirty-second Annual Convention and Exposition held by the National Cable Television Association in Houston, June 12 through 15.

Cable differs from network television technically in that both its sound and its image are carried through wire instead of over the air. Most often, the large cable systems rely on a combination of cable and "dish," a delivery system that uses a central "uplink" to broadcast to a transponder on a communications satellite that has been placed in geosynchronous orbit 22,300 miles above the equator. These satellites necklace the earth at intervals of no less than 15 degrees. They receive individual station signals and bounce them back to "downlinks" on earth. These downlinks are receiving dishes that either service individual homes directly or are the receiving centers of cable webs.

The cable industry in 1983 was worried that dish technology would make cable obsolete, that dish costs would soon plummet and manufacturers would be stamping them out like garbage can lids, thus sidestepping the cable process entirely. Consumers could be taking almost all television straight off a satellite.

While most people in television felt that direct-broadcast satellites would not become a strong force in the market until the late 1980s, Canada's Anik C-3 satellite was about to change that, moving the schedule up by years when it was launched in the space shuttle in mid-June. United Satellite Communications had leased transponders on Anik and was prepared to deliver service to homes in twenty-six states; some analysts were forecasting a $1 billion market for DBS by 1990.

And this was the major current concern for many of the people gathered in Houston.

Until recently, cable television had been widely perceived as an ungoverned and ungovernable industry that grew by erratic leaps and strange bounds—a nineteenth-century-style venture-capital business populated by get-rich-quick types, media pioneers, and network rejects, a phenomenon fueled by a bouillabaisse of hucksters, visionaries, and hacks. So it was perhaps appropriate that the conference was being held in Houston, a beautiful, disorganized, oppressive space-age frontier town where there are no zoning laws.

Instant slum, or glory tower by I. M. Pei—in Houston, build what thou wilt seems to be the whole of the law. Likewise, out at the ratty colossus of the Astrodome, where the big guns of cable had come to hawk their wares, Paul Klein's garter-snapping Playboy Channel booth seemed positively restrained. Just like a real TV channel, Playboy offered a complete menu of Playboy news, Playboy sitcoms, Playboy quiz shows, etc., all enlivened not so much with sex as with breasts. Each clip seemed to punctuate the screen with nipples, one on either side, all firm and erect. Everything else, the jokes, the info, the glitz, as in the magazine itself, was background in an ongoing ode to lactation. What the Playboy Channel seemed to be about was . . . great tits.

But that was scarcely all that was going on TV sexwise there at the Astrodome.

Witness the Pleasure Channel.

Which is to Playboy what Idi Amin is to Mr. T. On Pleasure's monitors, funky leather studs undulated with sex kittens from the twenty-fifth century, orifice-crazed musclemen grasped the tawny loins of jackbooted go-go girls, all in a ritzy, rocking, Bob-Fosse-choreographs-Gomorrah milieu. Slick, hard, glistening Technicolor TV sex. Explicit, but tasteful. The Pleasure Channel, in an "adult cable tier," was designed, according to literature handed out there at the booth, to "augment basic pay television services via Multiple Dish System Cable, Satellite TV, Private Cable, hotels and closed circuit systems."

It offered, on a seven-day basis, ten premiere films a month and no repeats within the first six months. Cable operators were promised that there would be several versions of each film, "enabling you to deal with the varying community censorship standards."

This information proffered by a beaming hive of Pleasure Channel "hostesses" there at the booth, all elegant and dewy-eyed, but with talon nails and lipstick that looked as red, bright, and hard as a taillight. Virgin nymphomaniacs.

Mother Angelica, foundress and chairman of the board of the Eternal Word Television Network, EWTN, was bivouacked a hundred yards away. Less than two years old, EWTN already reached more than one million homes. Doctor of Sacred Theology *honoris causa* from the College of Steubenville, former advertising staff member at the Timken Ball Bearing Works, this Franciscan nun had just secured a new satellite position on Satcom IIIR, Transponder 18. Her operating costs exceeded $1.5 million a month—transponder rental averages over $60,000 a month alone—and, because she offered her channel to cable services for free, all expenses had to be met by donations. She set up her station at the monastery she created twenty years ago, Our Lady of the Angels, in Birmingham, Alabama. Two hours of "family entertainment" and two hours of "spiritual growth" compose the typical EWTN evening. "We don't have the preaching type of program," she says, nor a political one. "I don't see Jesus involved in the political process." Because EWTN is the first Catholic television network, its expansion has been compared in importance to that of the parochial school system. Such are the potential powers of cable.

And Mother Angelica's operation was just one of a half-dozen "Christian broadcasters" displaying their wares in Houston. Chief among them was PTL Satellite Network, offering twenty-four-hour "inspirational" programming by the likes of Robert Schuller, Oral Roberts, Pat Boone, Rex Humbard, and Jerry Falwell. PTL Satellite programming is as slick as Rodeo Drive. Production values are state of the art and the "inspirational" programs seem to be delivered by men in the service of a loving, corporate, Republican Jesus. A Savior who endorses the middle-American status quo. Most of the ministers themselves radiate a sort of hamfisted prosperity, their clothes relentlessly new but slightly tacky, out of style, like aristocrats forced to buy their suits from JC Penney.

For the executives of the cold-cash, mainline outfits like Showtime and HBO, however, the bespoke tailors of Savile Row seemed barely adequate. And their hostesses were more like the no-nonsense Kitty on *Gunsmoke* than the nubile, fresh-from-the-dairy-case Playboy maidens who handed out leaflets and bantered with the customers ("Kenny, you tramp. I've seen you lallygagging throughout the bars and sewers of the whole Southwest").

These companies had real muscle. Showtime, for example, had nearly four million subscribers in over two thousand cable systems and it was doubling in size every two years. Owned by Viacom International, Showtime was created in 1976 to provide Viacom's cable systems in northern California with a pay television service. In 1978, Showtime went national via satellite, and went on to become the second-largest pay television service, behind Home Box Office. It was the first of the cable networks to make a network-style commitment to a series, when it ordered seven original episodes of *The Paper Chase*, a radical turning point for a cable network and a frightening portent for the networks.

But the biggest money was with Time Inc.'s Home Box Office. HBO was out in full force here in Texas. Only eleven years old, HBO had made a bigger profit the previous year than NBC. It was "a money machine, pure and simple." Cable analyst Jefferson Graham said, "HBO's growth into the dominant entertainment force that changed Hollywood has got to be the business story of the century." What he was talking about was the Time Inc. move into feature film production. "Our goal," advised Home Box Office President Frank Biondi, "is to have two HBO premier films a month, probably starting in mid '84. . . . We have created a marketplace that is going to be something like $600 million this year for feature film product, which not only is substantially greater than network license fees have ever been in the aggregate, but generally we buy all the motion pictures that are not X-rated, which is a clear difference from network TV."

The country's leading pay-cable channel, HBO was on the air 8,760 hours a year and was well on its way to becoming the leading financial backer in the entire motion picture business. In the spring of 1983, HBO had moved to create Tri-Star Pictures, a whole new "mega-studio," in partnership with CBS and Columbia Pictures. The merger was being hotly contested by competitors who claimed Home Box Office was seeking a monopoly position in the business.

HBO was also involved with the investment firm of E. F. Hutton. Together they created Silver Screen Partners, a conduit for investors willing to put up at least $5,000. If all went well, Silver Screen Partners could raise as much as $125 million for production of films over which HBO would have total control. One way or another, Home Box Office was now thought capable of pumping $750 million into film production in just a few years' time, equaling the fiscal outlays of the largest studios and putting HBO in a position to potentially account for more than 50 percent of *all* films produced in Hollywood.

Despite the formidable presence of HBO, despite even the well-displayed attributes of the sex channels, the real attention grabber at the convention was an ex–Ivy League southerner. It was Ted Turner, his eyes ticking from monitor to monitor, trailing a little chevron of assistants, sitting down, then bobbing up, exercising in a kind of manic ballet. His voice was high and quacking, much like that of somebody who's just inhaled helium. Turner zigged and zagged around his Cable News Network booth like a metal duck in a shooting gallery.

As CNN promo handouts waste no time telling you, Turner should be remembered less as a boozehound and womanizer than as the visionary who changed the face of television by "originating the SuperStation concept." His SuperStation, WTBS in Atlanta, "is a reworking of the traditional television network concept, in which one station acts as original programming supplier for a multiplicity of distant cable markets." WTBS made history when its signal was first beamed to cable systems coast to coast "via a transponder of RCA's Satcom-I satellite. Its signal is now beamed to more television households than any other cable programmer."

As an occasionally avowed presidential candidate, owner of the self-described "America's Team"—the Atlanta Braves—and creator of the world's first around-the-clock television news service (which he once modestly referred to as "the greatest achievement in the annals of journalism"), Turner had just appeared in an exciting interview in *Playboy* magazine. His talk with journalist Peter Ross Range had ended, in fact, with Turner attempting to destroy Range's tapes. This, after a wide-ranging series of discussions. Dismissing the magazine as "just sleaze on

Gordon Farr and Lynn Farr Brao
Veteran producers of The Love Boat, they developed the much maligned Fred Silverman project We Got It Made

Photo courtesy of MGM

Fred Silverman
The Michelangelo/P.T. Barnum of prime time. The only man in history to have, over a period of years, directed programming for all three networks, Fred Silverman was—with Paul Klein—the major shaping force in network television in America. At least until a couple of years ago, when his luck ran out. Since then Silverman has turned to television production, with intriguing results.

Tony Colvin (left) and Scott Gorden
They had a dream: to have all of America watch the shows they wrote. Tony gave up a $70,000-per-year bank job to be a gofer on Cheers, where he hoped to write scripts.

Grant Tinker
The most admired producer of television in Hollywood, Tinker found running a network—especially a third-place network like NBC—a chore and challenge of startling magnitude.

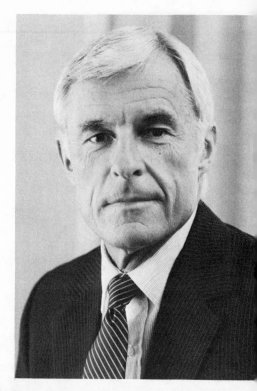

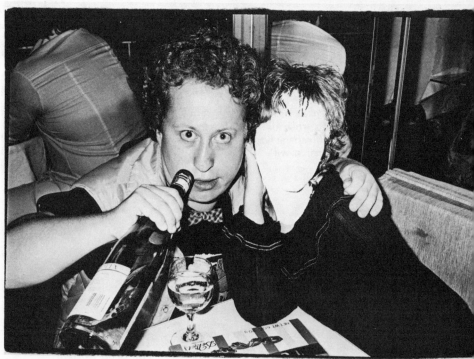

Corky Hubbert Corky Hubbert with a faceless fan. Living out of a backpack and dining at Ma Maison, guest star on Magnum, P.I. one week and a buck-an-hour extra in a Portland, Oregon, TV soap opera the next, Corky was determined to become the most famous midget in Christiandom by the time he was thirty-five.

Brandon Tartikoff
The first of the baby-boom programming chiefs. A protégé of Fred Silverman who was kept on by Grant Tinker as head of NBC Entertainment, Tartikoff battled the other networks and cancer simultaneously.

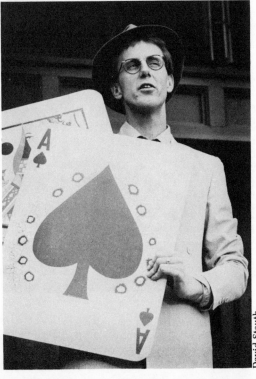

Harry Anderson
Former street magician and quasi-professional con man, Anderson was able to parlay his ghoulish sense of the ridiculous into frequent guest spots on both Saturday Night Live and Cheers and then, finally, into the starring role in the NBC sitcom Night Court.

Peter Grad
A former Wall Street financial executive, Grad was a power at Twentieth Century-Fox and a man responsible for developing a fat fraction of what America watched on prime-time television.

Gary Nardino
One of the top studio powers in Hollywood largely responsible for building Paramount Television into a giant in the late seventies

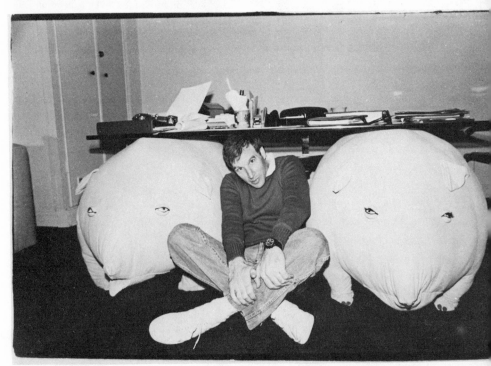

Allan Katz   A former producer of M*A*S*H, Rhoda, and the Cher show, Allan Katz was considered one of the brightest and most innovative writers in series television, a distinction he was willing to forgo given the chance to create and star in his own feature film, The Hunchback of UCLA.

Karen Salkin *Cable television's first true starlet. Her public access show, Karen's Restaurant Revue, made her the in-est item in L.A. during the summer of 1983. Her only problem with becoming a minor superstar was that she couldn't figure out how to make a buck.*

Larry Colton *A former professional baseball pitcher turned TV producer, Larry Colton came up with a plan to revolutionize prime-time television. It worked. Almost.*

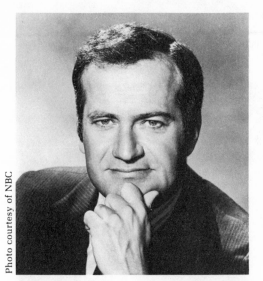

Steve Sohmer
NBC head of promotion, daytime programming, specials, and Saturday morning. A successful novelist at twenty-five, Sohmer went on to become one of the most original and astute promotion artists in television history, but even his inspired efforts weren't enough to float the network's fall schedule.

Jimmy Burrows
The power behind Cheers. A protégé of Jay Sandrich, Burrows fought tooth and nail to keep Cheers on the air and make it the best sitcom on television during the early eighties.

Hamilton Cloud
A young kingpin in comedy development at NBC. Part of the team that would take the heat over the collapse of the 1983–84 NBC development season.

George Wendt ▶
One of the most talented character actors in television, a vital atom in the Cheers molecule.

Danson ▼
·rs star Danson was an unas-
·ng young actor completely
·ised to find himself becoming
·jor sex symbol and one of tele-
n's most important leading

Shelley Long ▲
Star of Cheers. Unlike the more laid-back Ted Danson, Shelley was adept at setting goals and fulfilling them quickly. She was considered by many in the industry to be the next Mary Tyler Moore.

John Ratzenberger ◀
As the mailman Cliff on Cheers, Ratzenberger turned a working-class background and a lifetime of bit parts into fame and fortune. Then watched it almost slip away.

Ed Asner

Once America's favorite boss, then tarred by the press as a pinko subversive for helping send medical aid to Central America, Asner had to rehabilitate his image in 1983. It wasn't easy. Here he narrates a documentary, one of many Q-building activities that helped him once again power his way to the top of television.

Long regarded as a soft touch for a good cause, Ed Asner worked the telethon circuit as diligently as any star in Hollywood—even when practically no other star in town would help him with his political problems.

some pages and outdated information on others," Turner had allowed, "I don't need your . . . magazine. I've been on the cover of *Time*." Then Turner went on to address the question of inventing himself as head of a "fourth network." He had been trying to consume Metromedia, a huge independent broadcasting conglomerate, which, like Turner, was actively entertaining notions of creating a fourth free network. Though ultimately negotiations with Metromedia would collapse on the issue of who was to eat whom, Turner made few bones about the fact that he felt his stewardship of a network would be an excellent idea. Even a merger with CBS had crossed his mind, although he'd claimed a year earlier that CBS was a "cheap whorehouse" taken over "by the sleaze artists," and that its president, William Paley, was a "failure."

Turner explained to Range, "If I'm really up, I tend to let it all hang out. . . . You know, several years ago, I said that the network presidents were guilty of treason and all should be lined up and shot after court martial. . . . I said the worst enemies that the United States ever faced were not the Nazis and the Japanese in World War II, but were living among us today and running the three networks. . . . We're approaching the 21st Century with the most powerful communications force the world has ever seen. And it's being totally misused by three organizations that couldn't care less about what happens to the nation. . . . It's insane. . . . In prime-time, there are no choices. . . . only stupidity, sex and violence." Range asked Turner what he'd like to do about it and Turner told him, "I'd like to get my hands on a network. I'd like to be the big guy for a while."

Shortly thereafter, the equanimity of the interview evaporated. Flying thirty-five thousand feet over Colorado in an airliner, Turner allowed to Range that "I'm sick as hell of you." Then he snatched Range's tape recorder from his hand, smashed it on the cabin floor, heaved Range's camera bag into the aisle, and booted it against the cockpit door; cursing and yelling, he pitched the bag at Range's head while the interviewer was scooping up the remains of his tape recorder. Turner then adjourned to the airliner's lavatory, leaving his companion, Liz Wickersham, to console the shaken Range. "He's under a lot of pressure," she said. "He did the same thing to me once, getting on a boat in Greece. He kicked me in the shins."

Larry Colton's primary ambition in life was to make a living. But if he had to destroy network television, piece by piece, throughout the entire country, in order to do it . . . well, so be it.

"Larry," program director Steve Currie said, eyeing the bank of television monitors above his desk, "you realize this is probably the farthest out a network affiliate has ever gone on an independent project like this?"

"No, uh, actually, I didn't," Larry Colton said, hunched forward, elbows on his knees, on the seller's side of Currie's desk.

"But we're impressed, I mean *impressed*," Currie said, still watching the monitors. "I saw national-quality acting in those clips; I saw great dialogue, some real moves, Colton. You've come up with a superrevolutionary idea."

Colton said, "Thanks. Is my check ready yet?"

Currie took a deep breath, his whole torso expanding over the back of his chair. At thirty-seven already a past president of the National Association of Television Programing Executives, Currie wore a gray suit and a sort of Beachboy-Dutchboy haircut trimmed for executive action. He sighed, "No, sorry. I can't cut it 'til Howard gets back," Howard being general manager of Currie's station, the CBS affiliate KOIN in Portland—where Larry Colton, ladies' man, former pitcher for the Philadelphia Phillies, and freelance writer, had just completed a deal he believed would revolutionize prime-time television, through the creation of a nationwide interlocking net of "franchised soap operas." Like every other cable and nonnetwork programmer in the country— and there were hundreds of them now—Colton knew that the Big Three networks were suddenly vulnerable. So it was war. And Colton was taking no prisoners.

He had a simple four-part plan that was soon to get him written up everywhere from the front page of USA Today to The Wall Street Journal:

• Create a prototype for regionalized weekly soap opera, designed to preempt network broadcasting during its weakest hour or half-hour slot. All production would be done and/or paid for by local affiliates, though basic scripts and an "operations manual" were to be provided by Colton and company. Colton's notion was that regional broadcasting companies could form "affiliate molecules" in order to underwrite production costs, so that a single prime-time Colton soap opera might serve stations in Buffalo, Albany, Schenectady, and Troy.

• The revolutionary attraction was that each of these soaps would be tailor-made for each region and community, their fictional characters tied to real events. If the hog market in the Midwest was going down the toilet, a major character could be taking a beating in pork belly futures. If the Detroit vice squad had just had six officers arrested for acts of corruption, an actress playing a sensuous, blue-haired Grosse Pointe matron could be framed by the cops on a phony "heiress by the hour" call-girl beef. The *roman à clef* possibilities were endless. The names would have to be changed to protect the guilty, of course, *but everyone in town with the brains of a gerbil would know exactly who those guilty were.* It would thus be possible to out-*Dallas Dallas.* After all, why should viewers watch the sleaze-ball machinations of the network's cowboy-hatted fops when they could get the real dirt on their own friends and neighbors?

• Congruent with this, Colton and his two other partners, film director Tom Chamberlin and producer Evelyn Hamilton, had made an interesting discovery: that Portland—and so probably most every other burg of more than a million in population—had a community of actors capable of putting on at least *one* prime-time series of national quality. And by spreading the action over a three-city area, the burdens of talent could be further absorbed.

• Costs could also be reduced, Colton had discovered, by selling advertising not only around the show, but *in* it. He and Chamberlin had offered to shoot scenes at local businesses, in exchange for a "location fee." If the sexy, blue-haired matron was going to blow her brains out on account of the vice cops' framing her, no reason why she shouldn't do it at the Beef 'n' Brew. So far, almost a dozen local establishments had ponied up for the privilege. And why shouldn't the actors look like walking billboards for the local manufacturers? Portland's Nike and Jantzen had already signed on the dotted line.

"So," Colton said to Currie, "I get my check on Tuesday."

Currie nodded, spreading the papers containing Colton's plot outlines for *The Pillars of Portland* in a fan across his desk.

"Cuz I'm hurting, man," Colton said. "I gotta pay the rent, the wives, everything."

Currie smiled and opined that Colton was just his kind of independent TV producer: smart, hardworking, hungry, and cheap. "You're the wave of the future, Larry," he concluded, eyeing the outlines. "And like I said, we're impressed. All *Pillars* is going to need is a little tinkering here and there. More humor, sex, action, character development, and plot. Just the small stuff." He put both hands on the edge of his desk. "You know, people have talked about pulling this kind of stunt at the affiliate level for years, but I'm pretty sure this is the first time any affiliate's ever had the guts to produce something like this. Except maybe in Boston once."

Colton looked at Currie. "What happened there?"

"I don't remember."

"So," Colton said, "how do you want to construct the first show?"

"As a two-hour special. Doing it that way will have more hooks for the advertisers. We'll sell it on sizzle; it's easiest to sell prior to the ratings. We'll bump a network repeat theatrical. Hey, another thing: We talked to Seattle; they're interested."

"You mean in showing it up there?"

"Yeah," Currie said. "If we can syndicate even the first one within the region, we can really cut costs. By this time next year, you'll probably be able to buy us all." Currie went on to paint a brief, expert picture of the hopes, dreams, joys, fears, loves, and angst of Colton's characters—who would be franchised, Burger King style, across the country. Colton would become an amalgam of Neil Simon, Hugh Hefner, and Ray Kroc. It was a portrait so immediately vivid it seemed to hang in the air, neon between them.

"Just remember," he said in closing, "focus on the community."

They continued their banter as Colton ambled out of the office, conjuring up a Larry Colton who would be the king of preemption, a Larry Colton who would drive the networks crazy, replacing their weakest shows from sea to shining sea. A Larry Colton so powerful that Grant Tinker would have to order Brandon Tartikoff to have him hit. A Larry Colton who was such a threat to the networks that he'd probably end up facedown on the bottom of the Columbia River.

Leaving, Colton, nicknamed the Fun King because of his general blue-chip mentality, was ecstatic about the big picture and desperate about the little one. After almost two years of work on *Pillars*, he was finally getting his shot. But the rent on his house was, as usual, weeks overdue, and it was that kind of stuff that was really beginning to drive him crazy. Colton had spent most of his life as a winner, but lately it had been a long spell between victories.

Colton had grown up on the beach in L.A., in what is now Playa del Rey. Twenty years prior to realizing how to revolutionize prime-time television, he'd signed as a pitcher with the Philadelphia Phillies and married Hedy Lamarr's daughter— several months earlier than they had planned actually, in order to sidestep the draft notice Colton had received in the mail.

Instead, he graduated from Cal Berkeley as an English major and signed with the Phillies. After three years of pitching for a farm club, he was moved up to the majors. But his career as a ballplayer effectively ended the night Bobby Kennedy was shot in Los Angeles, which was also the night before Colton was scheduled to make his debut against the L.A. Dodgers. Colton and a friend watched the news about the shooting from the vantage of a bar in San Francisco. Leaving, Colton got into a fistfight with two other patrons, was blindsided, and hit the ground with such force that he dislocated his shoulder. He moved around the minors for the next couple of years, but his touch was gone, and Colton is remembered in baseball circles not for his talent but for his axiom "It's not whether you win or lose, but how good you look in your uniform."

* * *

Colton went on to teach at Portland's Adams High School—which during the early seventies was the most radical public school in the nation. "Which meant," he remembered, "that half the time, when you stood up in front of the class and said, 'Well, kids, today we're going to learn about past participles and the Crimean War,' they'd say, 'Fuck you, zipperhead, we're all going to the park to snort laundry detergent.' " Frustrated, Colton attempted a return to pro sports. But when you're thirty-five years old and your main dish is a forty-mile-per-hour fastball . . . Suffice it to say, he had problems. Later, however, Colton wrote a story about his return to the minors and sold it to the *Portland Oregonian*. And soon he was writing freelance for *Sports Illustrated* and had published a book, *Idol Time*, about professional basketball.

So, life was good. Larry Colton had what this country is supposed to be all about—athletic ability, brains, personality, and drive. No odd habits. No crazy temper. While he probably wouldn't have put Rhett Butler out of a job, he was a ladies' man, too, and had all the girlfriends any rational human being could want.

Still, there was a problem. No money. Here he was forty years old, twice divorced, the father of two daughters, and living alone in one of those flaky-paint neighborhoods where America hasn't quite kept its promise. Making money writing freelance, Colton had discovered, was about as easy as truck farming the moon, and he faced Christmas of 1980 with just enough cash to buy presents for his girls.

It was then that he got a call from a woman named Evelyn Hamilton. Ms. Hamilton had approached him before about turning a comic soap opera he had written for a local weekly paper into a TV show. Colton didn't know much about her. Except that she was a Junior Leaguer type, was living in a wealthy Portland suburb, and apparently had inherited some kind of estate. They pitched the idea to KGW at a time when NBC's ratings were at an all-time low, right after the departure of Fred Silverman in 1979. But KGW passed, and Colton forgot about the idea until Evelyn Hamilton called back twelve months later and said she would finance the experiment herself.

She offered him a thousand bucks to start a script, and the

next thing he knew, he was in a meeting with the Bird Lady—a female rock-promoter friend of Hamilton's who, the only other time he had seen her, was dressed head to toe in an outfit that Colton believed to be stylishly recycled chicken feathers. She exclaimed, "This can be national! I'll get you an appointment with Ted Turner for fifteen percent of the action!" Colton demurred.

Still, he went to work on the script. Evelyn Hamilton put up $10,000 to cover start-up costs. Interest at KGW, he was told, had been rekindled, because NBC programming had become so disastrous the station was on the verge of preempting several prime-time network shows and producing its own. So, within a week, Colton was writing a script designed to encompass four dozen separate actors. Staying awake half a week straight, he cranked out forty separate scenes. Creating characters, he went for the local archetypes: the hip lumber broker. The hip, bored, and luscious suburban housewife. The hook was psychotherapy. Evelyn Hamilton even arranged for the two of them to be analyzed by a psychiatrist.

The original budget for *Pillars* was to be $7,500 per half hour, a rate just slightly higher than what local stations were accustomed to paying for syndicated programming. And the idea that Colton would write generic scripts was an early one. His hip lumber broker in Portland could become a hip steel magnate in Pittsburgh or a hip cattle rancher in Houston.

KGW passed again. The local ABC affiliate wouldn't even talk to him. But Colton got to watch a bidding war take place between the Portland CBS affiliate, KOIN, and KPTV, which was owned by Chris-Craft and was the most profitable single independent station in the country.

KOIN made the top bid and put Colton in business. The problem was, he was still living on cleaning-lady wages—clearing, if he was lucky, $200 a week, which was about a two-hundredth of what somebody like *Cheers* producer Jim Burrows got weekly just to get out of bed in the morning.

But all of that, Larry Colton was positive, would soon change.

Cable TV's first true starlet, Karen Salkin, too hyped to giggle, watched the monitor in the *Tonight Show* greenroom as

Johnny Carson did the monologue. Why, Johnny wanted to know, did all football fields in Iowa have artificial turf? No takers. Carson, looking as tautly exercised as a jockey in his perfect sport coat and tie, said, "So the cheerleaders don't graze at half time." Then, how come Minnesotans didn't drink Kool-Aid? Not a peep from the audience. "Because," Johnny said, "they can't figure out how to get two quarts of water in one of those little paper envelopes. Now . . ." He squinted slightly at this next one. "Why do the women in Tennessee go in for breast feeding?" Ed McMahon, his own natty, baked-potato countenance tipped toward Johnny's desk, shook his head and shrugged. "Because," Carson said, "the milk cartons kept falling off their chests." What, Johnny wanted to know, was the difference between an Ohio girl and an elephant? Titters from the gallery. "Thirty pounds." How did you make up the difference? "Force-feed the elephant."

Then Chevy Chase came on, served a few tennis balls into the curtain, and plugged his new movie, *National Lampoon's Vacation*. Next was comedian David Sayre, then Johnny said, "Most cities, including Los Angeles, have what they call public access TV. My next guest has a show on there called *Karen's Restaurant Revue*. We thought you'd like to meet her. Would you welcome Karen Salkin."

Karen Salkin emerged from behind the curtain wearing a tight purple gown and sat down next to Johnny's desk.

Johnny said, "You really dressed up tonight, didn't you?"

"This was Cher's. I bought it at her garage sale." At that moment, in Hollywood, Karen Salkin was hipper than money. Her scrupulously unrehearsed restaurant show appeared weekly over Group W Cable.

She was cable television's first discovery.

"Originally the dress cost fifteen hundred dollars," she said, "but for me, seventy-five."

"Listen," Johnny said, "this public access. How do you get on?"

"You pay, Johnny." She continued, "I've been talking to you for years. You don't know about it. Every night before I go to sleep and when I wake up. You're not there. My boyfriend is. He wonders why I call him Johnny."

Taken aback, Carson wondered what it cost to be on public access.

"Thirty-five dollars."

"You pop for thirty-five dollars," Carson said.

"This is so exciting!"

One of Johnny's brows curved upward. "I come to work feeling that way every day. So you do this public access, reviewing restaurants."

"Yeah," Karen nodded. Mostly big eyes and nose, black hair, slender legs, and a mouth reflective with lipstick, Karen Salkin was campy-glamorous, what Batman might like for a girlfriend, her face equal parts Cher, Marlo Thomas, and Boy George. "Public access is really good for people who want to do something and no one else will give them a job. People would always say to me, 'You must be good for something,' but the question was, what? So my boyfriend figured it out."

"Well"—Johnny's lips pursed; his head went up and down—"he's obviously a man who would know. What are your qualifications as a restaurant critic?"

"I have a mother who never taught me to cook."

Carson bonked his desk with his palm. "What solid credentials!"

Karen clapped both hands to her mouth. "You're so funny!"

"That's why," Johnny intoned, "I get more than thirty-five dollars a week. What's your favorite place?"

"Well," she said, "my new favorite place is Back on Broadway in Santa Monica."

"What's it like?"

"I'm not much of an expert yet. It's food," she allowed; "that's all I know."

Johnny was incredulous. "How can you get away with reviewing a restaurant by just saying they've got food? *I could tell you that.* What kind of a review is it just to say they have food? I know they have food. What kind of food is it?"

"Good food," Karen clarified, "good food."

The daughter of Woody Allen's high school English teacher, Karen Salkin grew up in Brooklyn, hitchhiked to L.A. at eighteen, got hired to be cut into twelve parts as a magician's assistant, and had one leg fractured on stage as a ballerina and the other broken in an auto wreck during the shooting of Robert Altman's *Popeye* on Malta. But now things were really beginning to coagulate. The *L.A. Times* had run a full-page profile of her in its Sunday supplement. She was the lead item in *Peoplescapes* in the latest issue of *Los Angeles Magazine*. The *L. A. Herald-Examiner* had done a feature story on her, headlined SUCCESS SERVED ON A SILVER PLATTER; HER CAREER'S COOKING AFTER "KAREN'S RESTAURANT REVUE." And a month earlier, she'd been interviewed by *Time*. When two producers of *Entertainment Tonight* came down to Group W to do an interview with her, one said to the other, "She could end up either as the new Streisand, or the new Joan Rivers. But which?"

It was fitting, though, that Karen should rise out of cable TV, for her act was nothing you would ever in a million years find on network television. Salkin was what she called a monologuist. She simply got in front of a camera. And talked. For example: "You know what's really good—I'm shocked—Carney's on Sunset. . . . Anyway, it's a cute place. Just don't go there in high heels because there are spaces between the boards and you could fall down, which maybe I did and maybe I didn't. I felt great that day and nobody saw me except those burger people. What a waste of looking good." Occasionally, she was not so enthusiastic. "Joe Allen's. I promised I'd never say anything good about this next place because I'm honest. Anyway, the service! Aren't they slow? Aren't they mean? Everyone's walking around going, 'Why do I need to be nice to you? You're just a customer and I'm going to be a star someday.' "

Sometimes she'd tell the camera about her life. "My friends got me a stripper for my birthday. Like I really wanted a stripper. I was jealous of his underwear. They tied my legs to a chair so I couldn't leave and when it was over I said, 'Thank you. Never do this for me again.' It was the most awful thing. So toward the end of the party they told me the police were coming because we were making a lot of noise.

"Then this policeman comes in and he wants to talk to me because it's my party, and my friend Nancy is taking pictures and he is saying, 'Ma'am, put that down.' I said, 'Listen, we had a stripper here before and she thinks you're a stripper,' and I said, 'Nancy put that down!' And then he took me up against a wall and started to hit me with his club and I started crying and saying, 'I hope this is a joke.'

"And it was! It was another stripper! You think I wanted him to strip after that? Forty-five minutes they had to calm me down—no one had ever scared me so much in my life *and* he had the skinniest rear end I've ever seen, so ha-ha. . . .

"Anyhow, this was my burger show. Buns, get it?"

And she allowed to Linda Christian of the *Los Angeles Times:* "The other day my boyfriend Ray said, 'You have no respect for what I do. You don't respect acting.' And I felt so terrible and I was crying so hard and I said, 'Well, you don't respect celebrityhood.' Our argument just had to stop there because Ray started laughing.

"I like celebrityhood. I would like to stay in celebrityhood. I'd like to strive for some national celebrityhood.

"I want to be a star to act, and I want to act to be a star. If there was a way to be a star without acting, I'd probably do that."

For almost ten years, Salkin had worked off and on in Hollywood, trying to figure out a calling. Finally, her boyfriend, Ray Buktenica, convinced her to do the restaurant show, offering to pay for it himself.

After her first show, she got a call from a casting director from the office of Norman Lear. He was most impressed with her maiden effort on public access television, and did she want to come in and talk about a job?

Next, she got an invitation to join the Los Angeles Restaurant Writers' Association. After her fourth show, a Canadian production company asked to use her tapes as part of a pilot for a comedy series. Then she got a call from a casting director who wanted to introduce her to Garry Marshall, of *Happy Days* and *Laverne and Shirley. Time* magazine asked if it would be okay if a *Time* story about cable TV highlighted her. Then *Real People* wanted to air part of one of her shows.

So during one taping, Karen Salkin held up a sign that read IF YOU WANT TO SEE ME ON THE TONIGHT SHOW CONTACT SHIRLEY WOOD AT NBC. But she had already been noticed by Carson's people. The morning *before* this tape was aired, she got a call from a *Tonight Show* talent coordinator asking if she would like to appear as Johnny's guest.

Meanwhile, Rogers and Cowan, the most prestigious public relations firm in L.A., took her on as a client, under financial terms so lenient she got what amounted to a scholarship. At the Polo Lounge of the Beverly Hills Hotel, Rogers and Cowan executive Beth Herman talked about Karen's future. "Something's emanating from Karen; she's putting it out there, she's attracting it. . . . She opens her door in the morning to get her newspaper and *Entertainment Tonight* is there to film it. And, like the piece in the *L.A. Times*. People who've been in the business for twenty years can't command one, and for a writer to just go out and *find* someone like Karen is almost unheard of. But it was the trade ad that did it." Karen had run an ad with a photo of herself in *Variety* announcing her upcoming appearance on the Carson show. "I looked at it and all of a sudden things started falling into place. She may be a national personality by this time next year," Ms. Herman said. "A superstar within two to five years. I see her as a combination of Joan Rivers and Mort Sahl. What we'll want to do is start by getting her name in the trades. Constantly. *Variety* and the *Hollywood Reporter*, they're the industry bibles. Get her name in Army Archerd and Bob Osborne. For whatever reason. 'Paramount is considering Karen Salkin for so-and-so. . . . ' Something based on truth—it's never a lie— but whatever thread you have, use it to get the name in print, so that a film executive or producer sees that name over and over and over and over and over and over until it creates, like, a Karen Salkin consciousness. Who *is* Karen Salkin? You start with the smaller magazines like *California, L.A., Orange Coast,* and *Shape* and you build and build and build and build. So that your director, in order to distinguish her in his mind, has read about her eighteen times that week, has bumped into her at a party, seen her on *Good Morning America*. She has to be *planted*. . . ."

And what was the ultimate Rogers and Cowan could do for a client like Karen? "The ultimate thing," Beth Herman said, "would be to get her on the cover of *Time*."

It seemed like a dream. For almost twenty years John Ratzenberger had scraped to pay his bills as an actor, and now, before he'd even paid off all the money he'd borrowed over the years, there he was, grand marshal of a parade! In his hometown! There were his buddies—still factory workers in Bridgeport, Connecticut—lining the streets to watch him sail by in a convertible. There were the girls who'd refused to go out with him. There was Maureen, who'd let him copy her homework. And there *he* was, wearing a Napoleon-style hat, waving at the masses. Even the mayor was two cars back. What bliss! It was the last week of hiatus before Ratzenberger, and most other TV actors, would have to go back to work.

Unlike Ted Danson and Shelley Long, Ratzenberger had not been hotly pursued by the producers of Cheers when he'd gotten his part. In fact, he'd had to create his role. He'd come in for a general audition with Les and Glen Charles and Jim Burrows, who'd been looking for extras to people the bar, and said, "Hey, what if there's a bar regular who's a know-it-all? This guy works for the government—he's a mailman or something." Then Ratzenberger launched into a little improvisation, and when he left, the Charles brothers and Burrows were laughing. Ratzenberger knew he'd scored. He'd been to enough auditions in his forty-odd years to know that if they hadn't liked him, they'd have said, "That's funny." Ratzenberger was scheduled for six of the twenty-two shows, with a commitment for twenty-two the next season and co-star status. It was the first big break in his career, which until then had been composed mostly of numerous bit parts in movies. He loved the money and the regular work, and he particularly loved not having to go sell himself, every few weeks, again and again, to an endless string of indifferent casting directors.

As they cruised through his old working-class neighborhood, Ratzenberger spotted some buddies and yelled, "Hey, tell the mayor this street needs more trees!" When he looked back he could see the mayor smiling and nodding at the citizens' instructions. In the next block, he shouted to more old friends, "When the mayor comes by, tell him this street has too damn many trees!" In the next block, he roused the citizenry to demand more street lights, and in the block after, the abolition of street lights. The mayor was still smiling, but just barely. John Ratzenberger felt great. He felt in control. This parade alone, he thought, made twenty years of catsup sandwiches worth it.

That evening he sat in a neighborhood bar and a guy he'd known in high school sat down next to him. They had been apprentice blacksmiths together, after high school, and the friend still worked in a local factory. But Ratzenberger's buddy was not jealous. "When I see you up there, I feel like I'm up there, too. I've heard other guys here say the same thing." To Ratzenberger, the remark was thrilling.

8
Let the Word
Go Forth

Late June–Early July 1983

He was called Leaky Bill, because he was an unquenchable sentimentalist whose eyes would begin to float in brine at the mearest hint of sadness, and now he had a real corker to mull over, because he had just heard a rumor that both Brandon Tartikoff *and* his wife had cancer. The rumor, untrue, was enough to send *Tulsa Tribune* TV critic William Donaldson—Leaky Bill—into a paroxysm of despair. "One wonders how a *just God* could do such things." At the moment Leaky Bill was in Ronald Reagan's bedroom—at least the one Reagan tucks himself into when he visits L.A.—his feet propped up, a rum and Coke in his meaty grip, the guest of NBC, which had rented the presidential suite at the Century Plaza as part of its annual seduce-the-press extravaganza. In the adjoining living room, reporters from dozens of publications were making quick garbage out of a sumptuous buffet.

"Brandon is a dear, heroic young man," said Leaky Bill, whose manner of delivery fell somewhere between Richard Pryor's and Billy Graham's. "When I heard Brandon had . . . cancer . . . I wept," he uttered sonorously, pausing often for effect. "But this latest . . . outrage . . . is more than a mortal

224

man . . ." He finished his drink but not his thought. The door to the suite's living room cracked open and a roar of reporters' voices washed into the bedroom. "Shut that god-damned door," said Leaky Bill.

"Thank you, dear love." Leaky Bill shifted his considerable bulk; crammed into a short-sleeved dress shirt and expando-waist slacks, he looked like two scoops of ice cream in a one-scoop cone. Bill, who had been covering television for the past seventeen years, was reminiscing with a few cronies. One of them wandered around the suite, which resembled a Holiday Inn as conceived by King Tut, marveling at the telephone and micro-TV by the toilet, and the roomy balcony, lofty enough to offer a view of the ocean. "This is where Nancy Reagan spits," he said from the balcony, "just to watch it fall."

The press members, for the most part, were rather more impressed with the Century Plaza than the affiliates had been a couple of weeks before. A number of the affiliates were millionaires, while the reporters ran more along the lines of necktied peasants.

Until recently, the TV beat at most newspapers had been a traditional repository of pension candidates and editors' nephews, and the television industry was used to having the press kiss its ass softly, if sloppily.

No more. As television became more and more America's most pervasive conduit of cultural influence, the job of writing about it became a reporter's plum. Television offered the controversy of front-page writing, the color of sports reportage, and an opportunity for analysis and personal opinion rarely offered even on the editorial page. Now a newspaper was likely to have its best and most critical people covering TV.

And since much of the art of television is compromising art— making drama and comedy palatable to people more inclined to watch than read—it's not astonishing that many of the reporters here in the Ronald Reagan room weren't exactly in love with all the products of the medium—or all of the wealthy creators of those products.

To prove that theirs was an impartial, objective view of the four days of seduction that the network lavished upon them, most of the reporters were sending their expense bills to their newspapers, rather than to the network, which used to pick up the whole tab for everyone. NBC would pay the bill, if the reporter desired, but the network was less enthusiastic about grabbing the check than it had been prior to the 1970s, when a particular reporter had begun to abuse the privilege. The reporter had learned one year that he could get housekeeping to wash and press his clothes for him, so he ran up an appalling laundry bill. The next year, he was back—with two extra suitcases full of dirty clothes. And the year after that, he showed up at the hotel's dry cleaner's—with a full set of drapes from back home.

"The networks don't even give us a twenty-dollar bill to cover cab fare from the airport anymore," Leaky Bill was saying, "because they don't want anyone to think it's a bribe. My . . . God! Can you imagine, in this day and age, being able to buy *anyone* with twenty dollars? You couldn't even buy a politician for that. Back in the early days, when the newspapers didn't have such a . . . fetish . . . about us being prostitutes, Lucy Ball's press agent would send each of us a CARE package full of cheese and fruit . . . and other necessities. Then we would later be summoned to the great lady's home. Lucy would never talk to us in a hotel room, or at a studio or network; it had to be her house, out by the pool. Chasen's catered it. They'd move in, serve the bar, serve the food, and be gone. Her husband, Gary Morton, would be there. But the terrible thing . . . about Lucy and Gary . . . was that she always, publicly, upstaged him. You'd ask him a question, and she'd answer. That is as . . . e-*mas*-culating . . . as you can get." Then deep volleys of laughter rolled out of Leaky Bill's stomach, and his jowly, creviced, vein-exploded face, which looked rather like a relief map of Australia, crinkled and folded into utter mirth. Leaky Bill had been at his job too long to be terribly concerned about public relations gaffes, or whether or not he'd ever again be invited to Lucille Ball's house. "But I must say that Grant Tinker did the same thing to Mary Tyler Moore, when they were married," said Bill. "Once I was talking to him, and she said something, and he said, 'Mary! I'm talking!" She didn't say another word the whole evening. Still, I felt terrible grief when they divorced, because I love them both very much.

"Bob Hope used to also have Chasen's cater his functions, at his home in the Valley. Marvelous host! You'd be driving down an area of modest homes, then there would be an endless brick wall on your right, and that would be Hope's place. Hope lived by the golf course, and you had to hit golf balls, whether you wanted to or not. He always threatened to make us come all the way out to his place in Palm Springs. He said he knew Delores was in that bloody house out there—somewhere—because he'd been there. But just not lately. He told me that.

"You have wonderful! crazy! people in this business. But it's not what it used to be. TV," he said sadly, his eyes suddenly fogged in, "has dried people out.

"Whatever happened to people like . . . like . . ." He looked at his drink, as if the answer were there.

"Tallu," supplied one of Bill's old buddies.

"Tallulah Bankhead!" Leaky Bill's tongue celebrated the mellifluous name. "Tal . . . lu-lah! My favorite Tallulah story, absolutely my favorite Tallulah story ever: Tallu is invited to a wedding. And she is deep in her cups when she arrives at the wedding. And she is even deeper into the champagne when she gets to the bridal couple in the reception line. And Tallu doesn't say at all . . . what she's supposed to say, which is 'Congratulations.' She says, 'Commiserations.' They say, 'Commiserations!' She says, 'Yes, commiserations, dahlings; I've had you both . . . and there's not a good . . . fuck . . . between you!' Huh-huh-huh-huh-huh. Tallu was so impossible.

"When *The Big Show* was on radio, they sent a prissy little male reporter out to interview Tallu. God *knows* Tallu had been interviewed. While he was there, the phone rang and Tallu said, in her marvelous, deep voice, 'Hell-ooh?' And this little man got his inspiration for a new lead. He said, 'Mith Bankhead, have you ever been mithtaken for a man on the phone?' And she said, 'No, dahling, have you?' Huh-huh-huh-huh."

"Hollywood is such a dead-ass place now," said one of Bill's pals. Virtually everyone in the room sighed in unison. Then there was nothing but the swish and tinkle of ice in glasses. After a few moments, an NBC press aide opened the door to announce yet another carefully engineered event.

Bill joined about one hundred reporters in a room downstairs, where innumerable press conferences were held over the four-day tour. Waiting for the reporters were *We Got It Made* producers Gordon Farr and Lynn Farr Brao, director Alan Rafkin, blond star Teri Copley, and three other cast members. The interviewees sat opposite six rows of reporters, each row about a hundred feet long, and in front of a gaudy mirror–Christmas lights–poster mélange that represented their show. The program's publicists, one from MGM and one from NBC, fidgeted nervously in the background.

"If you don't mind my saying so," began a young male reporter, "the humor in your show seems to derive from misunderstanding, which is the absolute staple of *Three's Company*—the stupid, the moronic, idiotic, infantile . . ."

"But what do you *really* think?" interrupted Lynn Brao, which got a laugh from the reporters, mostly as a tension breaker.

"No, no," protested the reporter, "I'm talking about that show. I'm sure your show's going to be completely different." That got an even bigger laugh.

Another reporter asked for Teri Copley's thoughts. She stopped chewing her gum for a moment and said, "Ummmm." Then: 'Ummm." Then she giggled and everyone laughed.

"How did you get the role, Teri?" someone shouted.

"Through my agent." She paused, then giggled. Everyone laughed again. The crucifixion, it seemed, would be amicable.

Someone else said the show was like a cartoon, and Alan Rafkin took up the challenge. "I did two pilots last season, and the quote-unquote more aesthetic one—I see this happen every year—the more aesthetic one, the one you go to New York for, *pfft*, forget it. With this one, I felt like having some fun. And these actors are wonderful. Where we are now, with this offensive-defensive thing, well, that's all your business. You do what you want about that. I work on this show the way I work on *M*A*S*H*, and *Mary Tyler Moore*, and all the other 'good' shows. We do the best *god*-damn job we know how to do." Rafkin's voice was stretching tight as a wire.

"Easy, Al," said Lynn Brao.

"He's a Korean War vet and he's had open-heart surgery," said Gordon Farr, and everyone laughed again.

The rest of the group interviews, however, were a love-in. The reporters liked the new NBC shows much more than ABC's or CBS's, and they let the producers know it.

Earl Hamner and Tom Byrd explained why *Boone* looked so promising: It was "homespun entertainment" that would be shown during "*the* family hour, Monday at eight o'clock." Steven Bochco explained that *Bay City Blues* would overcome its male orientation by adding strong female characters. Michael Zinberg told the reporters that *The Yellow Rose* would be the *Hill Street Blues* of prime-time soaps. Chad Everett said he was attracted to *The Rousters* because "this business has gone to the independents, and Steve Cannell is the top one." Even Glen Larson, the producer of the silly drama *Manimal*, got off easy. The reporters asked him questions like "Will the Manimal ever become a pubic crab?"

The generally amicable aura prevailed partly because the American TV press saw NBC as the last bastion of quality programming, and also because the network would, often as not, return their calls. ABC often treated the reporters like behind-the-lines spies, while CBS was civil but close-lipped. Maybe this was just seduction, but the reporters didn't care.

Grant Tinker came to talk to them. He could have avoided it, since Tartikoff had already appeared and had run, once again, through his list of reasons why NBC was not going to finish in last place during the new season. But Tinker appeared anyway, to deliver a suavely unvarnished account of NBC's current predicament. "When I say we've finally got it going, I mean just barely. It's taken longer than we thought. . . . I'm not out to be number one. That's not my present goal. I'm out not to be number three. . . . I will settle for equality at the moment." As for his overall strategy: "I have come back down off, a little bit, that 'quality' kick I used to be on. . . . There has been some evolution in my attitude. I remember my grand statements on arrival; I hope you've forgotten some of them . . . . When you just have a little production company that can live on three, four, or five shows, you can really do just what interests you. However, with seven nights to fill, there just aren't enough people to make that kind of programming. . . . I think we have to push on and do, I won't say all quality things: I've learned another way to say it, which is to hit targets that we aim at, whatever level those targets are at. So if it's *The A-Team* or *Dukes of Hazzard* we're talking about, if those shows are executed the way they should be executed, nobody will confuse them with what we normally think of as quality."

If this particular bit of flat-footed double-talk rested uncomfortably with members of the assembled press, no one let on, and Tinker continued more linearly. "I don't think there's anything wrong with a mixed-bag schedule. As a network, we'd better have such, or we won't be competitive. It's not a crime to have something for everybody."

If the NBC chief was hedging a bit about the network's commitment to high-quality program content, he was almost perversely straightforward about the medium as a whole, stating that television "probably has a negative effect on kids," but while he felt guilty about it, NBC wasn't strong enough to tackle the problem at the moment. Asked what he himself watched, he replied, "News . . . reality things," and noted, somewhat parenthetically, "It's unfortunate that the time you spend watching television is the time you don't spend doing something else," a reflection tantamount to Chrysler President Lee Iacocca allowing that Americans probably should ride more bikes and fewer cars.

But then a resonant voice from the back of the room thundered, "Do you *need* an *A-Team?*" It was Leaky Bill.

Tinker began to respond that the program filled a certain public demand for action-adventure, that it had helped rescue the higher-demographic *Remington Steele*, and that . . .

But Bill wasn't buying. "Do you *need* an *A-Team?*" he boomed.

"Yes," Tinker said, befuddled. "Why shouldn't we have an *A-Team?*"

"Because," Leaky Bill declaimed, "it is *vi-o-lent* as hellll."

Tinker handed that one off to a vice-president of research who could prove with a recent study that TV violence had little effect upon the public. But Leaky Bill had lost interest.

Back at the lounge of the Century Plaza, Bill laid out his cards. "I did," he rumbled, "what no one else had the balls to do. I got the chairman of the network to admit that he needed his most despicable, lowbrow show. What does this mean? It means that NBC realizes quality *ain't* going to make them what they want to be. So now they pull back from quality, and lose their core viewers. Or they go ahead with loser shows, and get everyone fired. Or, lastly, they try to walk both sides of the . . . picket fence. And we know how well that works.

"Any way you slice it," explained Leaky Bill, who had observed almost two decades of network flux, "it means this: For NBC, this is the beginning." Pause for effect. "Of the end. Huh-huh-huh-huh-huh."

But, press-wise, NBC's effort was a smash. Jerry Coffey of the *Fort Worth Star-Telegram* went home and told his readers, "NBC . . . is the one network that is successfully pursuing and finding and scheduling superior series entertainment." Julianne Hastings of the *San Francisco Chronicle* concluded that "NBC has the most attractive prime time programming by far for next fall." And P. J. Bednarski of the *Chicago Sun-Times* went so far as "NBC [is] network television's last best hope."

Coast to coast, there was much more of the same, proving that the exercise had been worth the price of the drinks, or even a little dry cleaning.

"Watch a show fail to come together," said Ted Danson, idly, on the first Monday of production of the 1983–84 season, "before your very eyes."

Two postteenage girls, watching Danson from the *Cheers* audience area as he rehearsed on the stage floor below, giggled madly at his remark, and squeezed each other's arms. It was as if they were thirteen years old in 1964, at Yankee Stadium, and the Beatles had just jumped onto the stage. "My God, he's cute," whispered one of them, a pretty, pink-mouthed cherry-blonde in black Gloria Vanderbilt jeans. Her friend, who had casually lifted a script from a stack near the stage's wings, thumbed through it to see what scene would come next, and her eyes glazed over. She elbowed the blonde and pointed at the script. It said, in all-cap stage directions, SAM COMES OUT OF THE BEDROOM WITHOUT HIS PANTS. The blonde lurched and bit her lower lip, blemishing an incisor with lip gloss.

But then a small door at the other end of the cavernous stage creaked open, sunlight flooded inward, and an old man in a security uniform closed the door quietly behind him. "Megan," hissed the girl with the script. But the reddish-blonde was glued to her seat, her eyes fastened on the door that Danson would soon emerge from . . . with his pants off. The guard headed straight for them, and in moments they were gone.

It was the guard's fiftieth year at Paramount. Fritz Hawkes, seventy-eight, could practically smell set crashers. "Might be some problems here this year," he said. "Used to be the problems were at *Happy Days*."

On the stage floor, Ted Danson, in clingy blue underwear, his pants around his ankles, was saying to Jim Burrows, "I don't mean to cause trouble, but I just don't think there's enough of a reason for the characters to be having a fight."

Shelley Long, in a baggy sweatshirt, agreed. "He's right. Not enough motivation." Burrows nodded, but had them do the scene again, the way it had been written. Burrows, very quiet in most situations, gushed laughter at all the jokes, even after he'd heard them several times. Because not all the jokes were very funny, his laughter sometimes seemed inappropriate.

George Wendt, watching from the side of the stage, near an urn of coffee and a tray of doughnuts, vegetable sticks, bagels, and cheese, grimaced. "It feels really rough out there," he said. "Ted and Shelley's scenes are working, but the bar scenes are for shit. We're feeling like we're a well-oiled machine, but we're not. We're been away from each other for five months."

It was of crucial importance to the show that this be a strong, well-executed episode, because this would be the much-promoted first program of the new season, the show in which Ted Danson's and Shelley Long's characters finally, after a year of tease and grope, went to bed. If the episode disappointed the audience, the show would be in serious trouble, because gaining a new portion of fall viewers, just after the Emmy Awards were announced, would be crucial to the show's struggle for survival. This strategy was predicated, of course, on the assumption that the show would receive a healthy number of nominations. It was also based on the hope that the show, by the fall, would have added to its audience during summer, when it went against weaker rerun competition. That segment of the plan, thus far, seemed to be working. Going up against soft competition like a news documentary on McCarthyism, the already canceled *It Takes Two*, a failed ABC pilot called *Wishman*, and the movie *The Last Ninja, Cheers* had recently risen to near respectability in the ratings. But if it didn't hold its position in the fall, it would almost certainly be a casualty. And because it was one of NBC's flagship high-quality shows, its demise would spell the worst kind of news for Tartikoff's attempt at quality programming.

At four o'clock, Glen and Les Charles, both in Hawaiian shirts, came over from the office across the street, accompanied by two staff writers and about a dozen office workers, including Tony Colvin. The cast began a run-through of the whole show, holding their scripts for reference, as the producers and writers sat at stageside in orange canvas director's chairs. The Charles brothers, who had written the script, marked their copies as they went, scoring them as if they were freshman exams. Over the next two days, they would rewrite much of the script, even though it had already been through a thorough rewrite process, on the axiom that a script is an essentially unproven commodity until it issues from the actors' voices and makes people laugh.

Burrows continued to giggle and guffaw during the run-through, but with the added laughter of the other voices, his chuckles at the weaker jokes seemed more natural; the laughter made the jokes funny—a principle probably appreciated by the inventor of the television laugh track.

The stage phone rang and Tony, not presumptuous enough to sit in a director's chair, leaning against the rail that separated the stage from the audience seats, bounded after it. Scott Gorden—still Tony's writing partner—was standing right there, but Scott didn't move a muscle when the phone jangled. Because Scott had a new job—he was now Ted Danson's private assistant. In the show's first year, Danson had had neither secretary nor publicist, unlike his leading lady, Shelley Long. Danson had been leery of a Hollywood-style entourage. But he'd begun to feel as if Shelley's career was moving faster than his—so he hired a publicist and put Scott on his payroll. But Danson drove a hard bargain. He offered Scott $250 a week, $50 less than Scott had made on *Cheers*, on the grounds that in his new job, Scott would be working less than he had as a *Cheers* P.A. Scott held out for

$300. Scott had heard that Ted was making $25,000 a week from *Cheers* plus his pay for movies and for promoting Aramis cologne, and he figured, "What's fifty bucks a week to this guy?" But to Danson, $50 was something he could understand—the other monstrous sums seemed like just so many hieroglyphics on his accountant's ledgers.

So, every day for a week, Danson would offer Scott $10 more. Negotiations finally culminated in a victory for Scott. He began to run errands for Danson, read scripts that were submitted, and helped him to learn his lines. He still had time to write scripts with Tony, but their writing was going nowhere.

After the run-through, Tony and Scott sat on the stairs outside the *Cheers* office. The sun had dipped beneath Paramount's fortresslike wall, but still perspiration beaded on the faces of both Tony and Scott. "These guys," Tony said, jerking his head toward the producers' offices, "are never going to admit our scripts are good. 'Cause if they did that, then they'd feel guilty about treating us like peasants. We oughta quit."

"Hey, man," said Scott, "I just bought a house."

Glenn Miller, Jim Burrows's twenty-five-year-old assistant, who wanted to use his degree in art to become a set designer, bounded up the stairs with a box of Dixie cups. "Tony thinks we should quit," said Scott.

"I know the feeling," said Miller. But he didn't slow down. Then beautiful teen star Justine Bateman walked out of the *Family Ties* office, with that show's writer-producer, Michael Weithorn, who had been a schoolteacher only two years ago. Over the past year, Tony had watched while Weithorn had gotten the kinds of breaks that Tony wanted. Weithorn and Justine walked in different directions. Scott watched Justine, in shorts and a no-shoulder sweat shirt, walk away. "There's no business like show business," Scott said.

But Tony was watching Michael Weithorn, a tall, clean-cut twenty-six-year-old. "Fucking Weinhorn," said Tony.

On Tuesday, Ted Danson walked through the Paramount commissary on the way to his table. There were Big Faces in the room—Richard Benjamin, Alan King, Leonard Nimoy—but Danson was attracting most of the stolen glances. He was widely considered—by casting directors, other actors, producers, and talk-show talent coordinators—to be a possible breakout super-star of the eighties. He was, for example, on a list compiled by a national magazine of a half-dozen candidates who might re-place Sean Connery and Roger Moore in the James Bond mov-ies. Danson was in the most desired subgroup of all actors, the handsome thirty-five-year-olds with talent. As such, he was suitable as a leading man.

"Last year," said Danson, spearing a square of chicken from a salad, "was my year to grow up and take my place. The first year of *Cheers* was the first time my ass was definitely on the line. And if you haven't hung your ass out there, and let every-body take their shot at it, and lived through it, then you can't have that what-the-hell attitude, which is the attitude that an audience needs a leading man to have. It's not arrogance, really; it's balls—having the weight to carry it off.

"I didn't have that as late as last winter. I remember in Jan-uary, when they announced the Golden Globe nominations, the show got one, and Shelley got one, and I got nothing. The funny part is, I didn't even know what they were until I didn't get one, and then I was devastated, and trying to pretend I wasn't. Then I went to the awards show, and sat there with Shelley, to accept for the show if it won. And Shelley won, which turned the screw a couple of inches more. And she'd already landed a part in *Ir-reconcilable Differences*, while I was trying like crazy to find a movie for hiatus.

"Shelley used to go in and demand what she needed for herself. I always thought that was embarrassing. I figured, if people don't give me what I want, maybe I'm wrong to want it.

"Anyway, there I am in front of this crowd at the Golden Globes, trying to play the nice guy, and something just went . . . pop! Like—fuck this! And I got mad. Well, shit! This doesn't work anymore. The hell with this.

"So I went in to see Les and Glen and Jimmy. But I went in as Sam Malone, my character. And I said, in so many words, 'I've lost my bar! Here I had this great bar with great guys, and in walks this twit who does nothing but talk, talk, talk, and turn off the TV set. The show's not even supposed to be about this girl. Gimme a break! She's not even *putting out*! So fuck her. If she's not gonna put out, have her shut up. I've turned into this wimp! I'm supposed to be a cocksman, and all I'm doing is making adolescent sexual innuendos. I want my show back!'

"And they looked at me like 'Great! Where've you been?' "

On Thursday, the four camera operators and each of their assistants came in, along with a full complement of stand-ins for the actors. After rehearsing the first four of the show's seven scenes, the actors retired to a conference room above the stage, while the stand-ins went through the show, movement by movement.

Burrows had already figured out every piece of physical action in the show. On Tuesday, after another 4 P.M. run-through, the producers and writers had spent most of the night in the offices, doing a final rewrite, although lines would be altered up until the Friday night shoot. On Wednesday the actors had learned the new lines and had worked out all of their stage directions. Burrows had gone over every movement in minute detail, so that it could be photographed properly, and so that every possible piece of physical humor could be wrung out of the script.

By now, on Thursday afternoon, Burrows knew where all the actors would be for every line and what their gestures would be, but he had to relay this information to his camera, lighting, and sound men. For every shot, a cameraman's assistant stretched a tape measure from the camera to the actor's stand-in, to focus the shot perfectly in advance. Tape was stuck to the floor to mark the actors' positions. It was tedious work, and the stand-ins, wearing actors' names on placards tied to their necks, looked dead-eyed.

Upstairs, Rhea Perlman, who played Shelley Long's foil on the show, cuddled her newborn daughter while she practiced her lines with George Wendt. "Pretty rocky out there, huh?" said Perlman, referring to the many missed lines in the morning's rehearsal.

"We're money players," said Wendt. "It'll work by tomorrow night. Maybe."

Ted Danson was lying on a couch, massaging his own neck, his eyes cloudy and red. "I'm not ready," he said.

* * *

On Friday night, there was a crush to get into the VIP booth that overlooked the stage, as the audience, which had stood outside for over an hour, filed into their seats. But nobody was crowding Mr. and Mrs. Tartikoff.

Tartikoff and his wife were sitting in the first row of folding chairs, next to the pane of glass that separated the booth from the last, highest row of the audience. The people in the first two rows of the booth played to Tartikoff—their remarks, carefully clever, were for his benefit.

At the door to the booth, an independent filmmaker, who had just sold a movie to cable, told the booth's guardian, "You want me to stand? I don't stand, dear."

A four-piece band played rah-rah music and several versions of the *Cheers* theme song, as Earl Pomerantz, a staff writer, warmed up the audience by singing requested TV-western theme songs.

Dick Ebersol, the *Saturday Night Live* producer, wandered in and took a seat next to Brandon and Lilly Tartikoff. Next to Ebersol, a young casting agent was recounting to Tartikoff how a manager had convinced a thirty-year-old actress to accept an asinine role in a hare-brained sitcom: "He said, 'I just wanna say two words to you, sweetheart—Gwen . . . Verdon. Get your mug on TV or you'll be a forty-year-old that nobody *cares* about.' "

Hamilton Cloud, a thirty-year-old Yale friend of Tartikoff's who was now in charge of NBC's comedies, was sitting next to Lilly Tartikoff. He looked over his shoulder at the scuffle for seating in the VIP booth, murmured, "Show business," and returned his gaze to the $100,000 *Cheers* set, which, under a hundred brilliant lights, looked gleaming and real.

"God!" said Lilly Tartikoff. "Sitting up here is just like watching it on TV. Let's get seats downstairs, Brandon." Tartikoff stood, turned around, and found himself facing Danny DeVito, Rhea Perlman's husband and a former star of *Taxi*, which Tartikoff had canceled in May. DeVito was holding Lucy, their newborn. "There's somethin' we gotta discuss, Brandon," DeVito said sharply.

"There is?"

"You're damn right there is. Rhea carried Lucy during every

show of last season, right? Right. Well, we want residuals for the kid. If you don't come through we'll sue."

"It'll be a landmark case," said Tartikoff, leaving with his wife.

When Tartikoff left, Hamilton Cloud became the incarnation of the network. "I think you guys will beat ABC," the casting director said to him. "*Mr. Smith*—hey! What a breakout! And *Boone* looks tough; I think the USFL is going to hurt Monday night football. And I know you guys can blunt it with stunting."

"Could be," said Cloud, a fine-featured black man with skin the color of a suntan.

"Quiet on the set, please," shouted the second assistant director. "Sound."

"We have sound," came a voice.

"Speed," said the second A.D., and the cameras began to roll.

"We have speed."

"And . . . action!" said Jim Burrows.

The teaser, the scene that would play before the credits and theme song, began, and Burrows was still tittering at every laugh line. Later, when the show was aired, his laugh, a descending crescendo that started on a high pitch and tumbled quickly down, would be audible again and again on the sound track, as it is during every episode of *Cheers*.

Every scene was filmed more than once, some of them several times, but the audience remained enthusiastic. *Cheers* was already a minor cult phenomenon in media-infatuated L.A. Gaining entry to the shoot was difficult; each show was overbooked to ensure a full audience, and the crowd that managed to get in was full of true believers. The *Cheers* audience, which lined up outside the thick Paramount walls, was upscale in dress and accessory, in marked contrast to the *Happy Days* line around the corner.

People in the VIP booth clapped Hamilton Cloud on the back when there was a good joke, as if he were personally responsible for it. When the show was reaching its act break, where the commercials would be inserted, Danson's character sought to seal his and Shelley Long's newly acknowledged lust by inviting himself to her apartment. "But, Sam," she said, "it's so filthy."

"It doesn't have to be, if we care about each other," said

Danson, and the casting director hugged Cloud around the neck.

The twenty-two-minute show took over two hours to film. Then they shot almost the entire episode again in pickups, retakes of scenes that Burrows believed were flawed or that he needed from an angle he'd previously missed. By 10:30, they were down to the last scene, the audience was gone, and Tony brought an icy bottle of Heineken to Burrows, then two to Glen and Les Charles. A few minutes later, Brandon Tartikoff had one, then Hamilton Cloud had a bottle. Slowly, the bottles of beer filtered their way down the pecking order. Finally, by 10:45, Burrows said, "Cut. And we wrap," and there was small scattered applause.

"Ladies and gentlemen of the Screen Extras Guild," shouted second AD Brian Ellis, who had gone to a prestigious film school and was now making almost as much money as a waitress, "put ten forty-six on your cards, and good night."

The band jumped into the Cheers song, the sets were rolled back, and a crew of caterers began to wheel out tables of angel food cake, mountains of red strawberries snowcapped with whipped cream, tubs of imported beer in shattered glaciers of ice, rounds of cheese surrounded by croissants and scones and Danish pastry, and buckets of liver pâté and avocado dip.

Hamilton Cloud smeared butter on a scone and worried about his comedy lineup. Mama's Family, a high-decibel rural sitcom, was looking soft and an old reliable, Diff'rent Strokes, seemed to be running out of steam. Mr. Smith, thought Cloud, would be the big hit of the year, and hopefully would carry along Jennifer Slept Here, which Cloud thought looked promising in script form. Buffalo Bill, a summer tryout that had gained decent ratings, was being renewed for thirteen more episodes, which would push back the air date of Harry Anderson's Night Court. Several other comedies had also been developed since the close of the official pilot season, and they, too, seemed suddenly ahead of Night Court in the competition to get on the air. Among them were shows by comedy potentate Norman Lear, Mary Tyler Moore Show creator Jim Brooks, and former Saturday Night Live producer Lorne Michaels. Cheers was looking stronger, but if it finished in the lower 50 percent of all shows, it would be hard to reschedule it.

As midnight approached, the wrap party grew louder and

more crowded. Some of the cast and crew from *Taxi* arrived, including Carol Kane, who looked tiny and lost in a big sweat shirt. Tony's and Scott's wives appeared, and the two squired them around, introducing them to the actors and network executives.

"This kind of gig is where half the business in Hollywood gets done," said Tony, looking sharp in a jacket and slacks, proud of his wife in her party dress. "I may go over and put some ideas in Sam Simon's head," he said, nodding at the former *Taxi* producer, who was sitting with Carol Kane and talking about a new show he was trying to sell to NBC.

Then there was a hollow pop, and beer from a broken Dos Equis bottle began to flood in every direction. Tony looked at the bottle, then at his wife, then at Sam Simon, and finally, unwillingly, at a heavy gray mop that was being thrust into his face by his boss, Richard Villarino.

Villarino smiled a mild apology and pushed the mop's handle an inch from Tony's nose. Tony took the mop and began to saw it back and forth over the yellow liquid and the shards of glass. The people that had been watching glanced away.

"This chafes," muttered Tony. "This really chafes."

Scott, standing nearby, next to Ted Danson, said, "Pride cometh before a fall, Tony." But Tony gave no sign of having heard him.

"I think, in rejecting *Snoop*, Brandon Tartikoff was trying to say, 'I know television comedy better than Allan Katz.' He doesn't," Jay Sandrich said. Sandrich, former mentor to Jimmy Burrows and considered the preeminent director of sitcoms in Hollywood, sat on the lanai of his Manhattan Beach condominium. A dark, informal man at the middle of middle age, Sandrich had been the shaping force behind sitcoms ranging from *Get Smart* to *Soap*. "In my career, I've seen three or four pilots scotched by the networks that I knew would have been hits. *Snoop* was one of them. The irony is that the networks have no sense of showmanship, though Grant Tinker does. That's why a show like *Cheers* will ultimately make it, because Grant has stepped in personally to see that it's given every chance, because he's among the only people at any network that *knows* there's a real audience for quality out there. The other networks are so panicked about everything. They used to order thirty-nine shows of a series at a time when I started in TV, then it got down to twenty-two and thirteen and now they short-order like crazy. Four or six shows, that's all a lot of new shows get. The network people figure if they simply order a whole bunch of different shows, throw them all against the wall, something is bound to stick."

It was the middle of summer and Sandrich, who was scheduled to direct Harry Anderson in *Night Court* if and when that show was shot, looked up and down the beach. Unlike most of his counterparts, who lived in the manses of Brentwood and Beverly Hills, Sandrich lived in a neighborhood where you'd have to check many a denizen's green card to make sure he was an earthling.

Sandrich, who had directed the first several episodes of *The Mary Tyler Moore Show*, recalled, "When the show was first put together, the network wanted both Ed Asner and Valerie Harper recast. Can you imagine that, someone else as Lou Grant? It took Tinker and Allan Burns to stop that kind of thing time and time again. That's the way Tinker will succeed. He knows what is real. Once NBC starts building the nights of shows that people get a chance to really care about, NBC will succeed. In the meantime, though, I see more and more talent leaving the networks for film and cable, simply because they're tired of network people telling them that green is blue or that water runs uphill. It's often just not a very real environment."

The Hunchback looked like a cross between Alex Karras, Phyllis Diller, and Allan Katz. He was, in fact, the latter. Few had ever accused Katz of being faint at heart but, with the proposal for his screenplay *The Hunchback of UCLA*, the ex-dishwasher, delivery boy, and television repossessor had attempted the near impossible. He had stipulated that—if the contract was signed—he would be not only the writer, but the star as well. This from a man who'd never even been so much as an extra in a feature film and whose acting career to date was highlighted by the fact he'd done the voiceovers for the Screaming Yellow Zonkers ads.

But Katz was a known and admired quality in Hollywood.

He'd come to L.A. in 1970, thirty years old and enjoying the first flush of success. A year earlier he'd been an unemployed menial, a month earlier the toast of the Chicago ad world—his Screaming Yellow Zonkers, a box of popcorn and hype, had turned overnight into a monster success. George Schlatter, executive producer of *Laugh-In*, had hired him to write for the show and soon Katz was banging out sketches for the likes of Lily Tomlin and Ruth Buzzi. He met John Wayne and Goldie Hawn. Famous people called him on the phone. He made money. He bought cars. Even when he'd been broke in Chicago, he always had a nice car. It was a way of saying to people that he was different. Katz had had a classic Thunderbird when he was practically doing stoop labor, a Jaguar roadster when he was making $60 a week measuring eyeballs at the University of Chicago. Now, however, he could afford to up the ante. He bought a Corvette, then unloaded it because it had no trunk. He picked up a two-tone jade-green Bentley that burned mushroom clouds of oil and had absolutely no power, then sold it because he couldn't figure out if people thought he was important or insane. He bought another Corvette, which he sold when he noticed it didn't have a trunk either. Then there was a BMW. And another. Never mind that half these autos cost him a small fortune with their many and bizarre mechanical problems—cars were fun. And Katz could afford to replace an exotic busted valve or a shredded piece of leather upholstery. For Hollywood had opened its golden doors, and he was doing quite well, moving from *Laugh-In* to go on to write, then produce, some of the most successful sitcoms on television—*Rhoda* and *M*A*S*H* among them.

Still, there were problems. His marriage dissolved. And, as time passed, he began to grow restless producing other people's shows. He wanted to do something of his own. So Katz lit out. He quit TV and wrote a screenplay, *Skids*, about a bunch of winos who trick a big-city mayor into providing services for the poor. When that didn't sell, he did a pilot of Neil Simon's *The Goodbye Girl* that got ordered as a series, then canceled due to a catfight between Fred Silverman and the studio that produced it. Down to his last $80, he left his little manse in Hollywood Hills to move into a Connecticut boarding house, where he labored over the production of *Zapata*, a musical he'd written with Harry Nilsson.

Returning to L.A., Katz was determined to work on his own and he spent a year and a half writing and raising money for *The Hunchback of UCLA*, adamant that if the film was produced he would be the one to star in it. "Robin Williams' manager said Williams would be interested in starring as the Hunchback. I was flattered. But I knew that it would become his movie, not mine." So, an all-or-nothing approach. That would guarantee control and, besides, Katz was not at all opposed to becoming a movie star. And he was willing to forgo the financial advantages of having a big name behind the film. "I just don't care about money that much anymore. I practically paid to do *Zapata*, and if I've got a choice between getting nothing up front to do my work on the chance that it might succeed, and getting a million dollars to turn it over to somebody else, I'll take the nothing every time."

Katz had raised independent money for the movie and gone to Peter Grad to see if Twentieth Century would be interested in distributing *Hunchback*. Grad went to his boss, Harris Katleman, who liked the script so much he wanted to produce the film as a feature out of Twentieth Century Television. He told Katz, however, that if he wanted to star in *Hunchback* he would have to take a screen test. Katz was more than agreeable, and created a seven-minute *Hunchback* short. Grad loved it. Katleman loved it. Best of all, Twentieth Boss of Bosses Marvin Davis loved it. So much so that Katz was told afterward that Davis considered it "the funniest thing I've seen in twenty years."

So now, in his office, sitting behind the pink pigs and horizontal slab of glass that made his desk, Katz typed. Tapping out revisions for *Hunchback*. A screenplay about a startling metamorphosis, populated by characters who were variously suave and bumbling, confident and insecure, wordly wise and bumpkin foolish, handsome and gawky, athletic and awkward—characters that were, in short, various versions of Katz. The plot was sort of *Woody Allen Meets Beauty and the Beast*. It was funny. It was bittersweet. It was contemporary. It was a potential $10 million in Allan Katz's back pocket if *Hunchback* hit the youth market like *Animal House* or *Porky's*.

Reinhold Weege, producer of the in-limbo *Night Court*, did not want Harry Anderson to make any more appearances on

Cheers, since it violated Hollywood protocol for a star to do a walk-on guest shot on another show. Even if that star's show was still aeons away from actually producing rent money. Therefore, Harry Anderson, a star in theory, was back in the boon-. docks, rustling up cash. Specifically, he was in Ashland, Oregon, a psychedelic Swiss village of a town known primarily for its Shakespeare festival, and the marijuana it grows on nearby government timber land. Harry had made his very first onstage appearance here, in the mid-1970s, chucking a spear in one of the Bard's fight scenes. He had divorced his first wife and married Leslie here. Now he was returning, in a modicum of triumph, to headline a show of his own. The theme of the performance was "the Monster Midway . . . the Dark Circus"—it featured lots of good geek stuff and mind-bending weirdness, to show the home folks that he hadn't lost the common touch.

Harry had driven up from L.A. in Leslie's station wagon, carting the only indulgence he'd allowed himself to buy upon hearing that *Night Court* would at least make six shows. The purchase was a Decapitated Princess chair. With Leslie made up like Oriental royalty and positioned properly, her head appeared to float, disembodied, two feet above the chair's seat. It was one of the oldest props in magic, so antiquated that most of Harry's magician friends had never seen one. After tracking it down, he'd paid $5,000 for it.

He met Leslie at the Ashland-Medford airport, where she'd flown in with Eva, and they drove to their hotel, the Marc Antony, easily the tallest building in town. They had rented the top floor suite for $95 a night, twice the cost of the second most expensive room in town. They planned to entertain old friends in style.

"Hi, I'm Harry Anderson," he said to the old lady behind the desk. "We rented the suite."

"The suite?"

"On the top floor."

"Oh. That room. I guess we do call it a suite. Actually it's one big room. But it's got a huge bathroom. And a black-and-white TV. Free."

Harry carted his luggage over to the grate in front of the elevator and stared into an empty shaft.

"Never got the permit for it," the desk clerk explained.

"Oh!" Eva said. "What a beautiful hotel."

The next day they checked into a Best Western. "Nothing wrong with the Marc Antony," Harry told the room clerk at the new place, "that a match wouldn't fix." The room clerk did not smile.

On the day of the show, they went to a hot-dog stand near the outdoor theater for lunch. The guy sizzling the franks looked up and said, "Harry!"

Harry remembered the man, but not his name. "Hey!" said Harry. "How you been, man?" When Harry had left town, the guy had been running a prosperous health-food store.

"Not bad," said the hot-dog man. Then the guy began to tell Harry, in no uncertain terms, how sorry he felt for him—having to live in L.A., having to breathe all that smog, having to work in nightclubs, having to fight the traffic.

Harry apologized for himself and left.

The managers of the festival had given Harry two payment options—a flat $1,000 or a 70 percent cut of the proceeds from $6 tickets. He'd taken the cut.

Backstage, he watched as the theater filled all fourteen hundred of its seats. But after the show—which was great; the audience had come to laugh—he found out that the theater managers had given away hundreds of tickets to their friends and family, while turning away paying customers. "Never trust a nonprofit organization," he told Leslie. "They've all come up through the ranks by whining their way into money, and they'll give you the shaft every time."

Leslie and Eva went home, but Harry flew up to Seattle to perform at an event calling itself the World's Largest Pajama Party, which featured Playboy bunnies, a band, and Harry. The promoters told him the gig was a sure bet to get him into the *Guinness Book of World Records*, and offered him a cut of the gate or a flat fee.

"World's Largest Pajama Party, huh?" said Harry. "I think the Los Angeles VA Hospital has you beat. I'll take the flat fee." He got it in advance, as a cashier's check. That night, about 250 people trickled into the 4,500-seat auditorium.

The next night, he appeared as one third of the Comedy

Underground Hall of Fame, in a three-thousand-seat theater that was charging ten bucks a head. Again, he could take a fee or a cut. He found out that the two other members of this ostensible hall of fame had been playing a downtown club for the last two weeks at $5 a performance. He opted for the fee. They offered him a cashier's check, in advance. He wouldn't take it. He held out for the cash.

Late that night, after the show, he got drunk. Then he got a tattoo. From here to eternity, his left arm would say FUN.

"Truly," thought Corky Hubbert, a bright-pink can of guava nectar balanced on his bare mini-Buddha belly, the Hawaiian sun basting him in his own sweat, while Culture Club's incomparable new single "Church of the Poisoned Mind" wafted from his luxurious beach house, "truly, truly, *truly*, this must be paradise."

Corky Hubbert, the midget John Belushi whose pilot, *The Amazing Adventures of Sparky O'Connor*, had gone down the toilet at ABC, who had been reduced to living out of a paper bag, was back on top. All those old problems were ages away— a good six weeks away. Now Corky was lying out on the beach in Kailua, watching the translucent turquoise waves swell up and crash into bubbling slop, just lying there waiting for his big, fat, juicy Universal Studios paycheck to come in. It would be his compensation for having starred in an episode of *Magnum, P.I.*, the top show on television.

He hadn't just *appeared* in the show, or even co-starred— he had starred. That was what it said on the credits, for the whole world to see: "*Magnum, P.I.*, starring Corky Hubbert." What a coup. The whole show was based around him. His lines outnumbered Tom Selleck's twenty to one. Here he'd given up on L.A.—that's right, *given up*—had gone back up to Portland, and had been living off the meager earnings of the Cork Hubbert Cheap Theatrics workshop. He'd manufactured that gig to operate after-hours at the International Food Goddess restaurant. Then, lo and behold, he got a call from his manager, Paul Hiestand, and discovered that Paul and Cork's agent, Craig Wyckoff, had scored Cork the biggest TV job of his life—in fact, *the best job in TV*, moving the action in television's numero uno action-adventure.

What had happened was that Wyckoff subscribed to a "breakdown service," a list of parts currently available on episodic television. The casting director for *Magnum* was looking for someone to play a midget burglar–CIA agent–karate expert–friend of Ronald Reagan's. Corky fit that stereotype to a T. All his manager had to do was let the *Magnum* bigwigs take a gander at his *Fall Guy* episode and they were sold on the spot. No audition or anything. The only scary part was when Corky had to talk on the phone to the main casting guy and the guy—bolt out of the blue—asked for his measurements. When Corky's numbers suggested a fresh vastness to his stomach and rear, the guy said, "Holy shit! What are we hiring, a midget Orson Welles?"

"A midget John Belushi," Corky said. "But just because I've gained a little weight, there's no need to panic. I'm just a little more symmetrical, that's all."

"Yeah," the guy said, "about as symmetrical as a beach ball, it sounds to me."

But they sent him a first-class ticket to Honolulu and the shooting was fantastic, even though he had to work practically dawn to dusk eight days in a row. He got to drink beer with Tom Selleck in Selleck's office, and Larry Manetti, the guy who played Selleck's sidekick, even talked to Corky about the possibility of his and Corky's doing a *Magnum* spin-off. And the other actor, who played the English butler-aristocrat type guy, John Hillerman, was fantastic! One glance at a script and he had everything down, chapter and verse. Astonishing. The guy sucked up lines like a sponge.

Sure, there were some rough spots. Like remembering how to act. Corky had been out of the saddle so long it took him a little while to get used to memorizing his lines, which wasn't made easier by the fact that he didn't get a script until forty-eight hours before his first big scene. But the people were great. So nice. So professional. So . . . respectful. And so generous. Corky was back in the chips. And his hometown friend Annie, who was a scuba instructor, had married this Hawaiian doctor, and they'd given Corky use not only of their house in Kailua, but of their condo in Waikiki as well. So that as soon as he was ready to move out of the beautiful twelfth-floor suite the *Magnum* people had procured for him at the Colony Surf, with its stunning view of Diamond Head and top-drawer room service, he had not one but two spacious, luxurious digs from which to sortie from night spot to night spot—where he met scrumptiouslike barflies of the first order.

Corky took a sip of juice, rubbed some Bain de Soleil over his chest, and reflected on those recent days when he had been making a living as a sperm cell down at a club called Fool's Paradise, using his All-Purpose Costume Kit, which consisted of a sweat shirt with a hood on it. And creating characters like the Faggot Sheriff of Nottingham, then plying the crowds for quarters and dimes. What a downer.

But that was ancient history. Now he was back on track, headed for superstardom. He was pretty sure Larry Colton would create him a really flash part in *Pillars of Portland* when he got back to the West Coast. Corky had read about *Pillars*, even in Hawaii—there was a story in *USA Today* saying it was the wave of the future. A recurring role in *Pillars* as, like, the love interest, or something like that, shouldn't be too much for an actor of his stature to hope for. Whatever happened, though, Corky was damn sure that this time, for once, he wasn't going to screw it up. Absolutely! This time—NO MORE FUCK-UPS.

Back in L.A., early in the hot summer morning, Jack Valenti stood before a congressional subcommittee on television, an imposing figure in a silky suit the color of a ripe olive, his glossy mane combed straight back—just like the locks of his old boss, Lyndon B. Johnson. Jack Valenti was there to tell the congressmen that the networks were ravenous cannibals who sought to devour every last life-supporting asset in the television industry. If the networks' appetites went unchallenged, warned Valenti, if they were allowed to own all the syndicated reruns, television would become a monopolistic wasteland, devoid of the quality that only free competition could foster. The congressmen enveloped Valenti, a spokesman for the production industry, in fraternal gazes; they, too, had decided long ago that the networks should be stopped in their drive to invade rerunland, and were now simply drumming up support for a bill they'd written that would halt the invasion.

But Valenti wasn't about to let the fact that he was preaching to the converted impinge upon his opportunity for a bit of old-fashioned Texas oratory. So he implored the subcommittee, as if his life depended upon it, to stop the FCC from repealing the rule that kept the networks out of syndication. "If the rule is repealed," he drawled pontifically, "the three networks will totally! absolutely! completely! dominate all television programming in prime time and fringe times. No sane observer of the real world of television suggests otherwise!"

In fact, though, there was a bevy of network representatives on hand—men who appeared to have achieved at least the veneer of sanity—prepared to argue that television was an altogether different animal than it had been in 1970, when the Federal Communications Commission barred the networks from syndication, because of the rise of cable, pay, and independent stations. To imply that the networks now could hold a monopoly on television was ludicrous, they said. And the FCC, it appeared, totally agreed with them. The FCC, in the spirit of Reagan-inspired deregulation, was all set to unshackle the networks, if only the FCC could get past these congressmen's efforts to tell them how to run their own business. But the lawmakers, led by Hollywood Congressman Henry Waxman and inspired by lobbying efforts of the production studios, which were in jeopardy of losing an $800 million per year business, were not about to stand back and let the FCC give the networks this windfall.

"There is one question that must be answered," said Valenti, as TV cameras whirred and lights shone about him, "for that question stands supreme above all others. The question is, what is the public interest? The public interest is indisputably linked to the continuance of competition in the marketplace. Without competition, the market shrivels. The giants command the television landscape. Advertising costs rise. Alternative choices of free television entertainment shrink!"

Valenti was obviously just getting warmed up, to the obvious delight of the newspeople on hand. "But the real target of the networks," he said, gesturing grandly, "lies veiled and obscured by their thick, meandering files in this case. Their real target is the independent television station. According to FCC data, in 1972, two years after the rule was created, there were only seventy-three independent television stations in thirty-eight markets. In September of 1982, ten years later, the FCC reported the number of independents had soared to one hundred seventy-nine in eighty-six markets!" If the networks controlled the distribution of syndicated reruns, he said, they could shut off the supply to the independent stations, which need these shows desperately, as easily as they could turn off the faucets in an executive washroom.

For most of the day, representatives of the studios, independent TV stations, cable networks, pay TV companies, broadcast networks, and advertising agencies stood before the subcommittee, which was convened in a plush meeting room at the Museum of Science and Industry, and pleaded their views. The networks—in vivid contrast to their stances at their recent affiliates meetings—said that times were tough, business was bad, and the cable networks were capable of exterminating them. But no one in that room believed them.

The day after the hearing, Peter Grad was back in his office at Twentieth Century-Fox. "The way it looks now," he said, "the studios are in deep shit. I think that Mr. Reagan's bread is buttered by the networks. The worst part of it is, the public couldn't care less who owns the programs." That was too bad, believed Grad, because the public wanted high-quality shows, and the studios tended to plow their profits directly back into production. The $250 million that Twentieth had made from its *M*A*S*H* syndication sale, for example, was helping to subsidize new, as yet unprofitable shows, such as *AfterMASH*. But a network would be tempted to put the profits into its sundry other concerns.

Grad did not think, though, that the public would ever care who won the war over syndication. "The rest of America looks at Hollywood and they think of three things. Jews. Fags. And rich guys. They couldn't care less whose ox is gored."

"Entertainment execs get *potchkee* fever," Allan Katz said, walking through a studio dungeon at Twentieth. He reached out and grabbed the bars. "Do you know what that means in Yiddish? They can't keep their hands off anything they touch. They keep wanting to change your idea into their idea, and you keep trying to change their idea of your idea back into your idea. It's actually more complicated than it sounds."

Katz was showing an Englishman, George Ramsden, his part of America, and talking about his screenplay *The Hunchback of UCLA*, which, like his pilot *Snoop*, was now in limbo, having been transferred out of Peter Grad and Harris Katleman's hands to Twentieth's feature division. A switch that left Katz less than ecstatic. Grad and Katleman were his friends; they had championed *Hunchback* from the time he'd brought the project to the studio months before. It was Grad and Katleman who had sold Twentieth on the almost unheard-of notion of having a man who had never produced a feature being allowed to star in it as well. Despite the fact that Katz was not only an unknown actor, but scarcely an actor at all. Now the whole package was being put into the hands of people Katz didn't even know and into an arena he was only marginally familiar with. Film. "When *Hunchback* was with the TV division it was a go project. The feature division wants to take a breather. When the studios are doing TV, they're a lot looser. Because they get reimbursed by the networks, and don't have to stew about promo costs or distribution costs, because that's the network's responsibility. In a feature, costs jump incredibly, so they're much more reluctant to get enthusiastic." He walked out of the dungeon, around a corner, and into an opera house.

And if his experience coming up through the ranks in television was any clue, reselling *Hunchback* to a whole new crowd was going to be a chore. "You go to these meetings and it really takes from your dignity to have to go in and sell the same goddamned story to the same goddamned people over and over." He could just hear the film people saying, "Listen, what if we had a lot of the characters be drunk all the time—that worked in *Porky's*," or "Let's show some tits—that worked in *Porky's*," or "What if there's tons of swearing—that worked in *Porky's*."

"What I'm afraid of," he said, "is that there's a tendency of creative management to look at whatever the immediate past success is and try to repeat it. When, if you look at it, you realize that many of the real blockbusters have been total flukes. *Star Wars, Rocky, E.T.*—they started with no major stars and a new idea. What I'm afraid of is that these guys are going to look at the script and say, 'My God, you've got a beautiful co-ed making love to a *beast*. We can't have that. Make him more human; make him more conventional; make him this; make him that.' Pretty soon *Hunchback*'ll be the story of a love triangle between a co-ed, her boyfriend, and a guy that isn't very handsome."

Katz was prepared for the worst. What if they didn't like his script or changed their minds and didn't want him to play the lead? He had given up a lot to come to Twentieth Century, all for the purpose of shooting *Hunchback*. Just prior to signing with the stuido, Katz had been millimeters away from signing a deal with NBC for one of the most plum jobs in television—that of a producer for NBC studios, a position that carried with it a *guaranteed* thirteen-show series commitment. The only thing Twentieth Century had been able to do to eclipse this offer was to guarantee Katz they would make his movie. Now that seemed somewhat up in the air. And Katz figured that, if worse came to truly worst, he might have to go out and look for another backer.

It had been a difficult week. *Snoop* had been scheduled for a single airing by NBC and then rescheduled twice. No one from NBC had informed him beforehand and, in fact, he'd gotten the information from Jim Burrows, who'd heard it from somebody else. When it was finally aired it got panned by Dave Kaufman in *Variety* on curious grounds. Kaufman stated that the laughter was obviously canned—it wasn't; *Snoop* had been shot almost entirely before a live audience—and that the *Snoop* "newspeople" had laughed at the show's jokes—when the truth was, not one of those "newspeople" had cracked so much as the tiniest curl of a smile. "It makes you wonder," Katz said, "if he even saw the show." And even though Steve Allen, one of Katz's all-time heroes, had written Katz a letter, bolt out of the blue, saying that he loved the show, the *Snoop* experience continued to rankle. "The irony is, Steve's son was one of the execs that turned *Snoop* down at CBS."

For Katz, like many of all but the industry's most successful producers, studio life could be a luxury gulag. Either you were working your butt off in production—when producing *M*A*S*H* he had been responsible for twenty-four shows in forty weeks, and each script had to be rewritten an average of seven times—

or you sat around essentially doing nothing while talking yourself purple trying to get someone to bite on your next project.

Katz walked George through the opera house into an Art Deco dance hall. This was George's first look at America, he'd been in the country less than forty-eight hours, and the British bookseller was in hog heaven. For him, TV was what the U.S.A. was all about. "What will you do next, Allan?" he asked.

Good question. Katz had a chance to create and produce a half-hour sitcom for Charles Durning, one of the most respected character actors in the business. Durning had starred in the Twentieth pilot *Side by Side,* directed by Katz's friend Jay Sandrich, that spring. The show had been similar to the old Jackie Gleason *Honeymooners* series but hadn't jelled. Katz had been asked if he wanted to rework it and had suggested, instead, that he reinvent the show from scratch. ABC was the most formula prone of all the networks and sending Katz, whose love for formulas was about equal to his passion for tuberculosis, to that network's offices to pitch a totally new show was an interesting move on the part of Twentieth. Especially because the show Katz had in mind to sell ABC was astonishingly low on gimmicks and conventional hooks and heat. "When I go in to pitch them," he said, "what I'll do is just talk on my feet. When you pitch an idea for a series, you don't want to go in there with anything written down. Write something down and they have something to object to in writing. I'll just go in there and say, 'Let's do a show about a real family with real problems. None of this family-crisis-of-the-week stuff—the son accidentally robs a bank, Sis has to have an abortion, Grandma discovers she's an hermaphrodite. None of the stock characters—the battle-ax mother-inlaw, the kookie next-door neighbor, or the ditzy broad from work.' " Katz stood in the middle of the black-and-white checkerboard dance floor. George raised his gold horn-rimmed glasses off the bridge of his nose and, looking around, said, "Quite a set-up here, really."

Katz said, "It's a sound stage for *Johnny Dangerously.*"

"Do you think your chances are good?"

"Reasonably. Durning's terrific. I think he's a star. I think he can pull a show by himself. But I want to do a series people will identify with directly. Not so they'll say, 'Hey, I know a guy

like that,' but so they'll say, 'I *am* a guy like that.' It used to be,"
Katz said, stepping out of the Twentieth studio and into the sun,
"you could have a whole show based around *just makin' dinner.* Or you could do a *My Little Margie* thing in which a thirty-
five-year-old woman calls her father and says, 'I don't know what
to do, Daddykins; this man just asked me out on a date, and he
sent me flowers and I *lost* them!' But what I'm talking about,"
Katz said, stepping to the curb to allow a peach-colored Mercedes
coupe to ease by, "is a guy whose dreams are becoming more
and more modest. Who's forty-five, fifty, fifty-five, and who knows
that he's really not ever going to make it big. He's like a lot of
people, whose lives follow an increasingly narrow street, who's
basically just trying to provide for a better life for himself and
his family." Katz walked back toward his office. "I have the
feeling that ABC is going to have problems with promotion,
though. The networks always feel they need either a real high-
concept or a household name in order to hype a new show, and
I figure ABC will figure America's question is going to be 'What's
so special about this show?' Or 'Charles Durning? Who's Charles
Durning?' "

"When I was a kid," said Karen Salkin, in her humble Group
W studio, "I was in love with Paul McCartney. I had this big
color poster of him and I kissed it until his mouth turned green.
I got this plastic ring and told people that we were engaged. And
all my friends believed me! I said we were just using his regular
girlfriend, Jane Asher, for the newspapers. Listen"—she leaned
toward the cameras—"do you think this sounds crazy? Well, it's
cheaper than a shrink."

George Ramsden, son of a former member of the British
Parliament, stood with Allan Katz in the Group W public-access
control room, looking through the window as Salkin launched
into *Karen's Restaurant Revue.* George had asked Katz to show
him an example of "what was happening" in American TV at
the moment and Katz had offered him one of the medium's more
experimental wings.

Karen continued.

"I went to the Original Pantry for dinner last night with Mr.
X, my boyfriend. The decor wasn't all that great. Mr. X said he

felt like our house had flooded and we were eating at the Red Cross."

"Is this normal television fare?" Ramsden asked.

"Not entirely," Katz said.

"I know why they call it the Original Pantry," Karen said to the camera. "Because no one wants to copy it." She held her head in her hands. "It was so bad, it got me so depressed that I feel like taking my own life." Her eyes widened. "So I may go back there again tonight!

"I went to Viola's at the Beverly Center the other night. I thought it was going to be horrible, but it was gorgeous. My photographer showed up just as I was about to leave and insisted I dance punk with him. We did and he kept slapping my hair. Is that what punks do? I thought it was my father. Really, every man in my life: 'Your father used to push you around—that's *terrible*.' Then two months later: 'He should have *killed* you!' "

Ramsden asked, "Is she daft?"

"I want to mention one nice lady before I go," Karen said. "My gynecologist. She's getting married tomorrow. What kind of honeymoon would a gynecologist have? She'll probably be happy just to see a guy, for once!"

Katz folded his arms across his chest. He knew Karen from when her boyfriend, Ray, had worked at *Rhoda*.

A young, elegantly dressed black woman ducked between Ramsden and Katz, glanced at the monitors, and asked, "Is this for real?"

"I think so," Katz said.

On the monitors, Karen was speaking directly into the camera. "God, have I got the cramps. I'm a woman again." She was speaking to the girl on camera two. "Haven't you ever had your time before?" Karen picked up a piece of chocolate cheesecake off her desk. "You'll eat anything," she said, taking a bite. "You need food. I've been taking Extra Strength Tylenol every half hour for the last twenty-four hours. It'll be a pleasure to go home and talk to Mr. X after this. Mr. X only goes home and yells at me. No, he's very pleasant. I think. I don't remember. I haven't seen him for so long."

"I've never seen anything like that," the black woman said.

"How can she talk that way about menstruation?"

"It's America," Katz replied.

"I could sell this show like crazy." The woman introduced herself as a cable television sales rep. "She can't get sued for this? It's fabulous. I need to talk to her producer. Are you guys her producers?"

"No," Katz said, "we're her attorneys."

After the show, Katz took George into the studio and introduced him to Karen.

"Allan," she said, putting her hands over her mouth. "I haven't seen you since Christmas, when Ray and I were shopping, and I asked you, 'What do you get for a man who has everything?' and you said, 'Penicillin.' " Karen then explained to George that Katz had been producing *Rhoda* when her boyfriend, Ray, was a regular on the show and then told Katz that things had been going really fast for her, that the people from Andy Warhol's *Interview* magazine wanted her to be their official restaurant lady, and that she needed a little flat-footed advice. Katz sat down on one of the foldout chairs and asked her how she liked being on Johnny.

"I was excited," she said. "Numb, nervous. I told Ray a week or so before—I was *born* to do this. Before the show, we went to go to sleep and my head hit the pillow and I sat up going, 'Oh, my God, please let me be funny. Please let me be funny.' I spent two hours talking to myself about what if I'm not funny."

"How's it been going?"

"You know, like these people, this guy, sent me a treatment for a sitcom that was about a struggling actress who makes herself famous doing a restaurant review show on public access, and I was flattered but I told him, 'Hey, couldn't we make this a little less exactly like me?' I don't know, everybody—not everybody, but a lot of people—think I'm great, but I'm still not getting any *work* yet. A lot of bites, a lot of interviews, but . . ."

"I think," Katz said, "you've got to break out of this mold. Not that it isn't unique, but," he looked around at the minuscule set, "you're playing to a live audience of only a dozen people. Half of them are kids, who don't even get your single entendres, let alone your doubles."

"Well," Karen said, "I've been trying to think of new things.

Of having guests and stuff. But Ray says it's better with me alone. I had Moon Zappa on and she was great, but the show was blah, because I didn't know what to do with her. I mean, I'm hot now, everybody's talking about me, I'm getting all this free publicity, but, you know, *what next?* Maybe national cable."

"Well," Katz said, "I think what you do is terrific; it's a talent. But you're writing on your feet, just saying whatever comes to mind. It's too unstructured for network TV."

"The only other guest I was going to have," Karen said, "was Rick James. You know, the sexy singer. And I said I'd like to have him on because he was probably the best restaurant in town, which was smutty, but . . ."

"The thing is," Katz said, "you only create a half hour a week. And what you create is going directly on the air. Even the worst *garbage* on network TV is rewritten, rewritten, rewritten, crafted in a way people are not only used to, but entitled to."

"I guess the thing is"—Karen folded her hands in her lap— "I like to think, but I don't like to write."

"That could be part of the problem," Katz said.

"Another thing is, I'm afraid I can only play me."

Katz shrugged. "That's okay. John Wayne and Cary Grant made themselves legends by only playing me. However, they played themselves in a format that allowed them to be successful."

"Yeah," Karen agreed.

"Look," Katz said, "if you want me to tell you want I think, I will. I watch your show and I like it because it's different, it's original, it's contemporary, funny; you're fresh, attractive, all those good things. But I think you've got to start looking at some more conventional things. Like *acting.* Improv never works on TV. They've tried to do it. But audiences don't want hit-and-miss; they're not conditioned to it in television. They don't like people taking chances in front of a camera and failing."

"But," Karen said, "for some reason it works for me."

"That's because you have an eccentric kind of show and an eccentric personality. But what producer is going to let you free-associate on prime-time television at ten thousand dollars a minute? I mean, what are the other actors going to do?"

"But it works," Karen said. "It's working."

"Here. But you don't want this to be the end; you want it to be the beginning."

"I was thinking, I mean, maybe I could work as just a personality."

"Richard Simmons," Katz says, "is basically just a personality. But who would listen to him if he didn't have that exercise class? I mean, I'm not knocking your show. I watch it. You've got something. But if you want a network job, all I can say is that series television isn't going to change for you. Here you've come up with a format that's original, that works for you, that people love, but that can only go so far. It doesn't matter that you may have more talent than any number of the people in current series sitcoms, because they hit the marks. And the people at Rogers and Cowan, however well-intentioned, are not going to be able to reinvent that world for you."

Leaving, climbing back into George Ramsden's car, Ramsden asked, "Do you think she listened to you?"

"I don't know. She's an original, though. There's real talent there. But she's got to do something with it. What'd you think?"

"I think that we have absolutely nothing that *expansive* on English television. What did you think?"

"Success in the entertainment business is complicated. And, even in L.A., being totally off the wall is only one part of the package."

One hour after Karen's show ended, twenty blocks from the Group W studios, cable entrepreneur Norm Smith explained why he, unlike Karen Salkin, was having no trouble at all finding a proper niche in the video galaxy. His niche had found him. It was sex. And all he had to do now was find the best way to market it.

"What makes the Pleasure Channel unique in cable," Smith said, "is that it's different. Regardless of the delivery system— whether it's network, dish, cable, closed circuit, or whatever— if you turned on your set ten years ago, *almost* everything you now get on cable was already there, maybe not in the same release pattern, but there was sports, news, health, music, Dick Clark, *Shindig*. But you couldn't get sex. That's what makes us different."

Cable has become television's arena for the small business-man, for the wildcatter who can—with just a few dollars and an idea—get a crack at the life of banana daiquiris.

Witness Mr. Smith.

Gold watch. Pressed shirt. Rock jaw, precipice forehead—think of a slick, Neanderthal Billy Joel. Norman Smith, the young president of the Pleasure Channel, is clearly in command of his operation. More or less. Henchmen trot after cigarettes, Coors beer, and information. Secretaries play rattling concertos on electric typewriters under a poster that reads SHOGUN ASSAS-SINS: THE GREATEST TEAM IN THE HISTORY OF MASS SLAUGHTER, illustrated with Oriental men in robes at the cusp of going ber-serk with knives and swords. Next to it, another poster, titled BUBBLE GUM. Illustrated by a voluptuous teenage creature who wants to make love to you. BUBBLE GUM is subtitled PUT IT BE-TWEEN YOUR LIPS AND BLOW. WHAT GOES IN HARD COMES OUT SOFT AND WET.

The sense of single-minded, strongly directed commerce here would have been complete were it not for one of the secretaries' beating a T-shirted young executive over the head with a dildo the size of a baseball bat, right outside Smith's office. She kept it up until one of the other young execs yanked her aside with the words "Cool it. There's press in the office."

But Norman Smith didn't seem to mind. He was talking about his dream: "We'll get on the bird, the satellite transpon-der, sometime this fall. It's expensive, but right now the sky is overtranspondered. Only two thirds of the transponders are in use. We're the young independents. We're gonna give the Play-boy Channel a run for their money . . . even though I do busi-ness with them all the time. Because they need programming and I have programming."

He turned in his chair. "What I'm talking about is *Flash-dance* with real breasts. The misconception is that you can make ten adult films a weekend and they cost five thousand dollars each. Well, they don't. They're real films. They have a begin-ning, middle, and an end. They have good-looking women, good-looking men, a lot of new faces, and a lot of talent. You can make a top-shelf adult feature for a hundred fifty to two hundred thousand dollars. I'm not a cigar-smoking guy in the back room with a crooked nose saying, 'What have you got? Give it to me.' "

Smith took a call, holding the receiver between his shoulder and ear, leaving both hands free to pass out beers from a six-pack. He made a date to have breakfast with someone at the Polo Lounge, then hung up and continued. "It hasn't been easy for us to get in black ink. Our operation started without any money—it was just me when it started out—and now there are nine full-time employees. Now I can go out to raise the real money I need. I've closed one deal with a video cassette company three months ago, and I still need a half million!

"I project this to be a three- to four-million-net-per-year business after the end of the year. Building to five or six million within a two-year period.

"My goal is to create an entire second channel; the adult side will just be the locomotive. By that, I mean a second channel, as opposed to a primarily movie channel like HBO or Showtime. I'm talking about a twenty-four-hour-a-day secondary channel, featuring seven hours a day of adult programming, then a music tier, a drive-in theater tier with stuff like *Attack of the Killer Tomatoes*, maybe a *Kung Fu Theater*, maybe a little health care. That's the goal for the Pleasure Channel—the breakdown, in microcosm, of the entire cable universe. On one channel." Norm was going for the whole banana. "That's why the Pleasure Channel could be applicable to the low-power television market, the direct-broadcast satellite market, and could even be one of the five channels on the multiple-dish-system multichannel situation. And hey—probably would! Because of the fact that the adult side would be the locomotive. Sex sells suds.

"Face it," he said, lighting a cigarette, "the Playboy Channel is geared for cable operators. They, therefore, have to be more restrictive in their programming content, relative to how hot it can be. So our versions are not going to be hard core, but they will be hotter than anything else out there. The difference between Playboy and myself is that they are geared basically for the multiservice-operator cable people, whereas the Pleasure Channel is geared for *all* the delivery systems." Norm's cigarette smoke curled upward, conjuring visions of taverns, clubs, and motels offering his soft-X wares.

Norm Smith grew up on the West Side of Manhattan in what George Carlin refers to as White Harlem, went to the University of Virginia and NYU, was an economics major whose father was a banker. After working on Wall Street, he got a job at Paramount as a booker in the shipping room, "another word for male secretary." Then he went to work for New Line Cinema, selling theaters on films like *Pink Flamingoes, Jimi Hendrix Plays Berkeley, Sympathy for the Devil, The Seduction of Mimi, Reefer Madness, Cocaine Fiends, Sex Madness*—"basically working a youth-oriented market. There was no 'Midnight Movie' when I was selling *Pink Flamingo*. This was approximately ten years ago. I set up the promotions, the cross-plugs, the trailers, the radio spots, and the point-of-purchase displays. Which was great background for what I'm doing now."

Norm then scored the distribution rights to *Tunnel Vision* while that cult classic was still in postproduction, and went on to become marketing agent and distributor for *Attack of the Killer Tomatoes*. "Then I saw the handwriting on the wall, relative to the majors coming in and taking over all the independent-type products distributionwise, a la *Animal House*. I saw the video cassette business happening, got into that, then established my own video cassette company about a year ago." And what he saw in the cassette business was video sex, oodles of it—the demand was so striking that you'd have had to be sleepwalking not to see that what people *really* wanted on the tube was what shows like *Charlie's Angels* and *Three's Company* always promised and never delivered—naked slithering coitus.

"I sold my piece of the cassette company and got involved in the production side for presales. I was the exclusive agent for Marilyn Chambers's *Love You, Florence Nightingale*—which was sold to Playboy on an exclusive basis." Smith claimed to have written contracts totaling close to two million dollars during the last two and a half years for adult television, but started the Pleasure Channel because "I saw more money out there than in just my existing accounts." In other words, he found it possible to work both sides of the street, both as a distributor and as a programmer.

The Pleasure Channel's porn washes best in America's heartland, especially in the Bible Belt. A phenomenon he found easy to explain: "If you live in New York City, you can go see anything you want, up to live sex shows. If you live in a suburb of Livingston, Montana, you don't have that kind of choice. Also, because outwardly they're more restrictive there, when you get somebody in the privacy of their own homes, they'll pull the drapes, and the first thing they will put on is sex. To paraphrase Isaac Asimov, 'I've seen the future of video and it is sex.' And the American public is used to paying a premium for it. Whereas a copy of a *Star Wars* cassette is thirty-four to forty-nine dollars, a copy of *Talk Dirty to Me* is eighty-nine or ninety-nine dollars. The more hot, the more graphic the stuff is, the more it costs.

"The Mitchell brothers, Chuck Vincent, Anthony Spinelli, Howard Wenner—there are a half-dozen guys out there who really make top-flight stuff. No unions. A ten-day shoot. Half the people not even using their right names." As for his own prospects, Norman was pensive. "Because of conventions and travel, I'm on the road a third of the year and I probably put in fifty hours a week.

"I hope by the time I'm forty this will be a hobby. That's the goal."

9
Hype

July 1983

Summer was in full blaze—heat vibrated off the cement in NBC's vast, guarded parking lot—and, by the calendar, the fall debut of new shows was still seven weeks away. But Steve Sohmer's desk was a snowdrift of papers dealing with September, October, and November. He was preparing print ads for *Boone*, *Mr. Smith*, and *Manimal*, switching all the fall network promos over to the Be There theme, getting local stations on board for their own Be There promotions, and even planning hype for the November sweeps.

Sohmer had written a novel when he was twenty-five years old that *The New York Times* put on its "twenty best" list for 1966. He had, early in his career at CBS, been taken under the wing of William Paley and elevated to the top of the promotion department, where he managed to pump "Who shot J.R.?" into a worldwide anxiety. And now he had taken over *four departments* at NBC, and was considered within the industry to be within snatching distance of Brandon Tartikoff's job. Nothing within his aura moved slowly. Crouched behind his desk, just down the hall from Tinker and Tartikoff, a shiny black silk shirt stretched over his belly, a cigar the size of a cop's nightstick stuck in his rectangular jaw, Sohmer was central casting's idea of a *capo di Mafia.*

"Be There," said Sohmer, loudly, expansively, "is a *watershed promotion.* Television has changed—we're no longer living in the age of just three networks, where you come home and see what's on tonight. Now you go in there and demand a unique experience. If you don't get it, you say fuck 'em, you plug in your Atari, and away you go! The public doesn't watch networks anymore; it watches shows. You've got to *hit* people— with something like 'Tonight, Tip O'Neill visits *Cheers!* Be There!' The whole implication is that it's a once-in-a-lifetime event!"

Sohmer, head of NBC advertising and promotion, as well as daytime programming, children's programming, and specials, was convinced that NBC was the only network that was meeting the cable threat the way it had to be met—head on, claws bared, screaming. The other networks' promos, he thought, were "soft and irrelevant." ABC, apparently trying to move away from its image as the tits-and-pistols network, had created a series of mushy Americana promos, under the umbrella slogan That Special Feeling. Typical of their spots was a soft-focus, no-dialogue dreamscape featuring a kid bringing home an orphan puppy for dinner, then fondling it while the family watched TV. ABC seemed to be pushing television generically—along with families, food, and stray dogs. The hardest sell in the whole ad was Maureen McGovern crooning in the background, "It's that special time, yours and mine, good as it can be. It's That Special Feeling on ABC." CBS promos showed a paraplegic finishing a marathon in a wheelchair, and two deaf lovers signing their affection—ads so cloyingly exploitive that the network finally had to issue a statement proclaiming that the actors *really were* handicapped. The sound track had Richie Havens caroling, "A touch of laughs, a touch of tears, the friends you've loved for years . . . We've Got the Touch, America—you and CBS."

That type of slush was acceptable, Sohmer thought, for outfits that had a monopoly, like the phone company or the post office, but with practically everybody who owned an Erector set now jumping into the TV business, squishy promos were almost suicidal. CBS hadn't even had the sense to come right out and say, "We're number one."

"Anyone who thinks that We've Got the Touch says, 'We're number one,' is full of shit," said Sohmer, going to his office refrigerator for a diet cola. "Laying claim to the top spot can be a good campaign, but the only recent time anyone did a we're-number-one campaign was in 1976 when ABC had just taken over the top spot from CBS. You can't claim to be number one without getting an okay from the Nielsen Company, since it's their rating system, so what ABC did was to use a popular song of that time, 'Still the One.' It was an extremely shrewd campaign. The other of the three watershed campaigns, besides Still the One and Be There, was Looking Good Together, which I created at CBS. The idea was that the public shared the experience of television with the stars—if I wasn't father of that idea, I was at least godfather of it. In that one, a young man came home, rainy night, there's a party across the hall, he's not invited, he goes in and turns on his television set, and lo and behold, Loni Anderson materializes! In his bedroom! That was five years ago, and people still remember it."

NBC's new fall promos, Sohmer thought, would be equally as memorable. Each of them focused on a particular show, hyping the show as a can't-miss, buy-your-ticket-early event. They openly mocked the competition. The promo for Steve Cannell's *The Rousters*, for example, featured the stars of that show executing a model *Love Boat* with .44 Magnums, cackling gleefully when the thing turned into so much plastic confetti, as they shrilled, "Abandon ship, America! *Rousters* gonna sink *The Love Boat!*" Then the NBC peacock flashed its tail while the orchestra and chorus boomed, "Be what you wanna be, see what you wanna see, you can NBC there, Be There!" This chesty, take-no-prisoners tone made the other networks' promos seem leukemic, and Sohmer was thrilled with his work. "When I came to NBC, there were a lot of long faces around here. And a lot of desperation. And not a little despair. So I decided: I would reinvent network promotion."

Sohmer thought his new event-oriented promos would help boost NBC, if not into first place, at least into a dogfight for it. Some of the new shows, he thought, were a promoter's dream—he especially looked forward to sinking his teeth into *Mr. Smith*, *Jennifer Slept Here*, and *The Yellow Rose*. "If we can put a dent in *The Love Boat*," he said, "then *Yellow Rose* may have a chance, because *Fantasy Island,* which follows it, is over. As far as I'm concerned, it's a dead show. I have the same feeling about *Fantasy Island* that I had about *Happy Days* and *Laverne and Shirley* last year.

"The development season was great. Good property. Good concepts. Good casting. The development season at the other two networks must have been a disaster. I can tell you that if Mr. Paley were still at CBS, and they said they were playing three movie nights, they'd all be in the street. All!"

Not all the NBC shows would be easy to promote, though. "*St. Elsewhere* is tough. It's a very different kind of drama and the public can't quite get a handle on it, like *Hill Street* in its first year. And *First Camera*—the revamped *Monitor*—is a tough one. Because it's going against *60 Minutes.* People don't watch *60 Minutes* because it's a news show. They watch it because those guys are *stars.* The whole show is Mike Wallace. He's the box office there." *First Camera*'s witty and erudite Lloyd Dobbins, apparently, was not Steve Sohmer's idea of box office. Sohmer had promoted *60 Minutes* for five years, and knew that it was really a high-concept show—a cowboy show, white hats versus black hats—a formula Sohmer ably took advantage of. "While I was promoting that show, people would say, 'Who are they after this week?' That's why people would watch it."

High-concept shows, of course, were easier. The low-concept *Cheers* and *Taxi* had driven Sohmer nuts—"I burned up over a million dollars' worth of airtime during the World Series last year promoting *Cheers* and *Taxi,* and it didn't budge the ratings one share point." But high concepts could blow up in your face—right now—if the viewers didn't immediately like them. "*Co-ed Fever,* a show we tried at CBS, would have been the first television show about life in a co-ed dormitory. That's high concept. But it so happens that America hates college students. We played it after a two-and-a-half-hour *Rocky,* which did forty-five or fifty in its last half hour. Then *Co-ed Fever* came on and did about a twenty-five. It never played again."

Sohmer remembers his days at CBS with some ambivalence. "None of the stuff I've done here would ever have gotten on the air at CBS. None of it. Too conservative. My whole role is different here. Nobody looks at my work here; I do it and put it on the air. Over there, we had two ninety-minute meetings a week with a bunch of programming guys who didn't know a goddamn thing about promotion. When you get enough people looking at something, you get down to the lowest common denominator real fast. The only thing that gave us some spark over there was that I would shout as loud as they would! I never thought I had a career there. I had a job. I've been fired from every job I ever had, except that one, and I was sure I'd get fired over there. I had to quit twice a year."

Sohmer liked it better at NBC. "When Grant Tinker says, 'What do you think?' and you tell him, and then he *writes it down*—you thank God in heaven. Same thing with Brandon. When we sit down to have a creative meeting, all of a sudden he's not the president anymore; he's just another guy with ideas. That's how I am with my people. They do what they think is best, and if I don't approve, they tell me to go fuck myself and they do what they want. When you sit down at a table to work on a creative project, everybody is reduced to the level of copywriter. Otherwise, don't sit down." As evidence of the power of this approach, Sohmer pointed to the fact that virtually the entire executive team in the CBS promotion department had followed him to NBC—some taking lower salaries.

Sohmer considered Tinker one of the two smartest men he had ever worked with. The other was William Paley. "Paley taught me an awful lot about follow-through, and detail, and budget. Tinker is very different. Tinker's always looking for some radical breakthrough, some new and different idea."

Five miles west of Sohmer's office is a conglomeration of film-business cottage industries, overpriced houses, and a major studio—it is Universal City, the only town in America named after a movie lot. Here, press agent Cynthia Snyder was working on her own summertime promotions, on the third floor of a red-brick building that housed the American Stuntmen's Association, a hiring hall for the near suicidal. There was also a company here that would, for a fee, see that your particular soft drink or cigarette ended up in the hands of an actor during a movie scene. Down the hall from Cynthia Snyder Public Relations was the office of *Intro*, a classified-ads-for-the-lovelorn magazine that was in turmoil. The owner had recently placed her own ad and been so dazzled by its respondent that she'd sold him a controlling interest; she was now being booted out of the office by the new publisher. Snyder's own line of work is scarcely more secure. "The competition in public relations is incredible now," she said, pacing her office in Farmer John bib overalls and red spike heels. "Somebody will work for a biggie, like Rogers and Cowan, and as soon as they get a relationship with a star, they'll split him off and start their own company. But I'm spoiled by my clients. If I worked for a big name who called me up in the middle of the night, I'd probably kick him out the door."

Currently, Snyder had about ten major clients, including several actors from *Hill Street Blues*—Michael Warren, James B. Sikking, and Barbara Bosson. Sikking was due in shortly for a conference.

Before he got there, she wanted to clean off her desk, which meant finding out who produced the ads for *Gentlemen's Quarterly*, checking to see if Barbara Bosson would be free for a charity event, calling the talent coordinator at *Hour Magazine* to pitch *The Jeffersons'* Roxie Roker for the show, setting up a meeting with the Australian distributor of *Hill Street* to arrange a promotional tour, and calling a beauty pageant director in Australia about using Gordon Jump as MC. Jump, who'd played the middle-aged station owner in *WKRP in Cincinnati*, was more popular in Australia than he was in America, just as Sikking had greater celebrity in Canada. Furthermore, Jump could use the $5,000 plus expenses for two that the pageant would provide.

When actors are out of work, it's tough to pay a $300- or $400-per-month publicist's fee—or the $2,000 fee at a big agency like Rogers and Cowan. But publicity keeps their Qs above water level and sometimes provides a crucial break.

Not always, though. Snyder recalled when Eileen Barnett was scheduled to be killed off on *Days of Our Lives*. Snyder had managed to get *Tonight Show* talent coordinator Jim McCawley to come hear her sing, and McCawley had her set up for two songs on the show. Barnett was thrilled—the singing slot offered a golden chance to get her injured career moving in a different direction. "Rehearsal went beautifully," said Snyder. "She was on after Charlton Heston, an eighty-seven-year-old woman who did commercials, and some singing dogs. We were in the greenroom and oh, shit, time is running out. I was horrified! The singing dogs! How many numbers did they *have*? I mean, they were just barking—I don't care what you call it. And sure enough, she gets bumped. By the singing dogs! Until recently, she'd been practically unemployed."

James Sikking arrived with no fanfare and took one of only three chairs in Snyder's cubicle-with-a-receptionist office. Sikking played the adorably bloodthirsty Lieutenant Howard Hunter, a scholarly fascist who did things like beg his captain for the privilege of blowing away some lawbreakers because "my boys need the validation." Sikking wanted to arrange some golf tournaments for his next hiatus; after twenty years in the business, he still depended upon relatively low-key promotions to further his career. "Jim Sikking will never be on the cover of *People,*" Snyder had said of him before he arrived. "It's strictly a T-and-A publication. Neither will Barbara Bosson or Roxie Roker. Michael Warren might. Jim Sikking will never be the centerfold of *Playgirl.*"

For the first approximate decade of his career, Sikking earned about $300 per month. Then he landed three years in *General Hospital,* which paid him $24,000 a year to start, and topped out at $32,000. Even the *Hill Street* job was not the waterfall of cash that most prime-time starring roles are. "MTM is notoriously cheap," said Sikking, a calm, genial type whom Snyder considered one of her "top-five nice clients of all time." He continued, "They pay their writers very well, and rightly so, but have less respect for the actors. They are not overly generous. I'd be the same way, because I don't think the actors are that important." Sikking believed that, for the most part, talented writers would rise to the top in television, but that actors needed more than talent. "The reason I am here, and a hundred twenty-five kids that graduated with me from the university are not, is pure shithouse luck. It's pure luck. I will trade a pound of talent for an ounce of luck any day. It is luck."

"And promotion," said Snyder. Sikking dipped his head in acknowledgment.

"And perseverance," said Snyder.

"And perseverance," Sikking agreed.

"But one of these days," said Sikking, "*Hill Street Blues* will be gone, and I will go back to being a shit kicker. It'll be tough."

For Frank Bonner, another Snyder client, the end of the run has already arrived. Bonner, who played Herb Tarlek, the salesman in *WKRP*, was out now with his softball team, the Hollywood All-stars, a group of mostly unemployed TV actors playing benefits to keep themselves in the public eye. Bonner had been asked to read for a number of pilots, and had been cast in the ill-fated *Sutter's Bay*, with Harry Anderson's friend, Jay Johnson. Now he had another year to wait for a new pilot season to roll around. Bonner could have scrounged a decent living from public appearances as Herb Tarlek, but he was trying to avoid being typecast, mindful of the problems that his old friend Gary Burghoff was having. Burghoff, who played Radar O'Reilly for many years on *M*A*S*H*, had been making Arrow shirt commercials with Bonner one year, a star on *M*A*S*H* the next. But after he quit the series, and attempted a stint in nightclubs, he found that he'd been *swallowed up* by Radar—he could do a little cameo work in *Fantasy Island* or *Love Boat*, but to producers and casting directors, he was a ghost from the daily reruns.

"I think people are nuts to jump out of a series," Bonner said. "A series won't go on forever, and if you can ride it out, you can make your nut and sit in your nest." Bonner was finding out how difficult it was to reenter series television; it was almost as hard as getting in in the first place. And getting in had been almost impossible. Bonner, like every other cast member in *WKRP*—except for Jan Smithers and Richard Sanders—had been the second or third choice of the producers when the show had first been cast. The actor who'd first been offered Bonner's role, Rod McCary, had also had a chance to star in the pilot *Mother, Jugs and Speed*. "He literally tossed a coin," Bonner had said, "and it came up for the other show. He said, 'Good. I don't think *WKRP* is going to make it.' They aired *Mother, Jugs and Speed* once, and that was it."

So Bonner had gotten the part, and had modeled himself "after a used-car salesman that sold me a Mustang. He looked like he only bought two colors of pants, yellow and lime-green, because they both went so well with a white belt." *WKRP* was moved so frequently that even cast members would miss it, and then it was canceled, a move CBS's Harvey Shephard admitted had been a mistake. Shephard tried to reassemble the cast, but Loni Anderson was not interested and the show remained in its grave.

When he got back from his softball game, Frank Bonner would head out to appear at a rodeo in Wyoming—one of the places where fame and fortune, he believed, could be slowly and carefully reconstructed.

All through the midsummer week, the most famous actors in television dutifully trudged into a small, dusty warehouse in the fast-food and cheap-furniture section of Beverly Hills, near the corner of Pico and La Cienega Boulevard. Normally, it would have taken the lure of a taffeta-and-sequins awards ceremony or an antidisease dinner dance to funnel so many thirty-grand-a-week people into such a small area. But this shabby workspace held out the promise of a cover photo on *TV Guide*, found each week in 17,670,543 television-watching homes.

Toward the end of the week, the cast of *We Got It Made* was summoned to the warehouse. While Lynn Farr Brao and the publicists from MGM and NBC picked at fruit, cheese, coffee, and jug wine, the cast huddled in front of a cranberry-colored cardboard backdrop, a photographer squinting at them behind a Nikon. "Get much closer," he coaxed. Nineteen seventy-one Van Morrison issued from unseen speakers. "*No! Much closer. Teri, step out.*" Teri Copley was wearing a ruffled apron no bigger than a bib across her denim-squeezed thighs, and a pink cotton top laminated to her breasts. Her male co-stars, flanking her, smiled tightly behind coats of instant suntan. "Tom! A little higher— get more behind Teri." Soon he had them pyramided into the classic wedge offense, with Teri Copley's chest leading the way. "Great! Great! Great!" One of the publicists edged near and the photographer glared at him. "Please," he implored, "let them build a relationship!" The publicist tiptoed away.

Between costume changes, Teri Copley sat at the table with the fruit. Lynn Brao told her why Fred Silverman hadn't come to the NBC press tour. "He wanted the attention to be on you kids, and not on him and his past. Freddy just said, 'Tell the press I'm a big pain in the ass to work with, and all I do is scream, "More sex, more sex." ' "

Copley giggled. "Did you see the picture?" she asked. The picture was a shot in this week's *National Enquirer* of Teri in a low-cut, fringed buckskin shirt—sort of Pocahontas through the eyes of Larry Flynt.

"I thought you looked good."

"I thought I looked awful. I called my mom and told her, 'Don't you dare buy a copy.' "

"From now on," said the NBC publicist, "don't let anybody but NBC take your picture. Once someone shoots you, they own the negative."

"Any nudes out there, Teri?" asked the MGM publicist.

She shook her head.

"Good," he said.

"Sometimes it gets to be a real drag," said *National Enquirer* senior reporter Donna Rosenthal, "interviewing carhops turned TV stars, kids who've never been east of Bakersfield and never even heard of half the countries their shows are playing in."

Rosenthal sat in a café in west Los Angeles, ten minutes from where the *TV Guide* covers were being shot. About thirty-five, she had thick dark-blond hair and enough teeth to key a baby grand, though she only infrequently framed them with a smile and seemed to be under a strain. Perhaps because the *Enquirer* does not like its staff speaking to other journalists. Nearly as powerful a venue in the television press as *TV Guide*, the *Enquirer* is a vital outlaw force that doesn't even have an office in L.A., but depends upon largely anonymous staffers and a vast network of paid informants to brings its readers news they are likely to read absolutely nowhere else. Like the story about Jackie Gleason in the issue that was then on the stands. According to his ex-wife Beverly, the Great One came home one evening in 1973, slumped white-faced into an armchair, and told her that, thanks to President Nixon, he'd just been to Homestead Air Force Base, where "I've seen the bodies of alien beings from outer space.

"It's top secret," his wife quoted him as saying. "Only a few people know. But the President arranged for me to be escorted there and see them. . . . No one would tell me the full details, but a spacecraft has obviously crashed near here." Gleason, she said, described the experience this way: "When I arrived at the base, I was given a heavily armed military escort and driven to a building in a remote area. We had to pass a guard at the door, then were shown into a large room. And there were the aliens, lying on four separate tables. They were tiny, only about two feet tall—with small heads and disproportionately large ears. They must have been dead for some time because they'd been embalmed. It sounds incredible, but I swear it's true."

Rosenthal believed that such chronicling of the lives of entertainers for the *Enquirer*'s five to ten million readers gave this, the nation's largest newspaper, as much influence on the TV-watching habits of Americans as any other publication, if not more. "It's become very fashionable," she said, "for TV people to dump on the *Enquirer,* especially after the Carol Burnett libel suit against the paper. But they use us tremendously. We *made Dallas*'s Charlene Tilton. Eighty percent of our covers are TV oriented." For economic reasons. "TV stars are much bigger than movie stars on the newsstand. A Sly Stallone is much less salable than an Alan Alda. TV is an obvious hook for us, because the American public wastes so much time watching it."

Despite being Boswell to the Hollywood geek parade, Rosenthal was a writer of substance. She had cracked the John Belushi case, after tracking down the comedian's last companion, Kathy Smith, the sadly illustrious songwriter. Rosenthal spoke six languages and had covered wars and famines all over the Middle East. "The jump from Bangladesh to *Mork and Mindy* is very weird. You get twice as much money and twice as much praise and recognition covering Zsa Zsa's latest divorce as covering the plight of ten thousand starving kids in East Africa. It can be extremely depressing." A few of the big celebrities, like Bob Hope and Ed Asner, were gracious about dealing with the paper, but—by Rosenthal's lights—the majority were . . . a challenge. Even though, by her estimation, the paper *went easy* on the famous. "The *Enquirer* is very tame," she said, "compared to what really goes on here, like the whole gay, face-down-on-the-casting-couch scene. So many of these TV people are into completely bizarre sex scenes, drugs—you know, so-and-so, the clean-cut young talk show host who's so coked out the inside of his nose is made out of plastic. TV is amazing; it's America's biggest export. People in Europe and the third world see this country as a collection of J. R. Ewings and Archie Bunkers. It's all part of the Coca-colonization of the world. People don't realize the impact of the garbage turned out of this town."

Ben Stein takes another view, seeing national network and cable television as less the product of a clique of coke-snorting orgy addicts than the result of the bucklust and flimflam that is a reflection, to a large extent, of the hard-hussle atmosphere of the community in which most TV is created. "L.A. is the only town in the world where the women are tougher than the men, the children more frightening than the adults; where the old people are more frightening than the middle-aged, the rich more frightening than the poor, the whites more frightening than the blacks, and the Jews more frightening than the Gentiles. TV is a meritocracy where the merit is often very hard to find. Where much of the merit is in knowing the right people. It's very destructive to the human spirit."

A former Nixon White House aide, a former columnist for *The Wall Street Journal* turned now columnist for the *Los Angeles Herald-Examiner*, Ben Stein is author of a best-selling novel, *Ludes*, and a nonfiction book, *The View from Sunset Boulevard*—in which he made the now often quoted claim that television is controlled by a myopically leftist and anti-business clique—and was a creator of the satirical *Fernwood Tonight*.

Richard Nixon was commemorated here at Stein's home in the sunburned hills above L.A. by a framed picture of RN on the dining room wall behind his head. "The business of television is the business of making easy money. This is a business of orderly people, making a cookie-cutter product day in and day out. TV people are not wildly creative or emotionally alienated artistic types, but more generally are just small businessmen who've been put in a position to make a lot of money. If we're talking about the record business or the cocaine business, that's one thing. But TV—people in TV don't even stay up late."

Stein got up to fetch a Diet 7-Up from his refrigerator. "Easy money. The fact that you can get paid a million dollars in TV to do shit inspires you to do shit. There's no such thing as principle, only the deal. The network says, 'Sell your integrity to us for a dollar ninety-nine.' And you say, 'A dollar ninety-nine? Why would I do that?' And they say, 'Because a dollar ninety-nine is *only the down payment*. We're going to take your integrity, turn it upside down and make it look like a monkey's asshole.' And you say, 'Wow! You mean if I take the two dollars, I could end up a millionaire—just for that?' There's no such thing as integrity here. Until you get to be as rich as Norman Lear. Then you get your integrity back. I know, because I submit to that crap all the time."

Karen Salkin, who had promoted herself from anonymity to the position of darling to the stars, stared into the red light of the TV camera, about to begin another restaurant review show, stricken. "I have to tell you something. I may be sweating. I may look terrible. This morning I was happy. I just want to get this off my chest, because of what happened this morning."

Her eyes went wide. "I *am* an actress. And if anybody needs somebody on a show to do *this*, I can do it! So please don't anybody out there think: 'This is a good character; we'll get somebody else to do it.' I am the one!

"I'm upset. I'm shaky. See! If I can do this here, I can do it anywhere!"

That morning when Karen had come into the Group W offices to prepare for her show, a call was waiting from one of the producers of SCTV, the Canadian-based comedy series that had just jumped from NBC to cable. What a rush. Maybe this would be the big lift-off—a shot at the most ambitious comedy ensemble on TV. The producer came on the line and said he loved her show! She was *so* funny. And it just so happened that SCTV might need a character just like her. So could she send them some tapes, so they could see what they could develop?

She was ecstatic. Of course! She'd send the tapes immediately. She said she watched SCTV all the time, and really liked it, and would love a chance to re-create a part of her show on national television.

Well, the producer replied. Well. That wasn't exactly what they had in mind. What they wanted to do, the producer explained, was have somebody else play Karen Salkin. Somebody really good, of course.

What! Karen said. It wouldn't be me?

Not only wouldn't it be her, but it immediately became unclear whether or not she would even get paid. After all, she was just another public figure. He went on to fill her in on just exactly what he thought her rights were in this situation—that you could copyright work, but couldn't protect yourself from caricature.

She was incredulous. Actually staggered.

On the set, Karen put her hand down on a desk and knocked over a can of Coke. Cola urped from the can and snaked all over a sprawl of Group W paperwork. "I can't believe this," she said to no one. "They're going to steal me! They can, like, pirate your soul. Anytime they want. I can't believe this is happening. I really can't." She then called her agent, manager, and attorney. Those three had no trouble at all believing it.

Karen Salkin had suddenly come very close to promoting herself right out of business.

10
Summertime Blues

"The Hunchback is like religion: if you buy the basic prem- ise, you'll probably buy the rest. If I can get people to believe just one thing, that the Hunchback is *real* on the screen, then everything should pretty much fall in place."

Allan Katz's world had begun to turn around. Steve Tische and John Avnet, whose film *Risky Business* was making hundreds of thousands of dollars a day at that moment, had read the script of Katz's *Hunchback of UCLA* and had signed with Twentieth as producers of the film. With luck, *Hunchback* would make them all the kind of money during the summer of 1984 that *Risky Business* was hauling in now.

Too, Katz had just arranged financing for a play he had written, titled *Kaufman and Klein*. It was about two partners in the garment industry who bicker their way through a decade of business only to become romantically involved after one of them goes to Europe to have a sex-change operation. If all went well, *Kaufman and Klein* would be directed by Jay Sandrich and, if all went even better, it would star Ed Asner or Walter Matthau.

Katz was about to meet with the executives at ABC, to pitch them his idea for a low-concept sitcom for Charles Durning. Three solid bites: a movie, a play, and a series. Now if he could just get something reeled into the boat.

Lo, how the mighty have fallen! Corky Hubbert could not, for the love of God, understand it. Who could? There he'd been, cruising along at the height of his profession—getting his licks in *Magnum, P.I.* as a guest top-star, hanging out with Tom Selleck, America's top TV womanizer, a guy so suave he could probably vomit through a straw. Corky was really doing up the midget CIA karate-expert role—having Larry Whatshisface talking *spin-off* and getting paid a king's ransom just basically to play himself. And now, two weeks later, this. How the mighty can go straight down the shitter. And it was all happening in his hometown. Humiliation. Which might not, he figured angrily, have been that much of a coincidence. It was the Mark 6:4 Syndrome: "A prophet is not without honour, save in his own country."

Every time Corky tried to get anything done in Portland, his most favorite place on earth, he got screwed. Now that he'd wowed the *Magnum* folk with his Stanislavski-on-mushrooms Method acting, what happens? He comes back here, cannonballs down a flight of stairs, and breaks his leg. That's what happens. And right after that, Paul Bartel called to see if Cork wanted to co-star in his next movie. Paul Bartel was the guy who had directed *Eating Raoul*, Corky's favorite new classic film. So what was he going to do? He'd already accepted the part, but as soon as the Bartels people found out his leg was broken, well, he could kiss that opportunity goodbye. And, face it, parts like that weren't so easy to come by. It had been three years since he'd done *Under the Rainbow*, for God's sake. *No dream too big, no dreamer too small.* What in the world was he supposed to do?

Corky looked down at the cast that reached all the way up to his thigh. Every time he came back to this town, something went wrong. Like back in the days when he had been doing his comedy act. Portland clubs were full of all these too-hip-to-be-hip audiences, and he was always in trouble with the club owners. Like the time when he got the banana cream gook all over the rock band Quarterflash's equipment, when you could practically see things going off—pop pop pop—inside the club own-er's head. He went, "Don't . . . you . . . ever . . . throw . . . a . . . pie . . . in . . . here . . . again!"

But his return this time had taken the cake. Fresh in from Honolulu, Cork had run into *Pillars of Portland* director Tom Chamberlin at a party and the guy had leaned down, grinned into his face, and said, "Hi, Corky," like he was some kind of little kid or something. Corky said, "Hey, maybe I could help you out with *Pillars of Portland*," and Chamberlin made like he didn't even know what Corky was talking about. "What do I have to do," Corky thought, "beg for a part? What am I supposed to say—*'Attention K mart shoppers! There's a real, live profes-sional actor, Corky Hubbert, willing to work practically for free in this bozo production!'?"* Finally he just said, "Tom, look, if you could work out a part for me, I'll help you out." And Cham-berlin just grinned and looked puzzled and said, "Well, maybe in a crowd scene or something, Cork." Which got a nice laugh out of everybody. But Corky could tell the guy was *serious.*

Then he had fallen down the stairs at the house in which he was staying—where he sat now with his pal Little Gregory of Little Gregory and the Intolerables, the hottest Tex-Mex–reg-gae–funk–rock band in the county, watching rain bead on the windows. Here it was the end of July, and summer hadn't started yet in Oregon.

As it turned out, Corky had gotten a bit part in *Pillars* as a crippled midget baseball umpire. And then Chamberlin had made him stand around all day in the rain on this baseball diamond—which had God only knew *what* to do with the plot of the show. And finally, Chamberlin forgot about most of Corky's part. *For-got.* When cast and crew were breaking to go home for dinner, Corky had to hobble up to Chamberlin and say, "Hey, man, do I get to do my bit, or what?"

That's what was so scary. If Corky missed out on this Paul Bartel film, if he didn't keep his momentum going, he'd spend the rest of his life playing to the cheap seats, nickel-and-diming around, hat in hand, always wondering where the next dollar was going to come from.

"What are you going to do?" Little Gregory asked.

"There's only one thing to do," Corky replied.

"What?"

"Get me some scissors," Corky said, putting both hands on the cast that covered his leg. "We gotta cut this fucker off."

Ed Asner had two offices. One was on Sunset Boulevard, at the headquarters of the Screen Actors Guild, of which he was president. The other, where he spent most of his time, was in Universal City, across the street from a metal-skeleton skyscraper, thirty stories and rising, that proclaimed itself FUTURE HEADQUARTERS OF GETTY OIL COMPANY. Adjacent to the construction site was a billboard for *Jaws 3-D*—a three-dimensional great white, its nose sticking out over the roadway, with a mouth large enough to swallow the Liberty Bell. Someone had stuck a mannequin in the jaws, and hapless legs dangled from the carnivore's teeth. Asner's building, owned by producer George Lucas, also housed the Nautilus Tech fitness center, a bustling shrine for pilgrim actresses, female execs, and secretaries in pastel tights, their long, colored legs uniformly sweatered in the 95-degree heat by wooly leg-warmers.

Asner's outer office, perched above a giant, plant-filled atrium, was a jumble of paperwork and memorabilia, including a STARVE ONE FOR THE GIPPER bumper sticker, a picture of Asner and Nancy Reagan, and a blowup of a mock Steve Canyon cartoon, with the adventurous flyboy being told, "It's simple, Steve! Why don't you and your boys just get the fuck *out* of El Salvador." On the wall outside his inner office were photo collages from *The Mary Tyler Moore Show* and *Lou Grant*, and another cartoon, captioned, "My new boyfriend is wonderful. He treats me like an equal, but he also treats me like a woman. It's almost like dating Lou Grant."

The inner office was rather more formal, and starkly furnished. Awards papered the wall—a Tom Paine Award, a Woody Guthrie Humanitarian Award, a Paul Robeson Award, and evidence of Asner's seven Emmys and two Critics Circle Awards. There were a few books, a table, a couch—it was an office better suited to contemplation than organization.

Edward Asner sat on his couch, a cup of coffee in front of him, his knit sport shirt stretched into horizontal ridges by hard-muscled shoulders and chest. "I go to a therapist," he said. "And we discuss . . . difficult matters . . . and I fall asleep. He regarded this as a form of escape—of not wanting to confront particular problems circulating in my psyche and my neurons. Lately, though, he's come to regard it differently. As fatigue." Asner, like other actors, can work a couple of syllables for all their worth; he says "fuh-*teeeg*," making sure the *t* and *g* get just the right glottal occlusion, and that the drawn-out *eee* bespeaks weariness. He took a long pull on his inky coffee. "I haven't finished my reading for yesterday yet. I've not been getting to bed early enough. And I've been getting up too early." Over the past few days he had met with Senator Robert Dole, entertainment lobbyist Jack Valenti, and author Norman Cousins; he had appeared on the *CBS Morning News* and *Entertainment Tonight*. Within two days he would leave town to address the United Mine Workers, would address a stuntmen's safety committee meeting, and would attend the premiere of an expensive epic called *Krull*. Spliced into these commitments were lesser appointments, luncheons, paperwork, study, and a torrent of incoming phone calls.

This schedule came on the heels of a final thirty-six-hour negotiating session that formalized the guild's three-year contract. The talks, which had taken eight weeks, were Asner's trial by fire in the world of labor negotiation, but he was satisfied with his performance. There was no strike, as some studios had feared (Paramount had been careful to conclude its current filming before the end of the old contract), and the actors won a pact that would earn them an estimated $60 million more than they would have made under the old agreement.

Asner found that perhaps his greatest contribution to the proceedings was his reputation as a deep-dyed radical. "I was always in the background—the unknown quantity." The implicit message to management was: Look, our guy Asner is the loose cannon on your deck; you know he's a Commie sympathizer and a fire-breathing troublemaker, and everybody in town knows he won't be happy until he's walked over your faces into the White House—he's the flip side of Ronald Reagan—so, for Christ's sake, let's let sleeping dogs lie and wrap this thing up.

It was, however, only fitting that Asner should get some fear and intimidation value out of his rep, because for the better part of the last two years, *it* had been threatening to ruin *him*. Edward Asner had come close to becoming a walking anachronism over the last eighteen months. Not just a rad-lib in a conservative era, an actor between series, an idealist in a selfish time, but an honest-to-God unemployable, quasi-blacklisted actor, a financial disaster area, and a dethroned labor leader—all at once. And the whole nasty quicksand was still no more than one step away, waiting on either side of him. Ed Asner, once America's favorite father figure—the boss you dreamed of—was still a man with problems.

"For a brief while there, I felt . . . secure." Asner rolled the word off as if it were a magic incantation. "I felt . . . secure . . . enough to the point where I began to speak out on El Salvador. I had my series, a series that would probably continue for a goodly while, and it would not be easy to ignore me, either in hiring me or paying me. I had the name; I represented a kind of role. It looked like I would never have to worry about money again. And then—that all changed. It was very shocking, astounding—marveling—how swiftly it all happened."

What happened was that CBS, apparently alarmed at the controversy Asner was causing with his political views, dumped *Lou Grant,* which Asner partially owned and from which he drew an enormous salary. No one could prove blacklisting—but all of a sudden the months were going by with no offers. They called it a graylist. Concurrently, Asner was caught in an overextended position in the real estate devaluation of the early 1980s, and suddenly found himself with "what they euphemistically refer to as a money-flow problem." A group of conservative actors, led by Charlton Heston, tried to have him removed as president of SAG, and the press, which had previously championed him as its video archetype, began to squeeze him into the roles of bigmouthed actor, rich liberal, and pinko subversive.

The ferocity of the attacks was greater than Asner feared it would be. Having just won a landslide election as SAG's president, and having received virtually nothing but kind attention from the public, the press, the critics, and the industry for more than a decade, he thought he had some leeway to speak his mind.

Asner had become concerned about El Salvador when a Catholic nun who had been living there started to fill his ear with horror stories in 1980. He became "a fanatic" on the subject, according to *Lou Grant* co-star Robert Walden, to the degree that some of his friends thought he knew more about Central America than Secretary of State Alexander Haig did. He gave the nun the telephone numbers of some of his liberal friends, but succeeded only in irritating them. Jack Lemmon's publicist told Asner to cool it in the future with beseeching nuns. In order to get a little action, Asner began to support a fledgling group called Medical Aid to El Salvador, presenting it with a $25,000 check at a Washington, D.C., press conference. With him at the presentation were Papa Walton Ralph Waite and Dr. Johnny Fever—*WKRP*'s Howard Hesseman. At the press conference, Asner told a reporter that he wanted El Salvador to have a popularly elected, democratic government, whether it was communistic, socialistic, capitalistic, or whatever. And all hell broke loose.

There was a death threat. Jerry Falwell's Moral Majority issued a statement condemning Asner. The right-wing Minutemen organization mobilized its Committee of Ten Million in a "psychological warfare campaign" against Asner. Their bulletin *On Target*, which has the cross hairs of a gunsight in its logo, called for an anti-Asner write-in campaign to CBS, promising a boycott of *Lou Grant* advertisers. The somewhat more centrist Congress of Conservative Contributors also threatened a boycott. The *L.A. Herald-Examiner* editorialized that "the left-leaning, weak-kneed nellies such as Ed Asner are telling their lies, giving aid and comfort to Communists, trying to obscure the issue in El Salvador." Another article said that Asner's actions "bordered on treason."

Several *Lou Grant* advertisers started to get queasy. Kimberly-Clark, in particular, took a dim view of the proceedings: It had approximately $3.5 million worth of tissue and crepe-paper plants in El Salvador. Vidal Sassoon, Inc.—whose ad motto was "If you don't look good, we don't look good"—sent a letter to CBS Chairman William Paley imploring him to muzzle Asner, because Sassoon had gotten thirteen letters from consumers complaining about Asner's views. The Peter Paul Cadbury candy company had gotten even fewer complaints, but joined Kimberly-Clark in taking its ad dollars elsewhere. Other advertisers stood firm, however, and potential replacement advertisers were clamoring to get on the high-demographic show.

The high ratings of the fifth-year show, in the wake of the controversy, slipped only slightly, from a 32 to around a 27 share in the weeks after the Washington, D.C., press conference, still well ahead of the average show's ratings during that season.

But within three months, Lou Grant was canceled. Because of ratings, said CBS.

"During all the sound and fury back then," said Asner, who was still agitated about the cancellation, "during the talk about boycotts, and shooting me, and stringing me up from the nearest oak, I was contemplating a lawsuit. I talked to two actors who had been blacklisted in the fifties—they told me that while they were pursuing their suit, they did not work. They eventually dropped it. One of them summed up the whole situation beautifully. He said, 'When it comes to pushing the goods, forget the controversy.' I realized that was the dictum the network had followed."

It was not, however, an acceptable dictum for Ed Asner. He recanted none of his statements, softened none of his stances, even engaged Heston in debates that were dubbed "Star Wars." He beat back the attempt to throw him out of union office, then created further controversy by calling for mergers of SAG, the Screen Extras Guild, and the American Federation of Television and Radio Artists. He also pushed for SAG to openly endorse political candidates, an idea many union members opposed.

As the months after the cancellation dragged on, the effects of the graylist began to be felt. He said, in the summer of 1982, "Whether I could interest a network in another TV series at this point is doubtful." But he said he did believe he could get a role in a TV movie, a very humble assessment for a man who had been one of the biggest stars in series and miniseries television for over a decade. Eventually, he did get a TV movie role, and a role in a movie for cable TV—a type of work actors were still shunning because it did not pay particularly well. He also got a relatively low-paying role in the film Daniel. But business was not what it used to be.

Asner, on this late-July day, had just announced his bid for reelection to the SAG presidency. He would be opposed by a young black actor named J. D. Hall. There was speculation that a conservative candidate could win if Asner and Hall split the liberal vote, or that Asner might even lose a two-way race because of his spiderweb of controversies.

"Life for me," said Asner, "was always a slow, constant improvement. I came from New York to California on Memorial Day of 1961, and in the seven remaining months of that year, I made more money than I'd made in any year in New York.

"Everything is relative; each year kept getting better and better. That was at age thirty. Then, two years before *Mary Tyler Moore*, all of a sudden, my career went splat! I started going downhill, changing jobs and changing money. The reason for it? It just happened. I was losing things—possessions, jobs, money, whatever—and I did *not* know what would ever make it *right* again. That was a terribly nervous, terribly unhappy time.

"I never in a million years, ever, had thought I would be a big star. So not getting that was not what was bad. What was bad," he said, going again for his coffee cup, finding it empty, frowning, "was going backwards. That's what's bad—going *backwards*."

A message came in that Asner's manager did not want him to schedule an appearance on the afternoon *Queen for a Day*–type show *Fantasy*. Asner had helped a young woman cut through thickets of international red tape in order to be reunited with her adoptive daughter, a Taiwanese baby, and *Fantasy* had helped him do it. Now the show wanted to televise the reunion. But Asner's manager thought the show was shlocky, and that association with it would undermine Asner's attempts to reestablish himself as a star. Since the child was Chinese, it could even be misconstrued in certain hinterland homes that Asner was giving aid and comfort to some sort of Red Commie baby.

"When do they tape?" Asner asked one of his assistants.

"Friday. Ten forty-five," she said.

"Schedule it."

The mail came in, and Asner's assistant flipped through *Weekly Variety* until she came to the *Lou Grant* ad. MTM, trying to syndicate the show, had taken out a two-page spread, one half of which featured Asner's face, twice as big as life, beaming in the kindliest, most Norman Rockwell of smiles. The ad's headline, in bold red letters, trumpeted SOLD ON STRENGTH! The subheads said STRONG IN MASS APPEAL. STRONG IN YOUNG WOMEN. STRONG IN RERUNS. The ad admonished, "When every rating point means dollars, don't gamble. Make a sure investment in strength."

Asner, a profit participant in the show, stood to become a very well heeled capitalist from the syndication sale. The series had done 114 hours of programming, and now it was time, after all the years of work, to make a profit on it. There was only one problem—though *Lou Grant* was finally for sale, hardly anyone was buying.

Harry Anderson talked about his goals in life as he drove to an audition for *The $25,000 Pyramid*. If Harry impressed the show's producer, he would be scheduled to appear later in the summer.

"I want to get an education. I want to be respectable. I want to go to school and learn how to spot balls at a pool hall. I want to learn how to use a lathe so I'll have something to fall back on. Mostly, I want to get my deposit back from the phone company. What do you think Ma Bell nicks you on a deposit when you put down 'magician' as your place of employment? She wants seven hundred bucks, is what she wants. The day I get my phone deposit back—that's when I'll know I've made it. When that day comes, I'll enroll in junior college."

Actually, Harry wanted to read about the old magicians and about the history of magic, and spend more time with the guy he called "my guru," an eighty-nine-year-old magician named Dai Vernon. "He's the only one left of his generation," said Harry, pumping the gas pedal to stutter Leslie's beaten Volvo through a yellow light. "I don't think there are six magicians in the world who wouldn't consider him the best. He's got a column in *Genii*, the magicians' magazine, called *The Vernon Touch*, and the way he writes is, like, 'In the lovely mahogany bar in the Puff Room of the Lysol Hotel in Chicago, where I was giving card lessons to George Burns, Charlie Lindbergh, and Mary Todd Lincoln . . .' He's been around. He's done it all. He helped build the Brooklyn Bridge, and fell off it and broke just about every bone in his body, so now he walks like this"—Harry rocked back and forth over the steering wheel. "He knew Titanic Thompson, the greatest hustler of all time, who got off the *Titanic* by dressing as a woman. And he worked with Charlie Gondor and the Yellow Kid, the guys they wrote *The Sting* about; Vernon was the one who knew how to handle cards and dice. *The Sting* was taken almost verbatim from a nonfiction book about those guys called *The Big Con.* Vernon taught me the most basic stuff about magic, stuff most guys ignore. Such as don't run when nobody's chasing you. If you're doing a trick with a fake *Reader's Digest,* don't say, 'This is an ordinary *Reader's Digest.*' Nobody imagined it was anything else until you mentioned it. Shut the fuck up. Say what you need to and go home."

Harry wanted to get deeper into the lore of his craft, but first, in order to afford the sabbatical: "I want to make a whole lot of money with the least amount of work." Which meant, of course, *Night Court.* But *Night Court* was a fading dream. "They're hinting that it may go on in January, but I don't know if it'll ever get on. It's never been a real sure thing—when it's going to air, or how many we're going to do. So I don't want to arrest momentum in other areas, like *Saturday Night Live* or *Cheers,* because if *Night Court* gets a lousy slot, or if they just sit on it, then I'll have just been wasting time."

The $25,000 Pyramid, headquartered in a studio in downtown Hollywood, was one of only three game shows still on the air that actors could guest on to build their Qs. The networks now mostly thought game shows were soft programming. The genre's only consistent hit was *The Price Is Right,* which had long ago realized that housewives were proudest of their ability to shop frugally. Daytime legend had it that the average homemaker knew the exact prices of one hundred consumer items.

Harry auditioned his prowess in the cluttered office of an assistant producer, a young guy in a movie logo T-shirt. The game consisted of the celebrity's prompting the "civilian" to say a word by supplying synonyms or descriptive phrases. Harry was sailing along until the word he had to elicit was "mushroom."

"Dark," he said. "Musty. Grow."

"Basement?" groped the producer.

"Omelet," said Harry.

"Eggs?" Harry shook his head.

"Beefsteak."

"Breakfast?" The producer was holding a stopwatch, and time was ticking away. The producer was growing restive. But Harry *wanted* this exposure.

"Spongy. Tasteless."

The producer drew a complete blank.

Harry drummed his fingers on his tennis shoes and blurted, "Psilocybin. Magic! Trippy!"

"*Mushrooms*," said the producer, with complete confidence.

"My man!" said Harry. "Jeez, though, we couldn't say that on the air, could we?"

The producer shrugged. "We skew young," he said.

Later on, Harry drove down to Manhattan Beach, a young-money, avocado-and-cocaine surfside village peopled with the upscale swingers and swingettes who were always a good draw for Harry. Here he would play the Comedy and Magic Club, billed as "Harry Anderson: A Man Without Whom Life Would Have No Meaning." The subbilling went to "Turk Pipkin: A Man."

But Turk Pipkin, modest billing or not, got the biggest hand of the night at the very beginning of his opening act. He ceremoniously placed an Alka-Seltzer tab in a water-filled Coke bottle; then, signaling for a taped rendition of *Also Sprach Zarathustra*, he tore open a sealed Trojan condom. Topping bottle with device, he stood at attention as, majestically, the condom inflated to its fully erect position just as the *Space Odyssey* music hit its climactic chord.

None of Harry's stuff topped Pipkin's rubber.

Gary Nardino, who'd sold only two network shows this season, had asked Paramount to let him out of his contract. But the studio was balking. Paramount's TV business had increased 300 percent under Nardino, and it was in no hurry to let him go.

Nardino had been getting feelers, and he'd been tempted by them. Embassy Productions, Norman Lear's company, had offered him its presidency, and so had Tri-Star, the new corporation started by CBS, HBO, and Columbia. The Embassy offer looked especially good to Nardino, but Paramount held him to his contract.

Paramount then made him a counteroffer. Michael Eisner, Paramount's CEO, offered to let Nardino enter into a long-range deal with Paramount to do independent production of movies and television. For Nardino to take it would complete the classic career path of so many Hollywood moguls—from agent to agency head to studio division head to independent producer. If Nardino bit, his first job would be to serve as executive producer of Paramount's *Star Trek III*. He also could develop a project for *M*A*S*H*'s Loretta Swit.

Nardino went for it. He was out of the series rat race.

As soon as he made his decision, he felt a cloud of fatigue lift off him.

Ted Danson had the week off, because *Cheers* was in its short midsummer hiatus, but he and some of the other *Cheers* cast members were in New York at NBC's request. They were "surprise guest hosts" at an annual flatter-the-advertisers party, a free-booze-and-finger-food affair designed to show the advertisers and the agencies that a buck spent on NBC also bought at least a dime's worth of social cachet.

At the party, Danson recognized a few slightly familiar faces from the major New York ad agencies, such as the Ted Bates agency and Grey Advertising. They were the people who, several years before, had turned him away when he came in every few days to apply for a spot in a commercial or print ad. A frequent excuse was that he wasn't a regular size.

What Danson mostly beheld, though, were women—clusters of women, armies of women, whole galaxies of women, orbiting around him as if he were the natural gravitational center of the universe. They all looked their party best, with thin dress straps over bare shoulders, red lips and shadowed eyes, fancy shoes, and bright, inviting smiles. There they were—all arrayed in adoring concentric circles around him, as if he were Louis the Sun King and had only to crook his finger to summon a festival of earthly delights. Which was probably not far from the stark reality of it. One woman especially, a pretty strawberry-blond lady in a black velour dress, seemed in particular to regard him as too scrumptious for words.

Sexual adoration was new to Ted Danson. He hadn't even

been around girls much until he'd gone to college, having gone to a boys' prep school, Kent, in Connecticut. When he first arrived there, he was a 120-pound six-footer, given to wearing multiple layers of pullover sweaters as fake brawn. He was a basketball player rather than a womanizer (or girlizer)—actually Kent's MVP one year. Even after he put on a few pounds of dorm-food weight, though, he still couldn't cut it on the tough playgrounds of New York—once he joined a game of street ball, and the guy who was guarding him grabbed him by the collar, lifted him up, put him down on the sidelines, and went right on playing, without even calling a time-out.

By the time he arrived at Carnegie-Mellon University in Pittsburgh, he had begun to gain some confidence with women, but still felt more skinny than irresistible. At Carnegie-Mellon, one of the nation's top acting schools, which graduated several *Hill Street* cast members and *Hill Street* producer Steven Bochco, Danson found himself veering away from sexual, leading-man roles, and gravitating toward character parts, especially sensitive-young-man roles. When he started acting professionally in New York, he played male ingenues in comedies and light dramas. The experience of breaking in as an actor—the repeated rejections and the public recitations of his shortcomings—toughened Danson up, but also left him vulnerable to humiliation.

For example, there was the time he finally landed a part in a play destined for Broadway, a 1973 comedy called *Status Quo Vadis* that got a great out-of-town reaction. Danson's parents flew in for the premiere from Flagstaff, Arizona, where he'd grown up, and he took them to Sardi's before the play. Big Faces came up to their table and said, "I hear your play is great." After the opening, he brought his folks back to Sardi's for a party. Danson was standing beside the bar, his drink on the counter, when Rex Reed stood up to read a review by Clive Barnes. Barnes murdered the play. Danson heard a thump, turned around, and saw that a metal grill poised above the bar had been slammed down. The party was over. He couldn't even get his drink.

The next day, his parents dropped him off for the matinee. He walked up to the stage door and the doorman, who'd been palsy-walsy all week, said, "Where do you think you're going?"

"I work here," Danson said.

"Not anymore, you don't." The doorman pointed to a closing notice.

Danson had to run up the sidewalk after his mom and dad.

And now he was back in the same Forty-fifth and Broadway neighborhood, not just employed, but a star—what was more, a sex symbol—surrounded by a Carlsbad Caverns bat-flight of women.

Danson edged away from them and onto a down escalator, but the woman in the black velour dress with the strawberry-blond hair gave pursuit, and was just above him, trying to make eye contact, as he descended. Then her gaze landed on the silver-dollar-sized pink patch of scalp at the crown of his head.

"Oh," she sighed, crestfallen. "Oh!"

Danson looked up at her quizzically.

"You're losing your hair," she accused. "You've burst my bubble." He saw that she was dead serious. She pivoted and walked off.

Danson laughed, but for the rest of the day he found his hand continually creeping back to cover the top of his head.

The next week, as July drew to a close, Danson started to get fidgety about the Emmys. Within a few days, he, Shelley Long, and two stars each from ABC and CBS would announce the nominees at the Pasadena Civic Auditorium, where the winners would be revealed in September. Danson was starving for a nomination. He was sure that Shelley Long would get one, and he didn't want to be skunked again, as at the Golden Globes.

For the last few weeks, actors, producers, and studios had been taking out full-page and two-page ads in the *Hollywood Reporter* and *Variety*, pushing themselves and their shows as Emmy candidates. Winners would be selected by small panels, but nominees were chosen by a mass vote of peers. The ads for shows generally blazed with headline-typeface praise from the critics, but the actors' ads were just portraits with the single caption, "For Your Consideration"—becoming modesty so long as you failed to note that the actors themselves had arranged and paid for the ads. Some were spending huge sums. Marion Ross, the mom on *Happy Days*, was running a real no-guts no-glory campaign, presuming, perhaps, that for her it was now or never. Almost equally fervent were the producers of *M*A*S*H* and

Blood Feud, and also—in an almost touching display of faith in the powers of shameless promotion—the producers of *Solid Gold* and *Silver Spoons.*

Danson's ad had run that morning in *Variety.* It was a full page of him in a rugged wool-and-corduroy vest, his work shirt open just enough to allow a few chest hairs to see the light of day—and they were white, mature, leading-man chest hairs.

"I had a lot of qualms about the ad," said Danson, leaning against a wall at Stage 25, waiting for Burrows to end their lunch hour. "I thought it kind of smacked of, you know, all those people who don't stand a chance. But hell, it's good for me—my normal pattern would be to go, 'Well, let them come to me.' It's good to just say, 'Hey, I want to be a candidate.' "

It would be tough, Danson admitted, to stand there in front of the cameras while the nominations were read—particularly if he was reading his category himself. "I've got an idea," he said, brightening. "If my name's not on the list, I'll just read it off anyway. Fuck 'em. They'll be too embarrassed for me to say I lied.

"The worst part of these ceremonies," he said, "is playing the good loser. If nobody was watching me, I know exactly what I'd say: 'Fuck! this!' "

Danson figured his chances were fifty-fifty. There were only five nominations, and you had to assume that two of them would be sentimentally handed to comic aristocrats Carroll O'Connor and Alan Alda, up for the last time, and that Bob Newhart was probably good for the third. Fighting it out for the other two spots would be *Taxi's* Judd Hirsch, *Benson's* Robert Guillaume, *Buffalo Bill's* Dabney Coleman, *Three's Company's* John Ritter, and Danson.

Tony Colvin came in and posted last week's ratings on a bulletin board near the coffee urn, and Danson walked over to look.

"All right!" he said. "Twelfth-rated show! For the second straight week. If we keep that up and pull some Emmys, I might just stay employed."

"Tony!" It was Glenn Miller, *Cheers's* other PA, the aspiring set designer. "Go see Scott."

"What's Scott want?"

Miller just shrugged.

Scott, holding a telephone to his ear, waved Tony into his closet-sized office. He started to say something to Tony, but then his party came on the line. It was his wife.

"What's the best thing that could have happened?" Scott said into the phone.

Tony jerked erect. "Story editors?" he said.

Scott jerked his head up and down. "Sit down," he said into the phone. "Story editors!"

"Story editors!" Tony screamed, and Miller stuck his head in the door. "Story editors!" Tony said to him.

"Holy shit," said Miller. "Way to go!"

In just the last few days, Tony and Scott had submitted a script on spec for the new ABC comedy *Just Our Luck*, produced by Lorimar and MGM. The show was about a black genie. They were hoping to land a script assignment or, if they had unworldly luck, regular jobs as staff writers. But their agent had just called Scott and told him they'd gotten a job as story editors, a solid notch above staff writer.

Scott hung up and he and Tony bounced around the room, shaking hands, laughing like two girls who had just made cheerleader.

"What's the rate?" asked Miller.

"I don't care," said Tony. He ducked into his miniature office to call his wife, but couldn't get her. Scott dialed his brother, who wasn't there, and his mother, whose line was busy.

"Let's sing our song for the last time," said Tony to the small crowd of secretaries and PAs that now clotted the hallway. "We don't need no Xerox copies," he sang to the tune of Pink Floyd's "The Wall," "we don't need to get no lunch. Hey, Burrows, leave those PAs a-lo-hone!"

Miller, who had been grinning, suddenly went dead in the face. "Poor me," he said. Then he looked better. "Hey! When you guys gonna bring me over?"

Richard Villarino, Tony's supervisor, watched Tony and Scott shake hands with all the secretaries. In one phone call, his gofers had swept past him in Hollywood status, if not necessarily salary. "You're not making them kiss the ring yet?" he said. "Give 'em ten more minutes." Then he stepped back into his office.

"Go buy a bottle!" Miller shouted.

"Let them buy it," Tony said, nodding at Villarino's office.

"Yeah!" agreed Scott.

"Good luck," said Miller, and soon went back about his chores. It had seemed to him that Tony was perilously close to making a speech, thanking all the little people who'd helped to make it all possible—and Glenn Miller did not want to hear it.

Four hundred thousand dollars from Showtime. Four hundred thousand American dollars. Wasn't that what Currie said he planned to sting the cable company for rights to the first two-hour episode of *Pillars of Portland*? That would mean, Larry Colton calculated, $80,000 for himself, providing the deal went through. Eighty Gs. That was more money than he'd ever made pitching for the Phillies. Inordinately more than he'd ever gotten freelancing for *Sports Illustrated*. And one hell of a lot more money than he'd squeezed out of this project so far. While Allan Katz would doubtless score $20,000 a week for producing the new Charles Durning show if ABC okayed it, and while top-money guys like Jim Burrows were knocking down thirty and forty grand a week producing other network semihits, Colton—whose chores and responsibilities were exactly the same—was lucky to get heiress-producer Evelyn Hamilton to pony up $200 a week to keep him in typing paper and rent money.

Not even that, actually. His landlord was threatening to evict him if he didn't turn up with $700 by the end of the week, and Colton was strapped. Who was he going to turn to, anyway? It was kind of weird to have to hit up your parents for rent when you were forty-one years old. Even the rental car he was driving had to be subsidized by one of his ex-wives. He banged the steering wheel with the top of his hand. Well, that could change overnight if everything went halfway right the next few days. Eighty thousand dollars. A score like that would mean nice Christmas presents for his daughters—like maybe a couple of college educations.

Colton was on his way to *Pillars'* biggest location shoot thus far. The TV station wanted local color shoehorned into Colton's prime-time soap opera, so Colton was going to give it the most colorful local setting in the state—a small town taken over by a religious cult. Thousands of followers of the Bhagwan Shree

Rajneesh were encamped in a little town east of Portland, and Colton was going to shoot several scenes there, using the town as a backdrop.

He was enjoying his drive. Out of Portland, the highway had risen to the green fir-enshrouded base of Mount Hood, and then eased back down into the desert and prairie lands of central Oregon. Not many cars and a huge blue sky. Colton's car whipped by a herd of cattle, then vast yellow plains of wheat and an abandoned station wagon, red with rust, and the slumped carcass of an uninhabited farm building.

Finally, Antelope, Oregon, population ninety-five. It was a minor collection of white houses that appeared slightly shrunken and distorted by time, half lost in the shade of broadleaved trees. There were big new pieces of construction equipment here and there, however, and people dressed all in orange, purple, and red. "Maroonies."

Colton eased through the little town and then headed out toward Rancho Rajneesh, the main settlement, eighteen miles away over gravel roads. Yellow hills rose up on either side of the car. They were covered with juniper and sagebrush. The land became radical, rocky. High up on a mesa, a man in red stood with a walkie-talkie.

In the distance a car approached, feathering up dust. A Rolls-Royce, with a silver Lincoln Continental behind. "Here comes the main guy, the Bhagwan Shree Rajneesh. Or just plain God to his friends," Colton explained, scratching at his beard. "Every day he drives forty miles into Madras, turns around, and drives back. Never stops or anything." The Bhagwan passed, behind the wheel of the Rolls, a pretty young woman at his side. He had gray hair to his shoulders, a round brown face, large hooded eyes, and a beard. A guru's guru. His Holiness owned thirty Rolls-Royces, all for his personal use, and a quarter-million acres of land—Rancho Rajneesh had twice the land area of San Francisco. He'd dictated three thousand books and had three hundred and fifty thousand devoted followers worldwide.

His disciples considered the Bhagwan like Jesus, just a lot more middle class. His ministry had been acknowledged everywhere from *60 Minutes* to *Hustler*. The *60 Minutes* investigator, Ed Bradley, had shown a "documentary tape" featuring several of the Bhagwan's devotees beating the spleens out of each other in the name of "feeling therapy," and had then suggested that what the ranch was all about was sex, drugs, rock 'n' roll, and anti-Semitism. But Colton had written a balanced story about the guru and his crew in the Sunday *Oregonian*, and had gained permission to film on the ranch.

The question was, what to shoot? The story would not stop mutating, and Colton had found himself writing around the clock, turning out as many as five scenes a day. The plot kept metastasizing until Colton, who had cranked out a manuscript the size of *War and Peace*, had a showdown with director Tom Chamberlin at Colton's house. Chamberlin had come over to order up more scenes.

According to Colton, their set-to went like this: "This has got to stop," Colton said. "This whole thing should have been outlined from the beginning. We can't just do it willy-nilly and hope the thing falls together in the editing room. We gotta have structure; we gotta have plot. A script."

"Hey," Chamberlin said, "I couldn't care less about your plot ideas. I'm gonna shoot in two days. If you can't get the job done, then I'll take care of it myself."

"Look, asshole," Colton replied, "we've got a contract. The contract says I write the script."

"That contract means nothing to me. Your problem is, every time your rent comes due, you have a huge anxiety attack."

"Well," Colton said, "here's something for you to get anxious about. I called KOIN, and Steve Currie says unless I put together a conventional script and we quit all this anarchist Wonderland shit, you don't get any cameras to shoot with."

For a moment, Colton thought Chamberlin was going to literally explode. But there wasn't much he could do. So Colton spent the next several days outlining *Pillars* on big, sutured sheets of butcher paper that ended up wall-to-walling his living room and floor.

Meanwhile, though, the show had been gathering great public momentum. First a major Portland newsweekly had run a page-one story about *Pillars*, then the *Oregonian* had run a cover story in its Sunday magazine titled "Is Pillars Too Big for Portland?" Suddenly Colton was the belle of the ball at Rotary and

Kiwanis club luncheons, his tales of major-league baseball and incipient showbiz success a popular double bill in a town that was basically bored to death with itself. At the luncheons, Colton was almost certain that he alone was there for the food.

Just past the entrance to the ranch was a convoy of huge dump trucks, then the Bhagwan's giant man-made lake, then his airport, where the airliners of Air Rajneesh sat parked around their hangars. Dominating them was a hugh turboprop—once Howard Hughes's personal plane. From out of the desert appeared row after row of houses and tents. People dressed in red were everywhere. The ranch was the biggest and strangest commune in America and Colton, whose cast, crew, and camera trucks were due first thing in the morning, was damn well determined to make the most of it. If he played his cards right, the eighty grand would soon be piling up money-market rates. If he blew it, chance number two might be a long time in the making. Before letting him on the ranch itself—where he was soon to be greeted by the beautiful, nubile "Twinkies"—Colton was stopped. His car and body were searched for guns and bombs.

"We got an hour to do the crucial love scene, huh?" said Paul Nixon, star of *Pillars of Portland*. "Then don't worry. I can do sixty takes. My first wife used to call me the Minuteman."

Larry Colton, his breath flagging steam in the early morning air, said, "Was that before or after she threw you out?"

"Hey, look," Nixon said. "She was gorgeous. I figured her for a dream. We married in Nevada. I knew she drank, but after four days in Reno she still wanted to party. The crazy bitch was doing a quart of vodka a day."

Paul Nixon, who had recently been described as Oregon's Paul Newman in a local newspaper, stepped down out of one of the forty-foot *Pillars of Portland* Winnebagos and surveyed Rancho Rajneesh. He had just arrived with the rest of the cast and crew, after an all-night drive from Portland.

The arrival of the film crew this morning, though, was the least of local current events. Life here at the ranch had suddenly been complicated when the Bhagwan had announced that the cities of San Francisco, Los Angeles, New York, Bombay, and Tokyo were about to be obliterated and that it might be a good idea for people, his disciples particularly, to evacuate large metropolitan areas altogether. All morning the Rajneesh press relations trailer had been besieged by phone calls from newspapers, wire services, and television news departments, wanting to confirm rumors that upward of one hundred thousand of Bhagwan's followers had decamped the burgs of America and were, at this moment, converging on Rancho Rajneesh.

Handed the responsibilities of dealing with the holocaust hot line, as well as with the film crew, were the Twinkies, the Bhagwan's cadre of young sensuals. Dressed in the height of postapocalypse fashion—purple leg-warmers, pumpkin-colored blouses, burnt-sienna sweaters—the Twinkies were the commune's official hostesses. They were all very attractive and had adopted a sort of geisha–Dallas Cowboys Cheerleaders style that allowed them to juggle the end of the world and Colton's twenty-five-member cast and crew with equal aplomb.

Nixon sipped coffee. "When I was in Vietnam," he said to the horizon, "I was a door gunner on a helicopter. In the Marine Corps. One of those guys with a life expectancy of about thirteen seconds." Sound men and cameramen were moving camera equipment between the two big Winnebagos. "Never got a scratch. Three years in the marines, machine-gunning VC, and not so much as a Band-Aid. Then I come to Portland, enroll in this karate class, and they put me up against this ninety-five-pound, nine-year-old homosexual. On the level. He was already a black belt, the instructor's favorite. The guy that owned the place got his kicks watching this little fruitcake beat me half to death. They'd pair us in no-contact, and the kid would kick me right in the nuggies. I'd keel over, my knees hugging my chest, and say, 'Hey, little pal, what's coming down?' And he'd say, 'Your ath, marine! Fight me! Fight me on the street!' "

Nixon looked over the several pages of script containing his lines for the day. In *Pillars* he played a successful lumber broker faced with recession in the timber market, who had come to Rancho Rajneesh to offer the Maroonies a really great deal in either cocaine or plywood—take your pick. As Nixon read, time was running out. While *Pillars'* budget would barely pay the bar tab for a network prime-time production, there were still major costs involved; ultimately, they would come to about one thousand dollars for each minute finally aired.

Evelyn Hamilton, the heiress who had paid the show's start-up costs, stepped slowly down from one of the Winnebagos. She was thin, tall, and pale, and had the stern spacy look most recently in vogue during the Pilgrim era in Salem, Massachusetts. Like heiresses everywhere, Ms. Hamilton tended to exist above the fray, and since TV production is entirely fray, she seemed a little put out that there was not a great deal for her to do.

When it was time to shoot, Nixon was standing out alongside the road with about three hundred Maroonies watching the Bhagwan drive by in his Rolls, and the Bhagwan's eyes fixed on Nixon's own Paul Newman–like baby-blues and Nixon was transfixed. There was a camera pointed in his face and Tom Chamberlin was behind him pounding on his back, hissing, "Act! Act! Act!" But Nixon had been struck dumb, faced with the sight of . . . God?

Then they shot a nude scene with a whole busload of Maroonies frolicking around one of their homemade lakes—a sop to cable, which was ever eager to one-up the networks breast-wise. A sound man, down by the river's edge, walked in the early August desert sun, unaware that the boulders he was sitting on were tract housing for rattlesnakes. One of the serpents was coiled atop a flat stone, its tail a blur, when someone shouted, "Snake!" and the thing thrashed sideways off the rock and vanished twisting back into the debris.

It seemed as if they were getting the footage they'd come for, but Colton was growing more and more uncomfortable with the way the story was evolving. "The only thing that holds the thing together," he said, "is that all of these people are supposed to be in group therapy together. That's all we've got. Chamberlin seems to think that's all we need, that we can spin their various adventures around that alone. I'm not sure."

That night, prior to shooting a love scene in one of the rooms at the Hotel Rajneesh, Tom Chamberlin came into the suite and shouted, "Everybody out of here! I can't take this anymore!" And Colton, standing out front of the place, on a road under ten thousand stars, murmured, "The chickens are gonna come home to roost on this. This is just plain fucking crazy."

11
Struggle and Juggle

August 1983

It was as if Jesus Christ himself had chanted, "Lazarus, come forth." NBC had risen mysteriously from the A. C. Nielsen grave by August and climbed over ABC into the number two position. Shows that had been mush all year, like *Remington Steele*, *Cheers*, *Family Ties*, *Knight Rider*, and even *Facts of Life*, a half hour whose hidden premise seemed to be "Wanna do a chubby teenager?," were shooting regularly into the Top Twenty, where only *The A-Team* and *Hill Street* had trodden during the regular season. And NBC had actually won the first week of the July sweeps. The midsummer sweeps, riddled with reruns, are the least important of the quarterly contests, but still, *any* sweeps win sends tingles of joy up and down the ranks of the affiliates—and, it is hoped, keeps them from thinking about preempting network fare.

Maybe it was the thin air at these altitudes, but Brandon Tartikoff was beginning to feel . . . confident. Maybe Ann Jillian really was turning herself into teen America's new wet dream over at *Jennifer Slept Here*. Maybe Hamner, Larson, and Farr really *were* about to co-opt the whole market for warmth, violence, and sex. Best of all, maybe ABC and CBS really were as weak as he'd been claiming they were.

Of the new shows introduced by all three networks over the summer, only NBC's *Buffalo Bill* had been picked up for additional episodes. ABC had experimented with no less than five new summer shows. All dogs, and some outright embarrassments. *Eye on Hollywood*, for example, a straight recycling of a locally produced L.A. video magazine, gave five-part, presidential-type coverage to the likes of Pia Zadora.

Of course, NBC's ratings success was due partly to the fact that its shows were up against the lesser competition of reruns, including those of the serial shows, like *Dallas*, which always did badly the second time around because they were so dependent upon twists in their running plots. Still, the summer success was a very good omen, particularly for character comedies like *Cheers* and *Family Ties*, expected to grow slowly but now; it seemed, shooting up fast.

Even CBS's Bud Grant had been forced to admit that "NBC might be making some progress."

But even with all this—his first real taste of corporate triumph—Tartikoff was just a mite edgy. Nervous, you might say. In fact, he had decided to jump the premiere of *We Got It Made* to September 8, two weeks in advance of the "official" new-season opening date of September 26. Sohmer's promo—featuring a slow pan up Teri Copley's legs, from red spike heels, over vistas of black mesh stocking, to eventual near-crotch—was already running daily. The implied message in the voiceover was grab-'em-by-the-short-hairs simple: "Hi! I just got hired as a live-in maid by two guys. Do I service them both? Maybe even at the same time? To find out—Be There! September eighth!"

Tartikoff sensed big guns mustering just over the horizon. If his intelligence sources were correct, a third of all network shows would premiere early, jockeying for position. He decided on a preemptive strike for his all-new, all-white Friday night, so that the unproven *Mr. Smith*, *Jennifer*, *Manimal*, and *Love and Honor* could get in a few licks before having to face fresh episodes of the fearsome *Dukes of Hazzard*, *Dallas*, and *Falcon Crest*.

Or, as Tartikoff put it, "before they run into a brick wall."

* * *

The numerous transients and bag ladies of downtown Pasadena took no notice of the long cortege of limousines gliding through as quietly as the eastbound smog out of the San Fernando Valley. Maybe it was a Mafia funeral. So what?

But awaiting the limousines' passengers over by the dark hulk of the civic auditorium was a different kind of transient: the press, a hovering mass of publicists, and a clutch of representatives from the Academy of Television Arts and Sciences—the Emmy people.

All at once, strobes, microphones, Porta-Paks, glamour.

In a room upstairs, out of deference to press deadlines, the organizers wasted no time in getting to the good stuff. Ted Danson took the stage and announced the nominees for best actress in a comedy. When he got to Shelley Long's name, he smiled.

Somebody else read off best comedy series. *Cheers*, of course, was nominated, and so were NBC's *Buffalo Bill* and *Taxi*. For best drama series, NBC's *Fame, Hill Street*, and *St. Elsewhere* made it. ABC had been totally shut out for best drama and comedy.

Ted Danson waited in a chair off to the side, TV cameras studying his profile, as Shelley Long got up to read the best-comedy-actor category. Danson wanted to be cool about the whole thing, but his hands felt wet and his stomach was jitterbugging. Long read the names slowly and clearly; she'd gotten her start as an interview show host in Chicago, and had the Important Announcement tone down pat. "Alan Alda, for *M*A*S*H*," she said. "Dabney Coleman, for *Buffalo Bill*."

Alphabetical order. It had better be now, thought Danson, or it wouldn't happen. "Ted Danson, for *Cheers*"—she grinned at him. "Robert Guillaume, for *Benson*. And Judd Hirsch, for *Taxi*."

Almost immediately the entertainment reporters descended on Danson. "I've been preparing so hard to lose," Danson said to a crowd of them, "that all I can think to say is 'I was robbed.' "

Off to one side, *Entertainment Tonight* reporter Mary Hart was bug-eyed. "NBC got *what*?" she demanded of her assistant.

"One hundred thirty-three nominations."

"That's the story," said Hart.

ABC had received sixty-three nominations, CBS seventy-three; NBC had beaten each of them nearly two to one. An ABC publicist hurried out of the room to call Tartikoff's office. *Cheers* had gotten thirteen nominations, second only to *Hill Street's* seventeen. Jim Burrows had been nominated for directing, Glen and Les Charles for writing, and Rhea Perlman and Nick Colasanto for supporting roles. *Hill Street* had taken all five of the drama-writing nominations, and NBC comedies had won all five of the comedy-writing nominations. Tartikoff would later call this drubbing of his opponents "an endorsement of the quality effort we have tried to achieve this past year."

"The important thing," Danson was saying to a reporter, "is that this supports NBC. It makes them go, 'Hmmm, I guess we were right to stick with shows that weren't getting the ratings right off the bat.' " Danson was handed off like a football from one reporter to the next for "exclusive interviews." A disproportionate number of them seemed to be female—even in this professional situation, he was still the boy all the girls wanted to dance with.

Half an hour later, standing beside the elongated black Cadillac that would carry him, Shelley Long, and two publicists back to Burbank, Danson said, "My *God*, I'm glad that's over. I was nervous. Now I can enjoy the ride back. I won't even have to stop the driver every couple of miles because I have to throw up." He pantomimed losing his lunch down the shiny sleek side of the car.

Ted Danson married shortly after he got out of college. But that marriage, stressed by his attempts to break into professional acting, ended in divorce after five years. Then he met Casey Coates, a New York interior designer, and quickly fell in love. They met in *est*, the psychological support program, and he was struck by her strength and independence. She was almost ten years older and he enjoyed being with someone who was mature and experienced. Also, she was very pretty—she had shiny, cocoa-colored hair, smooth skin, and a model's face. Before long, he was married a second time.

They both wanted to have a baby, but were somewhat concerned about her being just past prime childbearing age. But the pregnancy went smoothly, until a month before the due date.

Suddenly Casey developed slightly high blood pressure. They considered having a cesarean, but Casey decided it wasn't necessary.

The labor was difficult. Casey had to be sedated near the end of it. Then, as delivery neared, her left arm developed spasms. Whenever she pushed, the arm contracted. She was unable to help herself onto the delivery table. The doctor told her to stop pushing and used forceps to deliver the child. Casey discovered she could not move. She'd had a stroke. It was Christmas Eve 1980.

Casey Danson lay in bed, still. While she was in intensive care, the doctor told Danson that it was quite possible that she would die. If not, she would probably be paralyzed and might not be able to speak. Even if not completely paralyzed, she would probably not be able to walk or use her left arm. Refusing to believe the prognosis, Danson began to take care of his new daughter, Kate, and to prepare for a play, *The Women's Room*. Even when she hadn't moved for two weeks, he would not accept the idea that he might lose his wife.

On January 7, Casey moved one of her toes. Danson heard about it over the phone, and he broke down completely. Since the stroke, he had been sleeping in her room at the hospital. Over the next few weeks, her speech came back. Still, she had no movement on her left side—she wasn't even aware that she *had* a left side. If Danson was on her left and said something, she would look to her right. She was also extremely depressed. It felt as if a huge weight were lying on top of her, or as if she were in an elephant's body but had only the strength of a bird. It was hard for her not to be able to hold her new baby. She had to concentrate for hours just to coax a few feeble movements out of her hand.

After two and a half months at the rehabilitation center of Cedars-Sinai Medical Center in Los Angeles, though, she was able to leave, using a walker.

Back home, she worked out with private therapists for a year and a half. She had to fight the constant inclination of her body to tuck back into fetal position. Getting frightened or upset triggered the tendency of her body to curl up. But she strengthened the muscles that had not been affected, and actually trained undamaged portions of her brain to perform new tasks. She began taking food supplements, swimming frequently, and working out with a system of pulleys and straps. In the end she was left with only minor impairment in her left foot and shoulder. Unless she tried to run, you could hardly tell she'd had a stroke.

But the pall that the stroke had spread over their lives would not go away. As Casey began to feel better, they began to fight. Usually over a minor matter—a forgotten errand, a difference of opinion—but inexorably the arguments led to the core source of tension: Why did you let it happen? Why didn't you get the doctor to intervene sooner? she would say. Why did you insist on a noncesarean birth? he would counter. And there were fights over what the stroke would do to their relationship: How can you love me when my body's all screwed up? They had had to work hard to put things back together.

When Danson and Shelly Long got back to Stage 25 at Paramount, there was a brief celebration. The thirteen nominations virtually ensured several Emmy wins, and these awards, presented during the fall's premiere week, would probably bring some curious new viewers to the show. If these viewers kept watching, Cheers would in all probability be over the hump.

In the Cheers offices, across the street from the stage, Harry Anderson was talking with Glen and Les Charles about appearing on the show when a secretary told them the show had gotten thirteen nominations. Harry congratulated the Charles brothers, but they only casually acknowledged him; they seemed utterly unsurprised.

"Anything else?" the pretty young waitress in the executive area of the Paramount commissary asked Tony.

"Thanks, we're fine," he said, basking in the glory of his first lunch on the bucks-up side of the studio eatery, ladling generous dollops of espagnole sauce onto a lovely parsley-bedded soufflé. Tony smiled dreamily at Scott and their manager, Jim Canchola, of the prestigious firm Shapiro, West, and Associates, as he hoisted his first forkful of Status Food into his mouth. Then he almost spit it all over the white tablecloth, and the gleaming silver, and the tan, exposed arm of Richard Benjamin at the next table. "That's . . . ketchup. No wonder the waitress was looking at me like I was from Outer Mongolia."

"Shake it off, Tony," Scott said, and redirected the conversation to his favorite subject, money. "What you're saying," he said to Canchola, "is that Lorimar is giving us two options on *Just Our Luck*—we can take twenty-five hundred dollars a week for a guaranteed ten weeks, or thirteen hundred fifty a week for a guaranteed season. Right? Well, I'd vote for the twenty-five hundred. If we don't belong after ten weeks, then we didn't belong in the first place. What about you, Tony?"

"I agree. Fuck security. What I don't feel comfortable with is their option on us at three grand a week for the second year. That locks us in to only a five-hundred-dollar raise."

"I'll try to do something about that," said Canchola.

"I also think," Scott said, "that we should keep in contact with the other people we have scripts with."

"Hey, no doubt about it," said Canchola. "I'm going to talk to them, let them know you're on *Luck*, let them know you're working, and see if we can't get you an assignment for something in their back nine."

Tony and Scott, on the eve of their entry into professional television writing, still needed all the work they could get. Of the $2,500 weekly story editors' salary, which they had to split, their agent got 10 percent, their manager got 15 percent, and about 40 percent went to taxes. After all the deductions, each of them would end up with $425, about the price of a tune-up and oil change down at the Mazda dealership.

"Waitress," Scott said, "we don't have any matches on this table." She went off to get some. They didn't have any matches because Canchola collected matches and already had them in his pocket. Now he would have more. Canchola beamed at Scott. It wasn't much of a favor, but Scott figured they'd better rack up all the points they could, while they still had possession of the ball.

A week later, Tony and Scott sat in their new office at MGM, waiting tensely for their first story conference. The office was hardly better than the little holes they'd been scrunched into at *Cheers*; it looked like a dorm room at a cow college. They had submitted their first story outline a couple of days before, and now it was judgment time. Because their show was at ABC, which liked to keep a chokehold on its projects, the outline had to be approved by the network. To calm his nerves, Scott banked wads of paper into a metal wastebasket. Every half hour for the last three hours, they had been told the meeting would start "in just a minute."

"I can't take it," Scott said. "I'm going to the bathroom."

Just after Scott left, a secretary called to say the meeting was starting.

"Where's your goddamn partner?" producer Bob Comfort said when Tony walked into the room. Both producers were there, with two staff writers.

"In the bathroom."

They started the meeting without Scott, but after a couple of pages of work, Comfort said, "He's been in there twenty minutes."

One of the writers said, "Is the guy *married*, if you know what I mean?"

"Yeah, but he's got a big bladder," Tony said, and just then Scott walked in.

"We were just talking about your bladder," said Comfort. "Have a seat. Now, first off, this scene where Steve McGarrett from *Hawaii Five-O* comes in—fuck it; Jack Lord's already rich enough, he doesn't need to be in anything, and it's just not plausible enough. Same thing with this Clint Eastwood scene—it's too farfetched."

"Too farfetched," Tony was thinking to himself, as he scribbled Comfort's directions. "We've got a two-thousand-year-old genie here, whose main job is to zap sexy chicks for an eight o'clock ABC show, for Christ's sake, and they're giving me too far-*fetched*."

The upshot of the meeting was, the network wanted a rewrite. Fast.

Later that afternoon, three miles to the north of MGM, Peter Grad was hurrying out of his Twentieth Century-Fox office, on his way to buy a house in Beverly Hills. As he left, a producer with an English accent stopped him to say that he'd just had lunch with Glen Larson, Grad's action-adventure genius, at Guido's, to discuss a new project. "Glen adored the restaurant," he said, "but I picked up."

"Of course."

"His pockets are lined with fishhooks."

In the hallway outside, a young producer, an Allan Katz protégé, exited a meeting with some of the people in Grad's development division—a successful meeting, judging by his grin. Outside, he had to step around the aftermath of a violent wreck. A Yamaha scooter, the type used by messengers on the studio lot, lay wadded beneath the front fender of a gray Mercedes, an ominous spot of blood painted onto the street. Two policemen were finishing their reports. The young producer leaped over the Yamaha and took a hard left, still grinning.

Grad, late to sign an expiring option at his realtor's, grimaced, slid into his car, and powered on an instant blast of frigid air. He was at a crucial stage of development for the 1984–85 season. Writers and producers from all over town, and particularly the writers under exclusive contract to Twentieth, like Katz and Larson, had begun in June to pitch new ideas to Grad for 1984–85 pilots. Out of about two hundred ideas, he had circulated about fifty to the networks, and had then gotten the go-ahead on approximately twenty scripts. Now those scripts, for which the network would pay $25,000 to $40,000 each, depending on the track record of the writer, were in first draft. The strength of these drafts could well determine whether Grad would be able to meet the monstrous mortgage payments on his new house a couple of years down the road. Grad still assessed his job security as "Two bad years in a row and you're out." He thought a script called *Jox,* about ex-athletes using their sports skills to fight crime, was just the sort of action-adventure that ABC and CBS seemed so thirsty for. NBC appeared to feel strong at eight o'clock, so it probably wouldn't want a show like that. But NBC was light in nine- and ten-o'clock product, so they might want a show he was developing about the first woman astronaut.

Grad, however, as he rushed to his Beverly Hills realtor, was angry that so much seasonal pressure even existed. "I just don't understand why we can't develop all year round. When a script comes in that looks good, hire the talent and make the pilot. There would be more talent available, and the networks wouldn't be faced with looking at thirty pilots at one time and having to pick a schedule."

In about a month, Grad knew, the networks would say, "We're out of money," and stop ordering scripts. "Then, along about November, they'll see a few of their scripts aren't working—or perhaps more than a few—and they'll start calling their favorite suppliers—Aaron Spelling, Glen Larson, Steve Cannell. And they'll say, 'Listen, none of our eight o'clocks are working. Can you give us something real quick? And can you get a box office name in it?' So then the producer has to pay everybody overtime, which is brutally expensive, and he has to call up Robert Mitchum, or whoever, and William Morris knows the producer's over a barrel and asks for big numbers, and maybe a package deal—'You can have Mitchum if you take along so-and-so.'

"Early in the development season you talk concepts, ideas. Which are cheap. Later, you talk names. Which are expensive. It's just not"—Grad pounded the steering wheel with the palm of his hand, inadvertently goosing a muted beep out of his horn—"*efficient.*

"In what other business would you conceivably order your entire year's inventory all at once? I think the network people are very bright. I also think they're terrible businessmen."

When Grad got to his realtor's office, the elevator was out of order. He was twenty pounds overweight and in no mood to haul his bulk up five flights of stairs; he was damned if he was going to detonate a ventricle just for the privilege of handing over a king's ransom to some do-nothing middleman of a property broker. On the other hand, however . . .

On the other hand, if you didn't buy in Beverly Hills, you were looking at rents that shot as high as $35,000 a month—and you couldn't even deduct them, like you could mortgage payments. At that very moment, there were no fewer than a dozen homes for rent in the area where Grad wanted to live, each with a monthly price tag of $8,500 or more. Granted, for thirty-five grand you got a lot of house—three or four master suites, a couple of guesthouses, and a tennis court. For $8,500 you got four or five bedrooms, a fireplace, a pool table, and a hell of a nice set of shrubs out front. If you wanted to talk affordable, though, in the $2,000-per-month range, you were looking at a "charming" one-bedroom guesthouse—in other words, glorified slave quarters—in the back of some *real* businessman's estate. At the moment, Grad was renting a place in Beverly Hills that most Americans would have been proud of—but it wasn't the kind of place where you'd bring Glen Larson for lunch.

Only about three men at each network were even in the quarter-million-a-year bracket, so his current home was fine for them, but producers like Larson and Spelling and Cannell—the men Grad had to traffic with—had nicer homes than his for their polo ponies.

So Grad gave in and mountaineered his way up to the realtor's office, breathing hard and cursing.

"Okay," Allan Katz said, "I can get you a guy to play the midget. And I know three guys who could play the giant. The only problem is, you'd have to tape 'em together." Katz was on the phone and in a good mood, talking to somebody who was trying to cast a show.

Katz had just found a director for *The Hunchback of UCLA,* after checking out almost two dozen possible candidates. Some had been too expensive. Some turned out to be weak in comedy. Some were tied up indefinitely. Finally, his new partners Tische and Avnet suggested he give the diminutive star of *Taxi* a shot. Katz thought, "Danny? DeVito? They must be kidding." But Jim Burrows told Katz that DeVito had directed a couple of episodes of *Taxi* and been excellent. Then Katz saw some film shorts DeVito had done and felt like the guy who first found gold at Sutter's Mill. DeVito's stuff was phenomenal. Katz started thinking, "How could this guy have gone undiscovered so long? If I tell him how great I think he really is, he'll say, 'Then maybe I'm too good to direct something starring you.' "

They met for the first time at Tische and Avnet's office. DeVito said, "I think *Hunchback* is wonderful, but I think it needs work."

Katz said, "So do I."

DeVito said, "I think it needs more reality."

Katz said, "So do I."

DeVito said, "You actually plan to star in this?"

Katz said, "Yeah."

DeVito smiled. The pact was sealed.

Katz, hanging up the phone, explained why starring in his own movie was an important career move for him. "Writers and producers, we have our share of power and pull, but it's the stars—the big stars—that have the real power. I mean, you've heard of doctors making house calls; well, Cher's *hospital* makes house calls. I couldn't believe it. When I was producing her show I went over to the studio one day and there was medical equipment in her dressing room. I asked, 'What's happening?' and was told she had some kind of ailment and didn't want to drive to a clinic and so, instead, asked them, 'Why don't you just bring the stuff by here?' I couldn't believe it. That's when I got my education."

However chipper, Katz was still concerned that Twentieth might pull the plug on *Hunchback* because of the unorthodoxy of the project and because he was now dealing with people who did not really know him personally. Also, he was eager to get a star for his play *Kaufman and Klein,* and was still waiting to see what ABC was going to do about his Charles Durning sitcom. Mostly, though, he was eager to get something of his own creation into production.

"I've produced successes, but I've never created one. I do not have a series on the air, and I have to start over every time one of these things falls through or doesn't materialize. You can go to the network, fight over a script, tell them you want a particular actress, and they say they want a celebrity instead. You persist and you get the person you want, it gets on the air, it becomes a hit, and the next thing you know the actress you fought for becomes a star and gets you booted from the show because of creative differences. The network can remove you from your own show unless you own it." Thus, the itch for a little public recognition and the punch that came with it.

"If I'm able to star in *Hunchback* and get some visibility, I can promise you a couple of things. I will use whatever leverage I can get to do TV, or another feature, or whatever, that delivers something to people. TV's idea of women's issues is to have Edy Williams raped in prime time; it turns problems into cartoons. I don't think there's been a show since *Lou Grant* that's really tried to deal in any regular depth with social issues. I'd love to do *Adams House.* Even now. So what if it was a ratings disaster

312 · THE SWEEPS

and only got ten million viewers? What if thirty thousand of those viewers had bad problems with depression and you were able to show how they could get help? What if on that same show you could show old people where to go if they needed help; what if you just had a character say to another character, 'Look, if you're old and lonely and don't have enough to eat, call Lutheran Family Services or Loaves and Fishes,' or whatever national social service was appropriate—if I knew I had helped a couple thousand people just by doing that, the whole show could go right down the toilet and it still would have been worth it. But prime-time TV doesn't want to take that responsibility. And I can't figure out why. There's lots of uncelebrated people out there who'd watch a lot more TV if it really offered them something to take away from the set instead of dumb jokes and fistfights. There are people out there who need help. I mean, you're talking to a guy who was one of them, who never made more than a hundred fifty dollars a week until he was twenty-nine years old." Were it not for the grace of God and a couple extra twists in his DNA helix that accounted for a wild imagination, Katz figured he could still be working nine to five mowing the Outer Drive in Chicago.

Still, he had a long way to go. The *Hunchback* project was taking one odd turn after another. He was now totally sold on Danny DeVito as director. He wondered what Twentieth Century was going to think about spending $10 million or so to produce and promote a film that (1) was written by a man (Katz) who had never written a feature before, (2) would star a man (Katz) who had never even acted in a feature before, and (3) would be directed by another man (DeVito) who had never directed a feature before.

It was going to be the turning point in his life when it came time to find out.

"We're gonna make ya a star, honey," Corky and his buddy Jay had told the Marilyn Monroe look-alike—and damned if she hadn't actually *believed* them. Celebrity was turning out to have its privileges, and instant credibility was certainly one of them. So there she was, naked with Corky in the bathtub, bobbing for apples, as it were, while Jay hovered over them with his cheap little pawn-shop Betamax camera, murmuring things like "Love it. Love it to death."

Still in Portland, still waiting for the goddamned rain to stop, still trying to get his poor smashed leg, prematurely freed from its cast, in shape for the Paul Bartel movie, Corky had time to kill. And what better way to kill it than to give a local scrumptious the big break in her career?

The scene was this: "Marilyn" was crouched naked below a multipeaked mountain range of bubbles, red mouth supposedly impaled on Corky's totem-pole-sized tower of power, giving Hollywood's top dwarf a hot-water hum-job. Then Corky's battle-ax "wife" would appear at the bathroom door, fresh (and surprisingly!) home from work, demanding to know what her husband wanted for dinner, etc., etc., etc., etc. Until Marilyn could stay submerged no longer, and would explode from the water and bubbles, her perfect water-balloon breasts sticking out all over the place. Corky's wife would get outraged and pummel Corky's curly head. Cut. End of epic.

She was really getting into her part down there—as Jay growled, "Be the character," from deep in his throat—rubbing those perfect, perfect, perfect breasts all over him, and pressing that naked slinky body all slick with suds against him until there was no way he would be able to stand erect without totally embarrassing himself.

So after the shot he had to sit in the damn tub as she got out and, ecstatic at the sight of her own naked body, began a crazy twirling dance right there in the bathroom. Corky was finally able to hop out of the tub and then—wham!—her long dancing legs smashed hard into his broken shin. He fell to the floor, grabbing at his leg, the world exploding all red and yellow and orange, moaning, "Jeez, Jeez, Jeez, Jeez," with the girl hugging and holding him and going, "Sorry, sorry, sorry, sorry." She was so passionately repentant that even as he lay sweating in agony Corky couldn't help but have visions of this delicious teen giving him a hot-water hum-job *for real.*

But it never happened.

Two weeks later, in New York, on the set of Bartel's *Not for Publication,* the leg was still throbbing like a son of a bitch. Corky was eating aspirin like M&Ms, and even Bartel asked him if he was feeling okay while Corky stood there behind one of the cameras, wondering what would happen if suddenly the damn thing just snapped, with bone jumping all bloody and splintery right out of his skin.

Other than that, *Not for Publication* was great. Since Corky was playing a chauffeur and didn't have a driver's license, Mayor Koch had *personally* gotten him a special one-week driving permit. Well, sort of personally, anyway. With some of the money he was making, Corky went out and rewardrobed himself. No more going to the laundromat three times a week to wash his only two pairs of socks. No way. Corky went down to this old place and bought dress shirts that were about sixty-five years old. They had been made especially for Rudolph Valentino or some other fop like that—the guy had died and the shirts had gone unworn. Until that very day, when Corky walked into the store and the owner recognized him from *Under the Rainbow*. Because Corky was one of the greatest midget actors in the universe, the guy sold him the shirts—all four of them!—for only $200, plus Corky's autograph. Corky inscribed the *Rainbow* "No dream too big, no dreamer too small" epitaph with his signature, and walked out of there feeling like Jay Gatsby, needing only a Daisy to show his shirts to.

Corky loved doing this movie, because it was about his forte—hard-hitting investigative journalism. It was just like the kind of stuff he used to do back when he wrote *Conspiracy Digest*. It reminded him of some of the headlines from his own top exposés, like IS SQUEAKY CLEAN? FROMME SET UP IN GERALD FORD ASSASSINATION ATTEMPT. And the people on the set were so nice—not like that putz in Portland who'd made him stand in the rain. The only thing that really bummed Corky out was that the *Fall Guy* people had called his agent, because Lee Majors wanted him to guest-star in another episode, and his agent had had to say sorry. Corky had simply become too busy and successful for network television. At least for the moment.

The houselights at the Lake Tahoe casino showroom dimmed to a somber twilight; the band crawled into an extraslow bridge to the next song—sure sign of a big finish. Country singer Mickey Gilley murmured into the microphone, "And now, good people,

I'd like to sing a song I learned from a man we still call the King," as the strains of Elvis's arrangement of "Dixie" became recognizable.

"Who's the King?" a woman from Iowa, crammed into a small table with couples from Washington, Oregon, and Utah, asked her husband.

"*El*-vis. You know. *Pres*-ley."

Gilley segued out of "Dixie" and into "The Battle Hymn of the Republic," pausing to say over the music, "If you think I'm doin' a little flag wavin', well, you're right. Cuz I love this country they happen to call America." Three backup singers in white, scooped-neck gowns moved in behind him and they brought matters to a fully orchestrated patriotic conclusion, ending with heads bowed as the curtain fell. One thousand people in Sunday-go-to-meeting clothes rose as one and called Gilley back for an encore, and Gilley plugged his appearance that evening on *Fantasy Island*, which, he said, "fulfilled a dream."

Then the curtain came down for good, and the audience went back to their uniform chicken-and-Stovetop-Stuffing dinners, for which they'd each paid $17.50. Harry Anderson walked out in front of the curtain in a baggy suit, a fedora, and black glasses. He had gotten virtually no billing and no introduction. Those in the audience who noticed him just stared. "Hey!" Harry accused. "I wasn't expecting you here, either."

That got a pretty good laugh, and he was off into his routine. Harry was doing the middle act for Mickey Gilley and Dottie West at Harrah's, a giant rectangle of concrete and neon on the south shore of crystalline Lake Tahoe, approximately four feet from California in the gambling state of Nevada. While Las Vegas tends to collect the midwestern, Los Angeles, and international crowd, the sister cities of Reno and Tahoe are playgrounds for small-town westerners and ranchers with skin as rough and brown as a baked potato's. Reno-Tahoe is the jerkwater entertainment capital of the universe—its crowd supports almost as many big-name country singers as Nashville—and, at the moment, that crowd was rather at a loss to know what exactly to make of Harry Anderson. But Harry won it over with a drop-the-pants gag and a tried-and-true cuckold joke, guaranteed to get a good nervous laugh from a middle-class, married audience: "Like my suit? It was a surprise present from my wife. It was on the chair in the bedroom when I came home last night. Came with a wallet, car keys, and everything."

Harry did ten minutes, while the next band was setting up behind the curtain, then beat a quick retreat to his room, to wait for the midnight show.

He dialed three cans of gin and tonic from the booze-filled vending machine near his bed, passed them around to a couple of buddies visiting from L.A., and sat down at a dull-red IBM Selectric typewriter. He was writing a screenplay treatment to kill the time in between shows. Called *Mastermind*, it was about the dwarf who hid inside the "automatic" chess player that toured Europe and America in the 1800s. "It was a standard kind of magician's device—it looked like it was all gears and levers inside, because the dwarf moved into one part when the other part was being shown, while mirrors provided an illusion of depth where there was no depth.

"Magicians' devices are generally simple—they trick people because people want to believe bullshit. Take a knife-throwing act. It works because the thrower flashes the knife in the light as he throws, and the eye sees an afterimage of that flash as it looks from the thrower to the target. So the eye thinks it has seen a knife fly through the air, while the knife's really up the thrower's sleeve. Then, when the person who's being thrown at activates the board behind him to pop the knife out, the audience is *sure* the knife's been thrown.

"But the real reason they're sure is because they *want* to be sure. Hell, why pay the admission if it's just a trick? If I'm in the audience, I want my bullshit! I *paid* for it. I want my lies! I want *magic!*

"Anyhow—the story is about the dwarf and the man who pitches him, and the Gypsy woman the pitchman marries."

The script was just one more insurance policy against the failure, or even possible nonexistence, of *Night Court*. Harry had been to New York in the past couple of weeks, to work in a nightclub and to see about further appearances on *Saturday Night Live*. "Those *SNL* appearances are as important to me in their own way as *Night Court*," he said. "Because if *Night Court* bombs, I don't have to proceed as an actor. I can just get deeper into the stage act."

The phone rang; it was Leslie and Eva. Eva wanted to talk to him personally. Her headline news was that she'd learned how to use the big potty.

"NBC has put us in a suicidal position," Earl Greenberg, producer of the daytime show *Fantasy,* said as he sipped from a Diet Coke, then opened a notebook and recorded the cola's consumption. "We're on at three o'clock against *General Hospital* and *Guiding Light. General Hospital* is like your top-rated show; *Guiding Light* is number three or four. This is really a prime-time show. If it were in prime time, we could do the fantasies we'd really like to. I'd have Leslie Uggams reuniting people in the Eiffel Tower; I'd have twenty-two people jumping out of an airplane; I'd have women going down the Nile like Cleopatra. I mean, there are only so many ways we can reunite people in a studio. We have to make at least *one* of the people surprised. I mean, the other day we had to reunite a woman that was disguised as a table."

Greenberg snapped the notebook shut. A former assistant district attorney for the State of Pennsylvania and a former NBC programming head, Greenberg was a balding man of less than forty who ran ten miles a day, recorded everything he consumed, and worked seven days a week. At the moment he was responsible for cranking out five hours of network TV a week—*Fantasy* was an hour show—more, in his estimation, than anyone else in daytime. And he was in trouble. *Fantasy,* which occupied a late-afternoon time slot that served as a vital lead-in to affiliate afternoon news, was getting killed in the ratings. Because daytime TV was such a huge profit center for the networks—ABC was said to be making $50 million a year off *General Hospital* alone—NBC was more than worried about *Fantasy's* performance. So Greenberg was happy to have such a popular— if controversial—figure as Ed Asner agree to make an appearance on his show.

At that moment, Asner—ultrarespectable in a banker's blue suit and accompanied by the two women who staffed his office in Universal City—was waiting in *Fantasy's* greenroom for the surprise appearance, in which he would reunite a mother and her adopted toddler.

Backstage, the mood was taut. Daytime chief Steve Sohmer had let it be known that only the most miraculous of interventions—say, a guest shot by the pope—would be sufficient to revive this program from its terminal condition. Based on the idea of helping ordinary (often poor) people achieve "fantasies"—like being able to feed their families or not die from lack of expensive medicine—the show had not struck the fancy of the greater American middle class. Last week, it had been twentieth out of twenty-five daytime shows. Since it was doomed anyway, NBC was feeding it bread and water; these days the fantasies the show was fulfilling were decidedly on the cost-effective side. Such as having Ann Jillian, today's guest host, fulfill her "fantasy" of teaching a couple from the audience a comedy routine.

Game show doyen Peter Marshall strutted onto the stage and bawled, "Hi, audience!" The audience boomed.

"How many of you watch *Fantasy?*" he prompted. This time he got a much smaller "Yay," and his grin frayed around the edges.

Marshall hustled Ann Jillian out to fulfill her fantasy of teaching ordinary people comedy, and she guffawed around the stage for many minutes with a couple from Muscatine, Iowa, who were wearing matching outfits.

Then a young couple, who had spent all their money and couldn't find work and were wondering, out loud, how in God's name they were ever going to get back on their feet, were presented with a minicomputer with home-finance capabilities. It would help them "straighten out their budget problems and get back on the right track!" The husband was presented with a handful of computer programs as the announcer plugged the donator of the gift, and was then hustled backstage with his new goodies. When he got behind the curtain, he surveyed the programs they'd given him. In one hand, he had "Super Cuda," and in the other, "Ape Craze."

Ed Asner was getting antsy. He had a full day's work waiting for him back at the office. For as long as Ed Asner could remember, he'd always had a full day's work waiting.

Ed Asner was the youngest of five children of a Lithuanian immigrant who could not read or write English. Asner's father had worked in the sweatshops of New York's garment district, then moved to Kansas City, where he became a junk dealer. When Asner was born, the family was living above a junkyard, across the street from an Armour meat-packing plant. One of Asner's earliest memories involved the packing plant, and he described it once to Pete Hamill for *New York* magazine: "A steer had somehow broken out, got away, and there were men chasing it. It ran into the junkyard, making panicky noises, knocking over metal, these men after it. And I was scared to death. I was about five, and I identified with that poor animal. They caught it and took it away to kill it.

"But years later, when I thought about it, I began to identify with the men who caught it. It was a trick. A way of protecting myself. It took years to get back to the original, and true, emotions."

The junkyard and slaughterhouse were in a part of town called the Bottoms, peopled mostly with Mexican and Indo-European immigrants. The Asners were the only Jewish family in the neighborhood, but there was not much discrimination. When they moved to a red brick house in a better neighborhood, anti-Semitism was directed at Asner more frequently, but was still subtle.

Asner was enrolled in a Hebrew school at a local synagogue, where a young, progressive rabbi encouraged the children to act out small dramas of religious history. In high school, Asner, whose arms and shoulders had become as thick as spools of coiled copper from hefting scrap metal in his father's junkyards, became an all-city tackle on the Wyandotte High football team. He also became feature page editor of the school newspaper, and had to be talked out of becoming a journalist by a teacher who felt it wasn't lucrative enough. Asner was restless in high school, and applied for jobs in exotic locales—Venezuela, the Amazon jungle. But he settled for the University of Chicago.

Encouraged by his roommates to look into drama, his first role was as the persecuted Thomas à Becket in T. S. Eliot's *Murder in the Cathedral*. He discovered that he loved to act, for a reason many actors give—because he liked his roles more than he liked himself. Even then, he says, he carried with him a sense of unworthiness.

After two years at the University of Chicago and two more in the Army, he helped found Second City with Mike Nichols and Elaine May. Twenty-six plays and four years later, he moved to New York, where he appeared in the theater but supported himself with a large dose of proletarian labor. He toiled as a cab driver, messenger boy, encyclopedia salesman, and steel mill worker. He married his wife, Nancy, in 1959, and headed for Hollywood two years later.

After almost a decade of scraping by in character roles, he caught the attention of Grant Tinker while playing a tough cop in a doomed pilot called *Doug Selby, D.A.* He was forty-one. Tinker asked him to audition for *The Mary Tyler Moore Show.* When he first read, he played Lou Grant as sober and earnest. The producers told him his rendition had been "very intelligent." In other words—not funny. They asked him to play Lou Grant as a buffoon. He did, and they gave him the job. He now believes they were merely "seeing if I would surrender my ego and take direction."

By the time *Mary Tyler Moore* and *Lou Grant* were finished, ten years later, the buffoon was a multimillionaire.

Now Asner, introduced by Peter Marshall, was shepherded to his stage-entry point by a very pretty woman, maybe twenty-one or twenty-two, in a blue-serge mock-police security-guard uniform, her auburn hair rolled under her hard-plastic brimmed cap. She was the backstage enforcer, a no-nonsense young overlord who was responsible for keeping law and order in the production facility. Even at her tender age, she had an assistant, a thirtyish guy in an equally imposing uniform.

Asner explained to the audience and camera that a thirty-five-year-old woman, Mary Reese, had contacted him because she'd heard that he had helped a Wisconsin couple adopt a child from El Salvador. Mary Reese had adopted a Taiwanese infant two years ago, a beautiful little girl named Omara, but immigration and visa statutes had made it impossible for her to be with her daughter. She'd run through her entire savings and was at a dead end until she prevailed upon Asner to wield some good old-fashioned celebrity clout. Asner had enlisted *Fantasy* and a local congressman to help, and now he was reading a letter from the legislator. "Dear Mary, I'm sorry I can't be with you. Your

living bureaucratic nightmare is over, and Omara is indeed for-
tunate to have you as her mother. David Dreier, U.S. Congress."
Mary Reese was starting to choke up, and when, with maximum
fanfare and a trill of the organ, little Omara was carried onstage,
the new mom was a shambles of happy sobs, hugging Asner and
the child, making no effort to compose herself.

Backstage, a sibilant sniffle sliced through the air. The head
of security—the tough little cookie—was crying. Her assistant
stole a look at her out of the corner of his eye. "Mind your own
beeswax," she hissed.

Then she said to everyone around her, "Lou Grant is *so*
bitchin'!"

Steve Cannell's new office was on the eighth floor of his new
building on Hollywood Boulevard. The building was a gemlike
cube, covered on the outside with massive sheets of green reflective glass. It contained facilities for a completely autono-
mous production company, everything from editing equipment
to the T-shirts and dolls of the *A-Team* fan club.

It was production time again in Los Angeles. Hiatus was over
for everyone now, and all across town the chains and gears and
levers of television series creation were back in motion. People
were going back to work—if you didn't have a job by now, it
was "wait 'til next year."

Cannell and his second-in-command, Jo Swerling, sat with
line producer Les Sheldon and a writer, Pat Hasburgh, mulling
over Cannell's new show for ABC, *Hardcastle and McCormick.*
They were going to be crashing a lot of cars and trucks in a day
or so.

"You know what the problem is on this one, don't you?"
Cannell asked, speaking over the scream of a Skil saw in the unfinished room next to his, where carpenters were still framing
doors and banging up plasterboard.

Les Sheldon, a cherubic man with frizzy, ash-colored hair,
blinked and shrugged. "I give."

"The bad guy's in the driver's seat."

Sheldon swallowed. "We can't do it any other way. As it
is, the good guys are shooting out tires and things."

"I think we'll be okay," Jo Swerling said, pulling absently

at the crease on his khaki slacks. About fifty, Swerling was a droll, fit man, tan and balding, whose presence was more that of a construction engineer than a television executive. "It's a good gag, a legitimate one. At this point, the important thing is to front-load the show with action."

"I dunno." Cannell made a wagging motion with his pipe. "This is a relationship show, right? Seventy-five percent relationship, twenty-five percent action."

Jo said, "Maybe we should load it fifty percent relationship, fifty percent action for the first few shows."

"Expensive," Cannell said.

The writer, who was younger than the other three, suggested, "What if we just flew a lot of cars? Do it all in one day."

Swerling said. "At one thousand dollars a pop? Every time a stunt guy jumps one of those cars into the air, you pay him a thousand bucks."

Sheldon nodded. "Yeah, but roll one and it costs you three thousand. As soon as a wheel touches a ramp it's three thousand dollars."

"And," Cannell said, "just because you do it all at once doesn't mean you've done it right."

Swerling said, "The first four shows, develop the action, get people hooked . . ."

". . And"—Cannell shook his head—"learn to love the characters in November. I'm not sure I buy that. What I see here is close to fifty thousand dollars a day out of pocket for second-unit work."

Sheldon sighed. "I've cut back as far as I can. I cut fourteen thousand dollars out today. This diesel thing with the truck is going to be a beautiful piece of work. Splendiferous. It's a phenomenal gag. I'd sure hate to lose it."

"I agree," Cannell said, "but I'd hate to see us trying to produce our shows from a soup line, too. After a point, I can't give you guys money I just don't have."

Cannell took a call from ABC and said, "Do me a small favor, will you? On *Hardcastle and McCormick*, let's not just cut a few scenes from one of the shows, slap on a voiceover, and call it a promo, okay? Take a look at what they're doing at NBC. Sohmer is really doing some beautiful promotions and any effort I could get like that from you guys, I would really appreciate."

Cannell faced the fall season optimistically, especially enthusiastic about the efforts being made at NBC for his show *The Rousters*. Putting the receiver back on its cradle, he said, "Sohmer is a real up-and-comer; he has such great ideas. Like when he came up with Mr. T doing the 'Magnum! You're a wimp!' I was skeptical, because there's been a real taboo in the networks against doing negative-sell against each other, but it was so funny. Like when T gets up there and says, 'Hey, Fonzie, yo Happy Days ah ovah!'—it was so bold, it really hung out there. He's got one for the fall. Mr. T on a lawn chair with sunglasses, laid back, saying, 'NBC has decided to let you know what's on Tuesday night. . . . Now on ABC they got something called *Just Our Luck* . . . and on CBS they got something called *Mississippi* . . . so this is just to let you know that when you're watching *The A-Team*, you ain't missing *nothin'*!'"

In addition to creating *The Rockford Files*, Steve Cannell also created a number of other classic action-adventure shows, including *Baretta*, and as television's most ambitious and successful young independent producer, he existed as a kind of space-age throwback to a wilder, more ungoverned time. He was the kind of guy who might go on safari. Cannell grew up in California, left the state for college, then returned to work within the studio system until he was almost forty. Then he struck out on his own.

Despite the phenomenal success of *A-Team*—syndication rights on the show were predicted to be worth *$2 million per episode*—Cannell made no claims to greatness. "I'm a little guy. I am not a big studio; I don't have a big economic reserve on hand. I'm totally on my own. I have no partners, no studio as banker." Nevertheless, Steven Cannell Productions was now almost totally autonomous, with everything from its own sound stage to cutting and screening rooms. "I won't shoot on a studio lot. MTM keeps their number of employees down because they use studio center crews; I don't do that, I payroll my own crews. Frankly, I think they're stupid. I think it probably costs them between three and four hundred thousand dollars a year to do that. It's easier, because studio center has all its own cameras, et cetera, all their own stuff. But it's a little easier for MTM to make money than it is for us, because they do three-camera sitcoms—even *Lou Grant* was mostly interiors. They don't risk getting killed by going outside. *Hill Street* was the first time they really went out in the streets. Action-adventure is just so much more expensive than sitcoms. With sitcoms you can stay within the license fees. That's why I think it's better to be completely independent. At MTM they go to the studio for most of their people and the studios charge them a commission for the service. I did that for a year and just got raped. So I stopped it. Everybody that works for me works for me. Otherwise, you can't even fire them—you can't get rid of them. I want to absolutely know that I'm running my own show."

An expensive prerogative. Said Michael Dubelko, Cannell's chief of business affairs, "*A-Team* may be a big hit, but we're still in the hole. You never make money on the first year of a series. You've got to write in a fifty-thousand-dollar loss per show, then try to recoup some of that immediately by selling foreign rights immediately at about twenty-five thousand dollars apiece. Still, you're facing a deficit. We're down about two million on *A-Team* after fourteen episodes." Cannell could, of course, have raised huge sums by selling as yet unmade episodes of *The A-Team* into syndication right then. "But you really pay if you start borrowing against the future like that," Dubelko said. "You're much better off if you can get people to defer fees and so on, juggle and struggle, until the merchandising and real full-blown syndication things start coming in two years or so down the road."

"You're gambling to get an *A-Team*," Cannell said, "to stay in business for a length of time until it's syndicated. I'll sacrifice my writing fees, my producing fees. There have been years where I've literally lived off my residuals. You go from gamble to gamble, hoping meanwhile that something like *The Rousters* or *Hardcastle and McCormick* will catch on and allow you to keep making TV, to meet your payrolls and stay in business."

"If you go through the little black book of Hollywood," said President of NBC Entertainment Brandon Tartikoff, "you're going to see that there are more producers out of work than there are those working. Those out of work go to breakfast together, and lunch, and dinner, and they're not going to say, 'I've lost my fast ball. What I see on television is better than the stuff I'm doing.' No, they're going to look for a scapegoat, someone to blame for the fact they've had to trade in their Rolls-Royce for a BMW. My job is to get high-rated, salable series on NBC. If I have zero hours from NBC Productions and a highly rated schedule, then I'm a hero."

Tartikoff was feeling some heat. Gary Goldberg, producer of NBC's *Family Ties* and one of the most respected creative talents in television, had just broken very important ice by joining with NBC in a co-production deal for thirteen episodes of an as yet unnamed series.

What this meant, in effect, was direct network participation in syndication profits if the series was a hit. And while federal law allowed the networks to produce two hours each of dramatic series themselves, a number of producers saw Goldberg's pact as the beginning of a full-court press by network executives who felt the Reagan administration was about to deregulate the industry and allow NBC, ABC, and CBS to reap not only advertising returns on all their programming, but full participation in syndication profits as well. Len Hill spoke for many producers in reaction to the Goldberg deal when he told the *LA. Times*, "The networks will be in a position of owning the field, and if you're not with them, you don't play." A theorem dismissed by Tartikoff as sour grapes from those who felt they weren't getting a big enough cut of the pie.

A point Hill was willing to concede. "Yeah, it's greed, damn straight. I want to make more money. But I don't know if that's to be frowned upon. If it is, maybe the whole thing should be socialized and run by a government bureaucracy."

Goldberg, for his part, was—mostly—just satisfied with his deal. "To me, working at NBC has been spectacular. It's like working with your friends. They are on our side, and I like to be on the side where Grant Tinker is. It comes down to that."

But the larger issue, what the Reagan administration was going to do about syndication, and whether it was going to allow network people like Tartikoff to become in effect producers as well as programmers, was, in TV land, the unresolved political question of the year.

*   *   *

It was the night of the Emmy nominations, but aside from Ed Asner, the people gathered around picnic tables at a Santa Monica political meeting could not have cared less. A girl with waxy buildup on her hair and a SAVE THE SEALS T-shirt stretched across her chest was telling her tablemates that the new co-chairperson of the group they belonged to "works from a real feminist perspective." Her significant other, in a matching-species shirt, agreed that "a lot of the negative forces have been cleaned out." A guy in a Tom Hayden T-shirt with a plastic plate of burned chicken and seared pineapple, procured from a smoking beachside barbecue pit, helped himself to a seat at the table. "Do you have an entity?" asked the seal lover, by way of introduction. It turned out that he had the biggest entity in all of southern California rad-lib politics—he worked for ex-Governor Jerry Brown. As such, he was more than welcome to break bread.

About 250 people of a similar persuasion were gathered on the sand about a mile north of the Santa Monica Pier for an event sponsored by a group called Santa Monica for Renters' Rights, or, more popularly, SMRR, pronounced *smur*. The group was represented in the state legislature by Tom Hayden, who was a sponsor of SMRR, along with Ed Asner. Asner was, at the moment, attempting to do something that Hayden's wife, Jane Fonda, had done with consummate skill a few years earlier—crawl out from under a rock of negative publicity and industry hatred, reemerging as rich, famous, popular, and powerful, while simultaneously keeping integrity intact. It was a hell of a juggle. Asner was making innumerable appearances like this, partially out of the goodness of his heart—his staff considered him a notoriously soft touch—and partially to build his own peculiarly structured Q. The job of rebuilding his popularity was essentially a lonely one—virtually the only public figures who had openly supported him of late were Fonda, Robert Redford, and *M*A*S*H*'s Mike Farrell.

Asner looked down at his chicken dinner. To eat or not to eat? His weight had always been a problem, but now, at fifty-three, taking it off wasn't as easy as it used to be. The fat, he thought, seemed to have little octopus suckers, attaching itself all over his body. It got to the point where the weight itself was sucking him down. He had been a jogger, but he'd hurt his knee, so he'd tried walking up hills. But walking just wasn't . . . *dramatic* enough. He liked *running* up hills, not plodding like a cow.

As a star, Asner had a host of problems that were totally the reverse of those faced by young actors like Harry Anderson and Corky Hubbert. He literally couldn't afford to work cheap. He was a corporation, a minor industry; he had people to pay all over the place.

Were he to do Allan Katz's play *Kaufman and Klein*, he could lose a fortune. Actually, what Asner was tempted to do of late was to simply go out, work his ass off, and make a hell of a pile of money. Then he could slap his enemies to death with wads of dollar bills.

While people were still eating, Asner walked to a podium that rested on sand; beside it were two fern-covered, environmentally compatible amplifiers. "I'd like to express my appreciation for this invitation to the directors of Smur," he began, glancing at three-by-five cards in his palm. "Smur—that sounds like something you find on your tongue after an all-night binge." His joke was met with only a murmur of laughter; a self-mocking sense of humor did not seem to be the long suit here. But Asner plunged on—"What's it stand for, Sadomasochists to Reelect Reagan?" He was wasting his jokes. So he began, mildly at first, to castigate them for getting their "butts whupped by the landlords" in a recent local referendum. "How does it feel to get your ass kicked?" he barked into the microphone. If he was expecting a 250-voice rejoinder of "Bad, Ed. Tell it!" he faced disappointment. "Well, I *know* how it feels to get your ass kicked, because I got mine kicked a year ago. After wallowing in Kierkegaardian self-pity, you figure out why you got your ass kicked. I hope you've reflected on the mistakes you've made—you failed to provide for your comrades. But I'm sure nobody has to remind you of the mistakes you've made." He concluded his brief address with mention of a planeload of supplies being sent "down south," and with a plea for them to beat their political enemies and reelect their friends. "Let's whip Reagan! And reelect Tom Hayden! Let's applaud our victories and admit our mistakes.

"Remember," he scolded them, shaking a finger in their faces, "you are a symbol. *Don't* screw the symbol up again."

At the back of the crowd a wisenheimer cut through the thin applause with a parody Mary Tyler Moore voice. "Aw, gee, Mr. Grant."

Miles north of Magic Mountain, just north of L.A. on an empty highway off Interstate 5, Steve Cannell's second unit for *The A-Team* and *Hardcastle and McCormick* was crashing the necessary cars. Cannell's crew of stunt men were absolutely vital to the fortunes of NBC. Unknown and moderately paid, these were the guys who pumped the flash into *The A-Team*, the network's only solid hit, and the backbone of NBC's strongest night. Though critics might cry about it, a well-done automobile accident was still what the greater part of America wanted from television.

Today's gag was this: The Coyote was chasing the bad guys, the Coyote being the automotive linchpin to Cannell's new series for ABC, *Hardcastle and McCormick*. It was a replica of a McLaren race car except that, instead of being powered by a monster V-8, its polished aluminum wheels were turned by a souped-up Volkswagen four-banger.

TV production for action-adventure shows is divided into two units, each with its own director and camera crew. First unit handles the actors; second unit, the gags—the stunts. The second unit for Cannell Productions went through a six-pack of cars a week. Cars mostly bought on the cheap from rent-a-car agencies. Big old cars from Detroit's Jurassic Period. Early seventies Fords, Mercs, and Chevies. Each usually good for at least four, six, or eight crashes.

There was a metal wagon train of equipment parked in a long semicircle about a mile above the shoot. A special-effects truck. A lunch wagon. Cop cars, real and fake. Trailers for the crew. A water truck and a backhoe to push dirt.

Hardcastle and McCormick starred Brian Keith, a muscular actor in some late decade of middle age who had the generic presence of a professional football coach. His stunt double was a man named Chuck Couch. Couch had been stunt coordinator for the long-running *Hawaii Five-O* and was looking toward retirement after a career launched in the forties. He had started in the circuses, working on the road, doing a high-wire walk, working on the trapeze with sway poles, and hanging by his teeth stories above the ground. "I worked with the Wallendas," he said, lying down on a bunk, head propped against the fiberboard paneling of one of the crew trailers. "The circus was less dangerous than TV stunt work because of its routine. You do the same thing every day for an entire season. I did a lot of things like working a hundred twenty-five feet up without a net. It's easier to get hurt in TV, but harder to get killed. Heights are my main thing. I did a fight going across the Grand Canyon on a cable car for Cornel Wilde, in *Edge of Eternity*. Hung by my teeth from a helicopter. Still, you shouldn't get killed in the picture business. In the circus, if a wind is up and a crowd is there, you pretty much gotta go up, regardless. In the picture business if there's a wind, they're gonna say, 'Aw, shit with this—let's do it tomorrow.' "

There was a noise, a swelling snarl from down the road. Something low and red whipped up from around the scrub-covered hills. The Coyote. It flashed by the equipment trucks, then pulled into the turnaround. The Coyote was as much a staple of prime-time network television as any starlet. Its job, essentially, was to precipitate car wrecks. That is, to be chased over and over by bad guys who were terrible drivers and who slammed their Fords and Chevies into stuff as regular as clockwork. Steve Cannell's stock in trade is good dialogue, justice, and fender benders. Cannell's characters—from Jim Rockford to Mr. T—are generally ass-kicking do-gooders, the affably disenfranchised who live on several sides of the law and will crash most anything to bring right to the world.

Which makes for an active second unit. This one worked for both *A-Team* and *Hardcastle*, though under different directors.

Today, the action was going like this: Eleven stunt men and one woman were representing the good guys, the bad guys, and sundry passersby. The good guys—Chuck Couch, a woman, and a young blond man doubling for Brian Keith's sidekick—were chasing the bad guys, who were driving like lunatics in a huge semi with a big boxcar of a trailer attached. The good guys were also being chased by some other bad guys in a veritable cornucopia of old Ford, Merc, and Chevy junkmobiles.

No one except stunt director Gary Combs seemed to know exactly what the geometry of all this was. Nonetheless, it all worked. One of the stunt men ran his old junker off the road and sent it plopping into a ravine at about forty miles an hour. Then Chuck Couch rolled up in the Coyote and pointed a gun at him. Then they blew one of the rear tires off a Ford. After which, stunt guys chased the semi and one of them hung out of one of the cars and waved a gun around. Between takes, the stunt men spent a lot of time standing around on the shoulder of the road, pitching rocks into the blue sky toward Interstate 5.

These were not young men. Almost no one in the *Hardcastle* second unit was under thirty-five. Yet, dressed mostly in Levi's, cutoffs, T-shirts, and tennis shoes, many did not look completely adult, but more like ancient boys. About three in the afternoon, one of these boys, named Tommy, climbed into one of the cars, an old blue Merc, in order to slam it into a pile of dirt. The inside of Tommy's car looked like a kid's jungle gym, with roll bars and protective piping everywhere.

Tommy's car and the semi drove up the road and disappeared around the corner. The cameras were positioned and Gary Combs barked into his walkie-talkie, "Okay, bring 'em on, bring 'em on!"

The truck loomed and boomed by, so close to the cameraman it blasted the baseball hat off his head. The hat spiraled up into the air. Tommy slammed the Merc into the dirt pile. Dust was everywhere. The cameraman swore. "I couldn't even see it! I couldn't even see it!"

Someone said, "Hey, what's with Tommy?"

Stunt people congealed around the car. The front end looked like it'd been hit with God's fist. Gary Combs ducked his head close to the window. Tommy emerged. Spidery blooms of blood wiggled from above his eye. His shirt was already blobbed with it. "It was my fault," he said. "I didn't harness up right."

The medic said, "Oh, Jesus."

Somebody else said, "There's a chunk of him out."

A V of flesh was missing from his eyebrow. The medic bandaged him up. "I'm okay," Tommy said, and minutes later he was back in the car.

The car and the semi took their positions around the corner

at the top of the hill and Gary Combs said into his walkie-talkie, "Okay, let's roll."

Suddenly, instead of the semi, a pickup came wheeling around the corner. "Oh, shit," Combs said. "Get that thing out of here!"

The semi boomed by again. Tommy's car hit the dirt behind it.

"Smoke!" somebody yelled.

"Fire!"

Tommy sat, wobble-headed, in the car. People ran toward him. But it was nothing. He was smiling.

"One more time," Combs said. The water truck wheeled up and sprayed the dirt around Tommy's car, to keep the dust down. Positions were again assumed. The truck flew by. Tommy slammed the old Mercury into the dirt pile; clods exploded everywhere. "That's a picture," the cameraman said. "Perfect." Tommy pulled himself up out of the car, his bandages tinted with blood.

12
September
Song

Ted Danson and John Ratzenberger, waiting on a Friday night in late September to start filming their last *Cheers* before the Emmy Awards, were hunkered above the stage, not far from the VIP booth, which was beginning to fill with people hungry for proximity to a hot show. Tickets were especially scarce now that *Cheers* had been authenticated by the Emmy nominations.

Ratzenberger was filling Danson in on the saga of his career—how he'd made the leap from being a dead-end working-class kid in industrial New England to being a semisuccessful young actor in merry old England.

"Friends of mine," he said, "they had a band, and they played in local bars, like in a hundred-mile radius of where we lived. Every now and then, when I didn't have anything to do, I'd help them lug the instruments and amplifiers, so I could get in and get free beer and dance. So this one place—I forget what the hell the name of the town was—we went in there, it's snowing outside, and I'm dancing the night away. I'm the white Ben Vereen. Towards the end I'm with this broad, and we're clamming each other's thighs with our crotches, just pressed ham. She says, 'You wanna come up to my place afterwards?'

332

"She lived right next to the bar in this old three-story place fixed into apartments, right next to the parking lot of the gin mill. Still snowing, and I'm lugging instruments for the guys, and I say, 'Take it easy, guys. I gotta spend the night with that broad over there.'

"They go, 'Hey, all right, okay, okay!' They take off in their van. It's like two thirty in the morning.

"So I walk upstairs, open the door, and I see her sitting in her bedroom doing something, her fingernails—I don't know. I take a step towards her, then I hear noises behind me, and I turn and I look in the kitchen, and there's like seven guys, standing in the kitchen. Now, all of a sudden, I think, 'Shit. There's a blizzard outside, I'm forty, fifty miles away from home, and my ride's left.' In normal circumstances you'd split, right? But I couldn't.

"I didn't know what the *function* was here, so to kind of get my bearings, I went into the kitchen, which was obviously the waiting room. I found out she had asked everybody up there, all these roosters hanging around the kitchen, smoking pot, and having a beer, trying to figure each other out. There were no two buddies up there. She had gone for like seven or eight *strangers*.

"Then they start talking about how tough they are. All of a sudden, they start saying, 'Yeah, I was in the Navy on this destroyer, and we pounded the hell out of Dem Qua Twa,' or some fucking thing. And I had never been in the service, and they're going all around the room talking tough army stuff. Here again I'm thinking, 'If I could steal a car, I would, just to get the fuck out of here.' But I can't.

"One of the meaner guys in the group says, 'What about you? Were you in the service?'

"And from somewhere, God helped me out and gave me the answer. I said, 'Yeah, I was in for a couple of weeks.'

" 'Couple of weeks? Whaddaya talking about?'

" 'Aw, well, they threw me out of the Special Forces. 'Cause I was too crazy.'

"There's just silence for quite a while, then a shuffle, and one by one, they start to sidle out. I got thrown out of the Berets 'cause I was *too fucking crazy!* They're all leaving.

"Now, it's like five o'clock in the morning, something like

that, and she's still in the bedroom. I say, 'All right, I might as well cash in my chips here, collect the door prize.' So I go in there, and we're shmoozing on her bed, my hand slips underneath the pillow as we're shmoozing, and there's cold steel. I pull it out; it's a forty-five-caliber automatic Colt. After going through the kitchen stuff, I'm like 'Whoa! What's this for?'

" 'Oh, it's for my husband. We're separated. Sometimes he comes by and goes a little insane.'

"Thank you, thank you. I get my stuff on—it's like five thirty, six o'clock in the morning—and hit the road. I hitchhike. Six o'clock in the morning in a blizzard. There's not a hell of a lot of traffic. It's not the *Autobahn*. Seven, eight o'clock, a third car comes by. It's a girl I recognize. She goes out with a buddy of mine. On the way to work or something. I get in the car. I'm freezing, cold, wet, I haven't eaten, hung over, all that shit. I musta looked a wreck. I mean just like a bum.

"So she's driving along into town and says, 'I thought you told me a couple of months ago you were going to England to be a great actor.'

"I say, 'Yeah, I just haven't gotten it together yet.'

"And she looks at me and says, 'You know, I *knew* you'd never get it together.' I was gone three days later."

Tonight was going to be Ratzenberger's big night—he and Rhea Perlman were the central characters in the show, which had been written by Heidi Perlman, Rhea's younger sister, now story editor at *Cheers*. Two years previously, Heidi, having watched her sister and brother-in-law Danny DeVito act in countless sitcoms, decided that she had the ability to write them. With absolutely no previous experience, she sat down at a table in her apartment, dredged up a script in a month's time, and took it to Jim Burrows. He wouldn't buy it, but encouraged her to try again. He bought the second one. Now she was in roughly the $100,000-per-year income bracket.

After Ratzenberger left to zip into his mailman's costume, Ted Danson groused about his part. "I don't drive the story tonight," Danson said. "I haven't had a script where I moved the story for a couple of weeks. I'm going to go talk to Jimmy and Les and Glen about it. Last year I'd have just pretended that it didn't bother me."

Danson was looking for a movie role for the spring hiatus. "People are asking about my availability. There's always TV movie stuff, but I'm not really going for that; I want to do films. What I need to do next is films, and there are starting to be rumblings. John Boorman has been in touch, and so has Ray Stark. But they're just asking. It means nothing, so far."

What Danson really wanted was not just to be one of the numbers in the hat when a movie role came up, but to be the name that made the movie happen, like Robert Redford, John Travolta, Dustin Hoffman, Burt Reynolds, Clint Eastwood, Richard Pryor, Eddie Murphy. Nine times out of ten, scripts and concepts and directors and producers are nothing more than cosmetic or utilitarian afterthoughts. In the beginning is *the name*. Danson knew that a popular TV show and just one hit movie could transform someone like him from mere actor to *name* in one Roman-candle burst of alchemy. It had happened to Travolta and Eastwood, and to Goldie Hawn. Just that year, Eddie Murphy had gone from being a popular player on a late-night show to superstar status with a $15 million movie contract, on the strength of two movies, *48 Hours* and *Trading Places*.

In Danson's favor were several factors—his age, his exposure on TV, his looks, and his talent for the work. He had been doing, he thought, the best work of his life in the past couple of years. The good period had started when he'd asked Casey how he'd been in an ABC *Movie of the Week*. She told him he'd been "weird."

"At first I was heartbroken. I said, 'Whaddaya mean, weird?' and she said, 'You're not being you.' Which translated to 'You're not being the way I love you.' All of a sudden I went, 'Whoa! Wait a second here! Here's somebody who loves me, and finds at least part of me absolutely charming, sexual, intelligent, et cetera, et cetera. So why not take a look at it through her eyes and see what she sees and go for that?' "

After that, Danson had tried to stop, as much as possible, relying upon tricks of characterization and delivery. "A trick is whatever you did once that worked. Acting is only good when it's right now. If you start removing yourself from that, you're like somebody who is being charming while they're not really there with you—they're off somewhere else, being charming by rote. It causes you to get that same sort of bad taste in your mouth." Danson had tried to abandon the "razzmatazz, catch-me-if-you-can kind of stuff. I tried to learn how to just be there, as myself, hanging out and letting people look at me, which is what they want to do. Standing still and letting someone look at you, to me, is agonizing, but it's what being a leading man is all about. You go about your business, aware that people are looking at you, but not allowing yourself to think about it. What acting really is, somebody once said, is pretending, while you're pretending you're not pretending."

A few yards away, off to the side of the stage, amid a stampede of orange canvas-backed director's chairs, John Ratzenberger, about to go onstage for the biggest TV appearance of his career, shifted nervously from foot to foot, teetering in front of Glen and Les Charles like a child in urgent need of a bathroom. "Glen. Les," he said gravely, his voice hollow, "I don't want to worry anybody. But I just took some acid."

"Don't worry," said Les Charles, "probably about half the audience is loaded, too."

"I'm gonna be okay. I am. I am. It wasn't the brown shit." Bogus LSD trip aside, when the Charles brothers left to go over a last-minute script change with Heidi Perlman, Ratzenberger sat down and launched a manic rap about his new videotape machine. "I waltzed in there and paid cash for that baby," he said. "Just threw the money down and said, 'Box that sucker up.' Man, I was raised not to buy that kind of shit. I'm, like, genetically incapable of buying something that expensive. That all changed, though, when the network hauled us back to New York for that advertisers' party. They put us up in these suites that went for, like, a grand a night, and chauffeured us around in limos. It gets to you. It gives you the *taste*. So I just sauntered into this store— the kinda place where I used to go in and they'd send Bruno the Floorwalker scurrying over to follow me around and go, 'Can I help you? Can I help you?' And the dude wanted seven-hundred-fifty dollars for this machine, so I say, 'I'll give you seven hundred. Cash.' He takes the seven. Tries to sell me a head cleaner for thirty dollars. I say, 'I can get that for ten bucks around the corner.' So he gives it to me for fifteen.

"It felt great. What it reminded me of was, once this buddy

of mine, his mom sent him twelve thousand dollars and said, 'Buy me a Cadillac for when I come to visit.' So we go out to this dealership, and they come up to us like 'What the fuck do you guys want?' Just treated us like dirt. My buddy says, 'I'm gettin' outta here.' So we drive down the street to this other place. My buddy crawls into this big Caddie and the salesman starts flying over and my buddy says, 'If you can have this piece o' shit ready in fifteen minutes, I'll give you cash for it.' Well, they *had* it ready. So we jump into it and cruise by the first place. And my buddy kamikazes the Caddie through their bushes. Across their lawn. Stops on a dime. Gives them the finger. And screams off.

"That was how buying that tape machine made me feel."

Ratzenberger got up to do a major scene. He did it well enough to get a hand not just from the audience but from the other actors, too. When he came back to his chair, he said, "I came off probation tonight."

Tony Colvin was home working, but Scott came to the shoot, fresh from a basketball game in a Mormon league. He liked the league, but didn't like praying before the games. Reveling in his new status as a nongofer, Scott held court among his former PA colleagues as they arranged the postshow buffet and shlepped tubs of iced beer up the stairs. From the stage, story editor Earl Pomerantz entertained the audience with TV-western theme songs as set changes were made. "We've got to pitch our rewrite on Monday," Scott said. "They liked the jokes, but they want it to be more believable. I'm not looking forward to it. If they don't like this rewrite, we're in deep shit.

"Earl Pomerantz, the guy that's singing out there, he wrote a script for *Phyllis*, and everybody was sitting around the table, and Cloris Leachman got up and said, 'Who . . . wrote . . . this . . . shit?' They say Earl Pomerantz has never been to table since.

"Your life's on the line there; you're listening for those laughs. If the big shots are there and they laugh, all of a sudden you might have an arm around you and some executive producer saying, 'Hey, pal, we've got a movie we're trying to script,' or 'Have you guys thought about putting a pilot together?' "

"Sounds awful," said one of the PAs. He didn't mean it.

Toward the end of the shoot, Ted Danson sat down in the

grove of orange canvas chairs and flipped through a copy of *USA Today* until he got to the entertainment section. His picture was on the same page twice, once with a story headlined "CHEERS" AND "BUFFALO BILL" VIE FOR BEST COMEDY AWARD, and once with a listing of actors who deserved to win. But his name was absent from the list of those who probably *would* win. Supplanting him was Alan Alda. Danson said, "If I win, in my acceptance speech I ought to first thank Alan Alda, for losing."

Though nervous, Danson was in a good mood. "I've stopped smoking! Both pot and cigarettes—I feel a lot better. If you smoke dope, you can do your work, but it fucks up your home life. We're trying to adopt an infant now, and I'm writing to an Episcopal priest, to see if he can help. I'm being much more active at home. We're also looking for a house to buy. When you work eighty percent of the time, and then smoke pot the twenty percent of the time you have off, it's a real fuck-you to the family." He laughed. "I sound like a reformed alcoholic." Danson tossed down the paper, and began to look around distractedly. "Anybody seen my wife?" he asked.

Hamilton Cloud sat in an empty audience seat to watch the end of the pickups and talked about the mood in Tartikoff's office. The network programmers were not saying it, but they expected to score big on Emmy night. Which would be of particular benefit to the quality newcomers, like *Yellow Rose, Love and Honor, Bay City Blues, Mr. Smith*, and *Boone*. It would also blunt the cable threat by demonstrating that the narrowcasters were not the only ones putting good stuff on the air.

"I think *Cheers* will win," Cloud said. "If anybody beats it, it'll be *Buffalo Bill*, believe it or not. I talked to friends on the judging panels, and they said that in the screening sessions, *Cheers* and *Buffalo Bill* were getting the biggest laughs. I think Alda will win. But Ted might. Dabney Coleman doesn't have a chance, because he'll be perceived as a theatrical actor doing a quick TV series. Shelley will win—easy. And I think Nick Colasanto will beat Eddie Murphy, because I think people feel like Murphy shouldn't be in the supporting-comedy-actor category, since he's not in a half hour. I think *Hill Street* will win a bunch, and maybe *St. Elsewhere*, too. I just hope the viewers don't think it's an NBC production, just because we're televising it, and two of our stars, Eddie Murphy and Joan Rivers, are hosting, and because we got a hundred thirty-three nominations."

Cloud had been asked to be a judge in the miniseries category, on the strength of his work several years ago on *Shōgun*. Cloud, who was thirty, had come to NBC four years earlier at the invitation of Brandon Tartikoff, whom he had met at Yale. Cloud had been working as a student intern at a New Haven radio station, and Tartikoff was a recent graduate employed at a local TV station. Tartikoff would come to the radio station to prepare promos for the afternoon movies. Tartikoff always seemed to have some new idea for a commercial that hadn't been done at that station before, and Cloud was the only staffer who consistently found a way of creating the effects that Tartikoff wanted.

Several years later, in Los Angeles, Cloud was producing a local-TV kids show, starring Sarah Purcell, later of *Real People*, when he heard that Tartikoff had been named the number two man on the West Coast at NBC. Cloud sent Tartikoff a congratulatory note, and Tartikoff responded with a job offer.

One of his first major assignments was to help shepherd *Shōgun* to success. After months of exhausting work, Cloud came home to his apartment on the first night of the miniseries and turned it on. Then he got a phone call and had to turn off the sound on his set. Suddenly he could hear the program coming at him from the apartment above him, and the apartment below him, and the apartments on both sides—and he knew he had helped pull off something big. He was still in his twenties when he became head of current comedy at the network.

Cloud had only the initial sampling of *We Got It Made* to go by, but he said he was "quietly optimistic" about the season. "When a show gets the kind of promotion *We Got It Made* received—a *lot* of promotion—it ought to get a good sampling. Personally, I thought the promotion was a little too suggestive, but apparently it worked. We'll just have to see how far it will drop—where it will level off. *Hardcastle and McCormick* on ABC got a lot of promo and a big sampling, too, but I think what you'll see there is that after about four shows, it will drop off. Because it doesn't have anything. Same thing might happen to *Hotel*—I heard it was boring."

Cloud thought that *Yellow Rose* would "clean up" against

Fantasy Island. "I think *Fantasy Island* will end up being a twenty- or twenty-one-share show, giving *Yellow Rose* a lot of momentum, maybe even enough to then move it straight against *Dallas,* which I think is starting to have a passé feel to it. But *The Rousters* will have a much tougher time, because I don't think *Love Boat* is that vulnerable. The kids will go to *Rousters,* but with the shifting-focus format *Love Boat* has, they can't really get stale. *Love Boat* could go on forever."

Cloud was worried about *Jennifer Slept Here.* Its competition, *Webster,* had had a Top Ten premiere, and Cloud had to admit it—the kid in *Webster* was the definition of cute. Still, *Jennifer* had sexy Ann Jillian, and as *We Got It Made* had so potently illustrated, sex sells, even better than cute.

As the pickups wound down, a PA appeared with a wet bottle of Heineken for Cloud. He pulled at it automatically, unenthusiastically. He had a distant look in his eyes, like that of somebody about to go to war.

Tony sat down with his Sunday night dinner—two massive jelly jars of white Gallo and a cold taco—and for the umpteenth time went over the rewrite for *Just Our Luck.* ABC wanted "plausibility" in the script about the two-thousand-year-old genie, so he was going to stuff facts into it until it read like the goddamned *World Book*—but funny, too. He put Hiroshima on the stereo and worked until 3 A.M.

At 9:30 on Monday he and Scott pulled into their parking spaces, and at 10 were sitting in the conference room with the producers and writers.

"Go for it," Scott said and Tony, his stomach jumping, started pitching the new story.

"Okay, the weatherman sees Shabu"—the genie—"so what Shabu and Keith do is make him think he's seen an alien, so they can keep Shabu's cover. For the alien, we get some strange old star. Somebody like Julius La Rosa."

"Too obscure," said one of the producers.

"I was thinking about Slim Whitman," Tony said. "Or somebody like him."

"No," said a producer. "This won't work unless you've got just the right guy as your alien." Tony felt the room tense—or maybe it was just him.

"Don Ho!" said Tony, forcing enthusiasm.

"Naw."

Tony thought he saw a look pass between the two ranking producers that said, "Who the hell hired this guy, anyway?"

"I've got your man," said one of the writers. Tony hoped to God he did. "Roy Orbison."

"Roy Orbison!" said one of the producers. "Perfect. He wears the shades, the satin jacket. Right? He's the Supreme Being!"

"Roy Orbison would be *excellent*," said Scott.

Now the meeting turned into a love-in. Tony had all the right moves—buttons, heat, inanity, silliness, and more credibility than the King James Bible. They were in.

When they got out the door, Scott, who had just turned twenty-six, said, "Who the fuck is Roy Orbison?"

Lining up Roy Orbison took only a matter of days. While it was still mid-September, the shooting of the episode began and Tony and Scott spent their lunch hour watching. The sound stage had been set up for a scene at a railroad crossing on a country road—a parody of the movie *Close Encounters of the Third Kind*. Tony looked around the mammoth MGM stage and thought to himself, "This whole thing is being used to produce *my work*." He felt like David O. Selznick. And when they did a run-through of the scene, the bystanders laughed at *his* jokes. That was even more exciting. "Twenty million people are going to laugh their damn heads off at *my* stuff." Walking out on the big money at First Interstate, swallowing his pride at *Cheers*, signing on at *Just Our Luck* with a short-term contract—all of a sudden, these moves, one and all, seemed to be the work of a shrewd and canny brain.

But when they got back to the office everyone was in a terrible mood. The NAACP, together with a group called the Black Anti-Defamation Coalition, had just declared war on *Just Our Luck*, publicly railing against "a black/white, master/slave relationship being sold as 'comedy' in 1983." The black genie had promised in the show to serve his master for "2,000 years or until your death, whichever comes first." The coalition stated, "We've already put in 400 years. Hopefully death (for this show) will come first."

The coalition had already forced NBC to make significant changes in a movie starring Gary Coleman. It had good public relations muscle and knew how to use it. It had also begun to privately lobby ABC to drop the show, or to at least let the group have advance screenings of its episodes.

The NAACP had sent word through its Hollywood president, Willis Edwards, that it, too, was ready to launch a public relations blitz. In a statement to the press, Edwards said that the *Just Our Luck* staff included "not one black writer. We thought this typifies the blatant and rampant racism that exists in the television and film industries." The NAACP was also threatening a boycott of the show's advertisers.

Suddenly Tony felt sick.

A narrow, lightly bearded boy in a red football jersey zigzagged dreamily through the crowd around Bob Radler's camera truck and told Jackson Browne he'd never heard of him because he was mostly deaf and didn't listen to much rock 'n' roll. "What is this?" he wanted to know.

"A video for MTV," Browne explained.

"Huh?" The boy scratched at his transparent beard.

"You know," Browne said, "Music with pictures. I'm doing it for my record album."

The kid nodded solemnly and then told Browne he was from Watts and that he used to carry a baseball bat, "with a spike banged through it" for protection, until he found Jesus Christ.

Browne smiled and the boy's mouth grew small. "You doubt I know the Lord?"

Browne allowed, "Not for a minute."

Ten feet overhead Radler stood on top of his camera truck under a big halo of gold hotel light. Walkie-talkie in hand, he directed the traffic of classic cars assembled for the shoot down the neon corridor of Fremont Street. The Las Vegas night was starless, but the casinos and their attendant restaurants, strip joints, and pawn shops shimmered, marquees offering $1,000,000 PAYOFFS, DOUBLE ODDS ON CRAPS, THE MOST FREQUENT SLOT PAYOFFS ON EARTH, 10¢ ICE CREAM, WHOLE PIZZA BAKED FRESH FOR 50¢, FREE BEER AND POP.

MTV videos—more often than not sexy, violent, hallucinogenic little fables hooked loosely around the rocker and his rock— had become, over the previous year, cable television's most distinctive and potentially profitable single ware, and had, all by themselves, rescued the sagging record industry. Or so the analysts thought—and Bob Radler agreed.

During the seventies Radler had made a name for himself producing films and commercials for clients ranging from cat litter makers to the CIA. But doing TV commercials drove him half crazy. He'd gotten his start as apprentice to a Marxist-Leninist ad producer in New York who did spots for major oil companies. "The guy I worked for was brilliant. He did this one public-service thing for the oil people in which we had all these little black kids playing in Central Park. In the punch line scene, one little boy climbs up this apple tree and shakes blossoms down on the face of his friend, who was lying on the grass. It was a beautiful photographic image, the blossoms brushing against his chocolate skin. The oil guys totally freaked out. They thought the scene represented *little Negroes defacing public property.*"

Radler got his first big break in TV commercials when he moved to Massachusetts. His first major commission was a promotional film for the Commonwealth of Massachusetts, which Radler had pitched to include "thirty-five of my favorite" things about the state. What were those things? Among them: a shipyard and a bagel factory and Leonard Nimoy. Who happened to be from Massachusetts. Seeing Radler's rough cut, the folks at the agency went through the roof. "Mr. Spock? Bagels? What is this? Where are the parades? The guys in Revolutionary War outfits?" Radler said those weren't his favorite things. "My God," one of the agency guys said, looking at the promo on the screen, "he's got the governor of the state walking around on the moon. What the fuck is this?"

"Actually," Radler said, "that's the governor at the state lunar pavilion, at the Apollo display."

"Listen," another agency guy said, "he's got somebody in here saying he likes California better! What state did you think you were promoting, anyway?"

"That was supposed to be a little humorous counterpoint," Radler allowed.

"Six months of work," said the ad man, "and he calls that humor."

"Who's the hippie singer?" the other agency man asked.

"Arlo Guthrie," Radler explained. "I got him to pitch in."

"Who the hell is *he?*" they wanted to know.

"They didn't know what to do," Radler said, "so first they sent the agency accountant to try to deal with me. Yeah, the accountant. Then their attorney. Then the creative director. 'You've got a car crashing in here,' he said. 'What kind of promo is that? When the governor sees this he's going to shit a battleship.' See, the whole thing was the governor's idea. So finally, the president of the agency came in to see it. This guy wearing herringbone slacks and carrying a gold cigarette case. He saw it and said, 'It's the weirdest thing I ever saw but, actually, I kind of like it.' Next they sent one of the state commissioners in to see it, and he said the same thing—that it really had some moments. But everyone was so scared of it they decided to arrange a private showing for the governor himself. Everybody files in. They roll the thing. A few seconds into it, the governor—Mike Dukakis—smiles. So everybody else smiles, right on down the line like the fall of a string of dominoes. Then the governor laughs, and everybody yuks it up, right on down the line. When it's over he says, 'I love it.' "

The promo turned out to be such a hit that Radler got a whole slew of TV commercials after that. But Radler was not what most clients expected. Certainly not some little unctuous blow-dry concoction in a Paul Stuart suit. No. Radler was big. Tall, too. He spent his spare time playing in a rock band, and if he did not look like a biker exactly or someone whose face has been made famous by the post office, he certainly stood out in the button-down world of Boston ad agencies. He felt he needed creative freedom to do his best work and found client involvement often tantamount to having Typhoid Mary bring serum for the plague. "Most often," he said, "I could listen to the suggestions and if they were good, well, great; if not, I'd just nod and do it my way and when the ad came out all was forgotten."

But still, there were problems. Like when he landed the cat litter account. "The product rep looked exactly like Lee Harvey Oswald. I mean, exactly. Enough to make you wonder if Ruby really got his man. The agency guy looked like Kojak and when they came into my office the day of the shoot, the first thing Kojak says is 'We should be doing this in Ha-lee-wud. I tell ya, we should be doing this in Ha-lee-wud.'" The gig was simple. One cat had to knock over the "Brand X" cat litter. The other had to lie down in the good stuff. "The problem is," Radler said, "you just can't direct cats. It's impossible." The first scene was relatively easy, though. Radler got a kitty to bat the Brand X stuff over by smearing cat shit on a crease of the bag. But none of the eight cats he'd assembled would lie down in the good cat litter. Or, for that matter, even relieve themselves in it. "It was a nightmare. My office never smelled the same afterwards." Radler tried everything. He tied a rope to one of the kitties' tails. At the suggestion of a vet he took an eyedropper and tried to force-feed another one blackberry brandy. One of the kitty owners dropped by the office and was mortified to see her beloved puss secured firmly to a table with gaffer's tape. Finally, after twelve hours in the office with eight agency people in the room trying to get ten seconds' worth of film, one of Radler's assistants said, "Dig it, I've got a joint of killer weed. Whaddaya say we try getting one of the little bastards stoned?" Desperate, Radler took the cat into an adjacent anteroom and began blowing smoke in its face. "It worked like a charm. The kitty turned into Sir Laurence Olivier right in front of my eyes. We got the shot done in about two takes. The ad turned out perfectly. But Lee Harvey Oswald, the product rep, went ape. I thought I'd never hear the end of it. 'You were stoned! The kitty was stoned!'"

Not long after that, Radler won a Clio Award for his ads for the Boston Bruins hockey team, and shortly thereafter—burned out on TV commercials—moved from Boston to Los Angeles, to do a feature film on the Doors. But that project collapsed after going through the hands of everyone from Brian De Palma to William Friedkin to John Travolta. Radler was then given charge of two MTV videos created around the heavy-metal band Blackfoot, and moved on to create, direct, and produce two more by Crosby, Stills, and Nash. Before long, he caught the attention of Jackson Browne and hired on to produce his video.

A mere four hours before Jackson Browne's encounter with the deaf boy in the street, Radler had lain staring up at the ceiling of his suite at the Golden Nugget, waiting for Browne to fly in from who knows where, wondering how in the world he was going to do what he'd promised: to create a complete video from scratch in less than three weeks. Bob Radler had a logger-from-Harlem style—plain untucked shirt over a black sleeveless T-shirt, tight jeans, and scuffed-up running shoes. He was tall, with a mop of curly black hair and the potent reserve of a bartender who'll allow you to make a complete obnoxious fool of yourself in his place at least twice before drop-kicking your ass. The video he was here to create was to be taken from the "Tender Is the Night" single off Browne's album *Lawyers in Love*. The problem was, the single had already been on the charts for weeks and as yet, no video. A previous producer had already tried to shoot the video, but the results had not pleased Browne—so here was Radler, ready to start from ground zero.

Radler had the script to the video scrawled on sheets torn from a legal pad and fanned out on his bed. He'd put the whole project together in four days with the help of his cameraman and co-director, Doug Ryan. Ryan was a short, wiry man decorated several times in Vietnam, who had previously been a cameraman on feature films, including *Apocalypse Now, Bronco Billy*, and *Scarface*.

By the summer of 1983, feature film people were flocking to MTV by the limoload. The top names and innovators. *Medium Cool* director Haskell Wexler had just finished a tape for Tom Waits. Bob Rafelson, director of *Five Easy Pieces* and the 1981 remake of *The Postman Always Rings Twice*, had just completed a $115,000 video for Lionel Richie. "When you're a filmmaker," Rafelson had said, "you get tired of waiting around for all the meetings and packaging to take place. All the deal making was driving me crazy." Rafelson, for one, saw film and video as very similar. "You create rapport by being honest with each other: You agree that your job is to make the star look good and you keep everything moving."

Which Radler was going to have to do very quickly. Browne's record company, Elektra, was sinking. The Elektra management considered Browne's *Lawyers in Love* LP the bucket that might bail them out. But albums are sold off singles and, more and more, singles are being sold off MTV. Their chance was slipping over the horizon, day by day. Hence, Radler.

The phone rang. He picked it up, said, "Yeah, yeah, sure, if we have to, okay, bye," and hung up. "The Teamsters," he explained. On top of his crew of twelve, Radler had also been forced to hire three Teamsters to "help" with production at about $25 per hour apiece. This was necessary in order to get the city of Las Vegas to allow his crew to film downtown. "The guy I'm dealing with as a negotiator," Radler said, on his way downstairs to meet Jackson Browne in the lobby, "told me his qualification for the job was that he was standing next to Joey Colombo when he got blown away in New York."

Browne was slight and precise. He wore a black-leather flight jacket and pegged pants, had poreless skin, ice-white teeth, and fine dark hair he probably could have sold. Despite the fact that his face was on the cover of the *Rolling Stone* magazine stocked in the lobby of the Nugget, Browne could still walk in front of the hotel and not be recognized. MTV would alter all that.

Shooting began at once. At a cost of about a thousand bucks per hour. Ryan shot Browne being spare-changed by young Indian winos under a sign that read A-GO-GO FOR LESS. Browne patted his pants and coat pockets and passed out quarters. "Who are ya?" one of the winos asked.

"Jackson Browne."

"Gladtameetcha," said the squat-faced Indian, putting out his hand. "I'm Charles Bronson."

Radler backed down the sidewalk as they did a shot of Browne striding toward the main drag, Radler hissing, "Don't look at the camera, don't look at the camera, don't look at the camera," to black extras who were weaving around Browne on their bicycles. Radler intended to make a "little movie" with this video, a four-minute film with images married to the plot of the song itself, that of two lovers fighting and losing each other to the attractions of a funky night in a gritty place like Vegas.

Toward this end Radler had commissioned a mob scene. Hundreds of people had been scouted up off the streets to walk around Browne as he was filmed approaching a line of beautiful classic cars from the fifties and sixties. An immense bile-yellow water truck sprayed fans of water all over the street to make it shine for the cameras. Everything became glittery, glowing, and reflective, from the hard sparkle of chrome wheels to the signs above the street advertising the world's largest taco and the most liberal twenty-one in the world. But after an hour or so, Browne told Radler, "These cars are an extension of my persona. They look too clean, too slick and midwestern."

Radler shook his head. "Look, don't worry. You're going to get what you want. You don't see what I'm shooting."

Browne looked up and down the street. "I don't know."

"Jackson, believe me, I promise you, we're going to get all the details we co-wrote—everything from the dirty alleys to the shiny hubcaps. It's not just going to be some glitz parade. But we've got to shoot for details. It's a jigsaw puzzle and we've got to shoot it piece by piece." Radler smiled with half his mouth. "Trust me."

Browne looked up at Radler, his head tilted to one side, his gaze narrowing.

Radler's smile held. "Jackson, don't worry, you'll get what you want."

Shooting continued past midnight and into the morning, but it never got "late." Things continue to continue in Las Vegas. Tireless battalions of senior citizens sat at slot machines inside casinos big as aircraft carriers, seeking the alignment of bells, bars, and berries, while outside Radler, Browne, and crew worked through setup after setup. The click, ring, clank, and splash of row after row of these machines made a sound as regular as a river or surf—backdrop to the blast of Browne's "Tender Is the Night," which played continuously, over and over, from stadium speakers on a pickup parked outside in the street.

Some of the old gals worked three slots at a crack. Every few minutes there was the crash and splatter of a jackpot, but the whoopees were sparse. Slot machiners to either side of the winner would lean to one side—as if all were passengers on the same swaying bus—to see who and how much, then it was back to the machines. Usually the winner was also deadpan, as if good fortune was only a chance distraction from the endless plunking.

The crew shut down near dawn. Some elevatored up to bed immediately. But Radler and a couple of others stayed up to check things out.

Here in the land where Mafia dons come to sun their scars in retirement, everything was possible. Hanker for an "around-the-world" or even a spanking? In Nevada the government says it's cool for the girls to do *anything*. Bet a hundred and win a thousand. But be careful. Or you may wake up in your room at the Golden Nugget watching *The Bugs Bunny/Roadrunner Show* and wishing you had grown up praying to different gods.

"You do it in the water. It's a ride."

"Boating," said Harry Anderson, hard at work on his Q on the stage of *The $25,000 Pyramid*. Behind his lectern, Dick Clark was genuinely enthusiastic.

The woman who was giving Harry clues shook her head, hard, like a poodle shaking off water.

"Californians do it."

"Go to the beach. Swim."

"They do it in wet suits."

"Snorkel. Scuba." Dick Clark looked at the timer.

"Waves," the woman commanded, her teeth not far from full clench.

"Uh. Umm." The buzzer blasted.

"Surf, Harry!" said Dick Clark. "Californians *surf* at the beach."

Harry smiled, but his contestant did not. Harry had just blown her chance to play for $25,000. She looked longingly at Vicki Lawrence, game player extraordinaire and Harry's celebrity opponent on the show, who now moved with her excited partner to the winner's circle. And sure enough, within two minutes, the contestant was twenty-five grand richer.

Harry didn't feel too downhearted, though, because over the five-game "week," which had been shot in one day, Harry had actually beaten Vicki Lawrence, point for point, and had advanced his partners to the big-money round more often.

As a brief contributor to *The $25,000 Pyramid*, Harry, who was anxious on this early fall day to pocket his $1,000 fee and hit the road for some real money, was a minor soldier in a big war: the fight to roust ABC out of first place in daytime, which had been ABC's fiefdom *every week* for the last four years. For over two hundred straight weeks, ABC had wiped out the competition, a phenomenon largely responsible for the network's being the most profitable of the three, netting over $300 million every year since 1978.

ABC had taken the lead in the late seventies by using daytime reruns of cotton-candy series, like *Happy Days, Laverne and Shirley*, and *Love Boat*. Allowed to run each of these female-oriented puffballs twice before ownership reverted to the studios, ABC used them to pump the morning ratings, which are of acute importance because day watchers, absorbed in housework, strongly avoid channel changing. They also develop loyal-to-the-death fixations on their favorite shows.

But when ABC grudgingly returned *Happy Days* and *Laverne and Shirley* to Paramount, its whole daytime schedule began to go soft. Breaking a long-standing tradition of daytime strategy, the network threw a soap opera into a morning, rather than afternoon, slot. It got murdered. Nothing else worked either.

While ABC was floundering, Tinker and Tartikoff put Sohmer in as head of daytime. Shortly after Sohmer began the job, he said, "If NBC daytime gets fixed, it's going to be because a bunch of people just got in there and ground their way out, day in and day out, fixed it, fixed it—this scene and that person—just stayed on it and stayed on it. That's the only way it gets done." The key to his strategy was to bring back game shows in the morning. In the 1970s, the money game show had been thought a programming dinosaur, bound for extinction in a future where crass materialism was going to be consigned to its proper place. Seventies hits like *The Dating Game* and *The Newlywed Game* stressed relationships more than acquisition, and were more like *cinéma vérité* soap operas than game shows. But Sohmer thought that the mothballing of game shows was a stupid screwup. They were hot. They were interactive. Just the stuff the cable-jaded, event-oriented TV watcher now demanded. Plus, they played on greed—and greed was eternal.

There was another aspect of game shows that made them dear to Steve Sohmer's heart: You could crank them out for peanuts. You could put together a lavish production, with a big payoff, for as little as $30,000 in above-the-line costs. That compared most favorably with a prime-time half hour, which would run a bare minimum of $250,000, or a soap opera, which would have twenty to thirty actors drawing a salary for a full fifty-two-week year. And daytime hosts came cheap; the dollar king of daytime, Bob Barker, was only drawing as much as a third banana on a long-running prime-time series, about $500,000 per year. The secondary stars, like Bill Cullen or Wink Martindale, were making less than $200,000, while the journeyman hosts below them were in the $2,500-per-week range—a salary that would be an insult to even the most insignificant prime-time actor. Plus, the Writers Guild did not include question creators as writers, so the unprotected daytime writers were averaging about $400 per week, about 500 to 1,000 percent less than most prime-time writers. Finally, even the prizes were cheap. Most of them were donated in exchange for a promotional plug, and the smaller prizes, like a package of hot dogs, were "fee items," which netted the show several thousand dollars per program. Only about 30 to 40 percent of the total giveaway came out of the prize budget of the show, mostly for straight cash. The game shows, which could be owned directly by the networks and which reeked of generous, unstinted expenditure every time some housewife shrieked over a new dishwasher or fur coat, were low-cost conduits through which poured an ocean, just an ocean, of black ink.

Not far from where Harry was taping, We Got It Made geared up to shoot another episode. The mood on the set was good. If the ratings held, the mood would get even better.

Director Alan Rafkin, in a sweater and jeans that showed off his fifty-year-old-but-still-marine-tough body, courted good luck by predicting doom. "This piece of shit?" he was saying to a cameraman and a propman. "This piece of shit won't even make thirteen episodes. They'll find a way out of the contract, and I'll be looking for another job by Christmas."

"Me, too," said the cameraman, trying to get into the spirit of it.

Rafkin stared at him. "I wouldn't exactly call boil sucking a job."

Valerie Bertinelli, star of One Day at a Time, which Rafkin had directed for five years, walked into the stage at KTLA with Shelley Fabares, also of One Day, who had been a child star on The Donna Reed Show. Rafkin whistled "Johnny Angel," which Fabares had sung in what seemed another lifetime, as he hugged and kissed both women. Both of them were rather distracted; it was a tough week at *One Day at a Time.* Mackenzie Phillips had just been fired—again—and the trade papers said the firing had happened for the same reason as the first one, because of her use of drugs.

Fred Silverman, like Rafkin in a sweater, but one that was baggy everywhere but the belly, came in and trudged upstairs to the control booth. Silverman had just recovered from a serious viral infection, and had also found out that he had diabetes.

The audience, a group of marines from Camp Pendleton, filed in and Rafkin began the warm-up. "We really appreciate your being here so much, when you could be out taking drugs." Rafkin introduced Bertinelli, wearing a teal-blue off-the-shoulder sweatshirt, and the marines went ape. When Teri Copley came out, they grunted like cavemen and stamped their feet. Gordon Farr, watching from behind a camera, smiled appreciatively as the marines roared enthusiasm right through to the end of the taping. It was a *We Got It Made* kind of audience.

In fact, if the A. C. Nielsen Company was even remotely accurate, it was a *We Got It Made* kind of world. But then Fred Silverman, now picking at the nonsugared items on a catered buffet table, had known all along that this was a *We Got It Made* kind of world.

Ed Asner wended his way through a gauntlet of politicos to the stage at the Screen Actors Guild meeting; on both sides of him were long tables manned by pamphleteers and advocates—many of them working for Asner's opponent in the upcoming union election. Asner greeted friend and foe cordially, but many of those he greeted, all of whom were actors, were intimidated—or, more precisely, star struck. These actors usually saw union leaders like Ed Asner and Charlton Heston where the folks back in Peoria saw them—on the silver screen. They did not often press the flesh with the Big Faces for a simple reason—most actors don't act. At any given moment a full eighty-five percent of them are doing everything to earn a living *except* acting. In an average week, there are about one hundred available acting jobs in Los Angeles, and fifty-five thousand SAG members vying for them. This unemployment rate had become Ed Asner's albatross in his bid for a second term as SAG president—the membership wanted America's favorite boss to see that all of them had at least a chance to work, and there just wasn't much Asner could realistically hope to do about it.

Consequently, another candidate had arisen, a struggling young black actor named J. D. Hall—a SAG board member but still one of the boys—who had promised his poverty-stricken colleagues that happy days could be here again, if only they would rid themselves of their current rich and famous representative. Hall was working the foyer, handing out leaflets and shaking hands, as Asner made his way to the speaker's platform. Hall had just finished appealing for support from a number of Asner's big-name, big-money enemies—God knew Asner had enough of them—and if even a few of them came through, J. D. Hall would soon find himself running a well-greased campaign machine. He would be the most enviable of political entities—the bucks-up people's man.

"I think I have an excellent chance of winning," said Hall, who was the first black presidential candidate in the union's fifty-year history. "I have a strong program for reducing unemployment, and Ed doesn't." And there was, as always, the ghost of El Salvador: "I think Ed has been a little naïve about not understanding the political ramifications of his support for the rebels." Lastly, Hall believed that the rank and file was sick of being represented by industry aristocrats.

But onstage, as the meeting came to order, Asner asserted that being a superstar was not all it was cracked up to be. "Many people see celebrities as wealthy, carefree sorts," he said, "oblivious to the problems of the great unwashed herd. It's not true. I checked with my business manager yesterday, and found out that in order to take care of my family, I not only have to die— but die accidentally." The joke got a forced laugh from the unwashed herd: tough toenails, Ed. And from there, enthusiasm waned.

In the back of the auditorium, J. D. Hall smiled wanly.

"This is insanity," Bob Radler said. The subject was asses. The assistant cameraman had filmed a couple of the female extras wiggling their tails down the sidewalk in front of one of the casinos. Nothing even R rated, but Jackson Browne's girlfriend, Daryl Hannah, had objected.

Hannah was a ropy-limbed girl of about twenty-two who had

a beautiful face, both strong and unfocused, framed by a long flag of straight blond hair. She was a hot item, having already starred in the kiddy sex feature *Summer Lovers* and in *Blade Runner*. Moreover, she had another three features, *The Pope of Greenwich Village*, *Splash*, and *Reckless*, awaiting release. And she objected that the filming of the two sets of buttocks was "exploitive."

She'd complained to Browne and Browne had complained to Radler. What had begun as a spontaneous gesture on the part of two extras and a cameraman now threatened the life of the project. A problem absolutely endemic to the new medium of music video, where two creative forces—the musician and the filmmaker—must share the helm. Radler's style allowed for latitude: Develop a linear story line, film each ingredient essential to it, and then take care to pick up any and all loose "moments" that occur around the camera work. So that, as "Tender Is the Night" was being shot scene by scripted scene, Doug Ryan was likewise shooting much of the satellite anarchy taking place around the scene. Totally spontaneous bits Radler could sliver into his rough cut in the editing room. "The best stuff," he'd said earlier, "is not what you plan, but what you find."

Unfortunately, Browne didn't seem to know what his producer was looking for. "Let's just forget about the butts," he said to Radler. "Daryl feels she's been exploited this way before and she doesn't want it to happen again." Hannah stood in the background, looking at once glamorous and unfinished, fashionable in ripped jeans and knit gloves, lovely and jointless.

Radler shook his head. "I thought we'd discussed all this kind of imagery before and agreed to just go for anything. We're budgeted to shoot at forty to one." This meant that only one shot out of every forty recorded would make it into the video.

"As for the girls"—he looked down the street to where the two young women in designer jeans were fidgeting with two guys who might have been their boyfriends—"I don't think the extras felt exploited at all. It was their idea."

Browne bit his lip. "We're not making a Jordache jean ad here, we're making a video about my song."

"Jackson, you wanted a video about the American night. Butts in designer jeans are just another part of it."

Radler was now angry and a little frightened. He did not

want to offend Browne, but knew that when a director loses control of a shoot his ass is 85 percent grass. Radler was an expert at managing people. An hour earlier, he had corralled over a hundred tourists in front of the Golden Nugget and made them all instant extras in a crowd scene, then picked out locals among them to interact with Browne—black kids on bicycles, winos, hipsters—coaching the hoi poloi into instant thespianhood while also keeping the crowd in line and directing his crew. Part Otto Preminger, part sheep dog. And, despite his unbuttoned style, Radler's plan for the "Tender" video depended on a rather complicated and precise series of shots. If he lost his ability to call those shots, he was afraid he would lose everything.

So nothing was resolved. Radler called for a break. Browne and Hannah went off to get something to eat. Browne's friend Buddha approached Radler. "Why don't we just forget about this thing? You guys can decide about it in the editing room."

Radler explained that wasn't the point. The point was, Radler said, he couldn't guarantee the quality of his service to Browne if he couldn't maintain control of his set. Suddenly, Radler was out of his element. Visions of Lee Harvey Oswald and the cat litter account. The very thing he'd come to music video to escape. So far he'd been lucky. The Crosby, Stills, and Nash videos had been a snap. He'd simply lit the rock legends "like Mount Rushmore" and let them belt it out. With Blackfoot it had, in ways, been even easier. "They were this southern heavy-metal bar band. We shot two complete videos in three and a half days for almost no money. The lead singer, Ricky Medlock, was fantastic." Medlock, who looked like the son of Marlon Brando and Alice Cooper, proved to be a workhorse and Radler had him crank out his two tunes against a backdrop of pimps, thugs, and topless dancers Radler dug up on the spot in the sleaziest section of Hollywood. And, despite the fact that Radler's fiancée was nearly raped by a bunch of Mexican lowriders in the process, the Blackfoot shoot had been a piece of cake.

Now he was in different waters. Browne was a superstar and a man not accustomed to delegating creative authority. Radler found himself in the position of an attorney whose client suddenly wanted to take his place cross-examining people on the witness stand.

Shooting resumed, Radler realizing he was going to have to

serve as both mentor and minion. Browne had already talked about wanting to throw himself into video, being the first to put out an entire video album if possible, and Radler knew now that he was going to be the hands on this project, not the head. And that he was going to have to make the transition from team owner to team coach very fast. He was determined to keep "Tender" under budget—shooting costs were running at $1,000-plus an hour. He was also determined to keep his client happy, so long as he didn't have to play Stepin Fetchit in order to do it. Despite his success over the summer, Radler wasn't yet the genre of producer who had a fleet of polo ponies and a deck of gold Master Cards back at the manse in L.A. No. And, despite the fact that he had another job almost lined up with a punk band fronted by a cute blond girl who made interesting squeaking sounds and wore transparent Plexiglas over her breasts, Radler thought "Tender" was a great song and could still make a classic video if he just kept a wrinkle in it and gave Browne some room to maneuver. Besides, what was he going to do—quit? Cook the footage? Hardly.

The cutting process was beginning to concern him now, however. "Editing is the most psychologically personal thing I do. I can handle all the ideas and instructions in the world, but I have to do the actual editing, the 'finding' of the film within the footage, myself." What was going to happen when they sat down in the editing room in New York? Radler hadn't the foggiest.

Still, he had confidence, both in "Tender" and in his chances of making a solid name for himself in this, the hottest new venue in recent TV history. Bob Radler had seen the future, and it was even more commercial than sex. . . . It was MTV.

For at least two decades Americans had been watching three camera, one-set sitcoms wrapped by ads that were the slickest film products since Sergei Eisenstein. Ads that cost up to twenty times more per minute than the shows they supported. Ads that had created the most visually sophisticated audience on earth: the American TV viewer.

For such an audience, Radler knew, the shows being churned out by the networks were . . . *predead*. Toes in the air before the first day of taping, before the first silly concept had passed jowl-to-jowl across a tray of margaritas at Chasen's.

And now, marvelous, a TV channel that was . . . *all ads.*

Real, honest-to-God black widow spiders had spun webs all around the alley where Radler was shooting, and one of the motorcycle cops he had hired to escort the production company was gleefully plying them with Mace, making the arachnids, in his words, "do the chicken." That is, to go into spasms as they died.

It was as if he'd been offered the world's largest orgasm and then told there was only one catch: that he'd have to wait for it, that this orgasm might take anywhere from two weeks to forever to occur. By any conventional yardstick, Allan Katz was a man who had everything. Talent, health, money, a plausible backhand, the love of a beautiful woman—actress/comedian Catherine Bergstrom—and enough friends to fill a cathedral. But now everything in his life was focused on getting *The Hunchback of UCLA* into production and nothing was happening. He found himself a kept man, paid a king's ransom by the studio to simply "be patient." This for a project he had every reason to believe was to have gone into production months ago. Katz was frustrated and confused. It had been almost a year since he'd done the screen test and made his deal with Twentieth Century-Fox and, despite the fact that he'd lined up Tische and Avnet to produce the film, he found himself having to go into meetings with Twentieth execs just to see if they liked the story. "It feels like the movie is back in development again," he said. "I have a contract, I thought I was holding the cards, but I talk to some of these people and it's like I'm pitching the idea for the first time." Twentieth Century was playing its own cards close to the vest. No one was saying much one way or the other and Katz was worried that his film was going to get axed simply because it was too unconventional.

"You know," he said, "sometimes I think it's better when your dreams are far away and unattainable and you can just sit around and complain that no one will give you a chance. When someone finally says, 'Okay, we'll give you a shot . . . maybe,' that's when it really gets uncomfortable. Because you can almost taste it. But it's like biting air."

Regardless, Katz was willing to compromise on *Hunchback*

only to a point. He was determined to star in the film. "Before I came to Fox I spent a lot of time looking for independent financing and one day I got a call from a money raiser–producer who said, 'Look, I talked to some people at Warner Brothers who said they loved the Hunchback and there would be a ninety-nine percent chance you could play the lead if you agreed to sign with them,' and I said, 'No thanks, I don't like the odds.' " Also, Katz was now totally convinced that Danny DeVito was the perfect man to direct, inexperienced or no. "He says all the right things. Not politically—in fact, there's a lot of the stuff in the script he wants changed. But he just has one great idea after another." And Katz was certain that, working together, they could conjure up a creative total that would be more than the sum of its parts. But would Twentieth buy DeVito, an untested feature director charged with steering an untested feature actor through a production whose costs would equal that of a good, used skyscraper?

Maybe. "The thing about films," Katz reasoned, "is that while the risks are greater, it's actually a much more adventurous medium than TV. Once they get behind a project they're much more willing to take chances." And it didn't hurt a bit that Katz had Tische and Avnet behind him. Their film Risky Business had grossed almost $75 million for the summer of 1983, and they made no bones about the fact they saw the same kind of potential in Hunchback.

Meanwhile, Katz had finally gone to ABC to pitch his idea for a low-concept Charles Durning sitcom. The first meeting had not been precipitous. Katz had trooped to the network with Harris Katleman and Peter Grad in tow to see Lew Erlicht, who had just been named head of ABC Entertainment, and Tony Thompoulos, the president of ABC itself. "I pitched the Durning character as a family man with traditional values and integrity and a lot of humor and compassion. But I kept getting the feeling that they were waiting for the other shoe to drop, for me to say, "And, in addition to that, his brother is President of the United States.' " A second meeting with Erlicht, however, had gone much better. Katz went into the show in detail and Erlicht was impressed to the point that Katz had reason to believe ABC would bite. Still, Durning had a thirteen-show commitment and once ABC said yes it was committed not just to a $750,000 pilot but to a full half-season investment. At least $5 million. And at the moment everything, absolutely everything, was still up in the air.

Time for affirmative action. On the home front, Katz's car—his Porsche Targa—was throwing oil so bad he had decided to replace it. Consequently, he'd just shelled out $31,000 for a stop-sign-red 1979 Ferrari 308 GTS. It sat, looking like a metal shark designed to break the sound barrier, in front of his office at Twentieth. Representing, for the time being at least, one of his problems solved.

13
The Beginning
of the End

October 1983

You did not have to be Roger Ebert or Gene Siskel to tell the difference between the two tapes—the sharp eye of a film critic wasn't necessary. The video Bob Radler had made for Jackson Browne was, well, simple. Clean. Clear. Literal. The video Radler kept for himself was not. Though it was only what Radler termed an exercise cut, designed to display the effects of matting and alternate cuts, his work was rich, subtle, and exciting. And useless. Except as evidence of four solid weeks of work.

After completing shooting of over two miles of film for Browne's "Tender" MTV video, Radler had edited it down to a five-hundred-foot rough cut within a week. Browne was on tour in the Northeast; at his request, Radler flew to New York, checked into a Sheraton, and presented the nearly completed rough cut to his client. Who hated it.

"Wait," Radler said, "perhaps 'hate' is too strong a word. Let's just say that his view of what should appear on tape and mine were somewhat apart." Browne felt the editing had been rushed and was more concerned about the quality of the work than meeting a specific MTV "world premiere" deadline. As for Radler, "The question was, who's going to make the art? That's

360

the whole thing in MTV. Is it going to be the producer or the musician? Jackson wanted to do it himself. That was his right. He wanted one video; I would have liked another. But it was his money, his neck on the line. The worst thing I could do was try to force him to make one thing when he wanted another. The whole thing would have got sausaged." Radler was pleased to get the extra time in any case and the next day they spent thirteen hours together editing at a rented cutting room in Manhattan. A surreal experience. Browne spent hours and hours in the on-line (the final video edit) trying things and changing his mind. At times it seemed he couldn't decide what he wanted; scenes were either too happy, or too sad, or finally too both. A week's worth of work was disassembled before Radler's eyes and there were times when Radler could not remember what the hell he had planned originally. Again, Radler got scared. Convinced that Browne certainly knew how to undo a video, but not at all certain he could put one back together. "I told him, 'Jackson, I think maybe you better let me handle this.' " But no. "For the first time in my life, I let a client, against what I considered his better interests, recut something already finished. I ended up just nodding and smiling a lot." Still, Radler conceded, "There's no question he did a good job. It's music. It's Jackson. It's a big hit, it's a nice credit for me. It's just not mine, that's all."

Marc Feidelson, one of the top media analysts in America, threw his hands high in the air. "There are a lotta gold rushes in this world and cable television is one of them. A lotta people rushed into cable that had no business being there. Greed." Feidelson's fingers wiggled and grasped at the air over his desk. "They thought it was a way to make a million. Now the whole industry is caught between a rock and lotsa hard places." Feidelson, media director for Dailey & Associates Advertising, of Los Angeles, was one of the advertising industry's primary conduits for ad dollars. Every year over $100 million worth of sponsors' money went through Feidelson's hands. "The network share will bottom out at about sixty percent. In fact, cable has now got to get very, very, very big—very, very, very quickly—to create terrific network erosion beyond where we are now, to expand to a bigger chunk of geography, bigger than the thirty-five percent it's at now." And that, Feidelson said, was not likely. The gold rush, as far as he was concerned, was over—for several reasons, not the least of which was the simple cost of stringing up cable. Cable start-up costs simply had not been absorbed by subscribers. America had discovered, in just the last year or two, that you could only watch *Rocky III* so many times before you started grimacing at your $29.95 monthly cable bill. Cable had done well in the rural areas, but in the cities, where a number of free independent stations were available, cable was on the ropes. There had even been talk of putting ads on some of the pay stations—a possibly self-destructive move—in order to keep some struggling companies from going bankrupt.

The newly elected chairman of Warner Amex, Drew Lewis, summarized the situation: "I'll tell you what has happened to us and to everybody else who has obtained major-city franchises. We've promised too much. In Milwaukee the contract we got called for four dollars ninety-five cents for basic service. We break even at eleven dollars. We have to keep our costs down, and we have to get our rates up. If not, I'm afraid we're not going to be in business."

It was four years since cable profits had peaked. By 1983, three of the largest cable companies, Storer Cable, Warner Amex Cable, and Rogers Cablesystems, lost money on cable. Time Inc.'s *TV-Cable Week* had just folded, losing $47 million. The Entertainment Channel had died, the Satellite News Channel had been gobbled up by Ted Turner, and Showtime and the Movie Channel had been forced to merge. Even HBO was having problems. During its latest fiscal year, HBO had gained about thirty percent fewer new subscribers than in the previous year. While Home Box Office and its sister corporation Cinemax had accounted for $168 million in profits over the last year, HBO's rate of subscribers who did not renew was a high 36 percent. Subscriber increase had clearly leveled out.

The shakeout in the cable industry in 1983 was, of course good news for the networks—particularly NBC, the network most vulnerable to cable both financially and in its emphasis on demographics. The first assault force of cable Visigoths had, it appeared by September, broken against the walls of Rome.

* * *

"Well," calculated James Patrick Devaney, president of the freshly formed JPD Television Network, "the networks may survive cable, but let's see if they survive low-power television."

LP-TV is the latest and most potentially explosive force on the television horizon, an infant colossus that may threaten the health and even survival of network, independent, cable, and dish TV within the next half-decade.

The principle behind LP-TV is simple. All programming originates from one or more "network" programming uplinks. That is to say, from a source much like Ted Turner's Super-Station in Atlanta. This programming is beamed to a satellite, then down to the low-power stations themselves, each consisting of not much more than two guys and a tape machine, who then fan the signal out in a twenty-five-mile radius. Free. No charge to the consumer. This is what makes LP-TV so dangerous to cable—anyone can get it, and for nothing. LP-TV is wholly advertiser supported. And LP-TV's overhead is almost nothing compared to that of network affiliate and traditional independent stations. Thus, LP-TV's ad rates are lower.

"Essentially what you have here," Jim Devaney said, "is radio with pictures." What Devaney himself had, as of early fall, was a real jump on the market. He was the sole programmer for American LP-TV, and his goal was to create a string of LP-TV affiliates that would rival the networks. The FCC was prepared to license three thousand LP-TV stations within the next decade, almost five times the combined total of NBC, ABC, and CBS affiliates.

Presently, Devaney, whose offices were in the sunny, posh recesses of a ranch-style office building in Rolling Hills, the richest town in California, had only ten LP-TV affiliates signed to contract. He had promises of another 35 by Christmas, and the prospects of 100 by April and 150 by the fall of 1984.

Devaney, with a background in affiliate television, got interested in LP-TV in 1982 after going to the second annual LP-TV convention and discovering the fetal industry was "all technology, but no programming." LP-TV had been in the works for a decade, but had been banned until recently by the Federal Communications Commission, largely as a result of strong lobbying by the rest of the television industry. Like so much that is evolving in television in the 1980s, LP-TV is essentially a product of deregulation and emerging technology. But now that the entrepreneurs have been freed of regulation, "a guy," said Devaney, "could set up a whole station in his garage for seventy thousand dollars. The basic broadcast package consists only of a receiving dish, two video recorder-players, a small switching unit, a transmitter that can fit in a clothes closet, and, for production purposes, two cameras, a couple of mikes, and a lighting package. Some of them are going in the back of radio stations and sharing radio towers."

Devaney saw a luminous future. "Down the road, LP-TV will screw up lots of major-market TV stations—because it will penetrate their Nielsen book. You could put fifteen LP-TV stations in Los Angeles alone. Think what that might do to the local independents. The major ones are averaging only a five percent share of the audience—screw that up even marginally and they're off the board. We'll sell those fifteen stations to advertisers as a package. We can sap advertisers from the networks. LP-TV can sell to local advertisers that could never afford to be on TV before. Monster ratings won't be a necessity, because ad rates will be so inexpensive."

With literally thousands of LP-TV stations across the country, Devaney would be able to offer programming to each station at a fraction of the capital outlay charged to network affiliates and, at the same time, collect considerable profits himself. "LP-TV stations have a much greater chance to survive, as opposed to cable stations, because low power is so much less capital intensive. There are simply no wires to lay." And a single station, such as the one planned for the World Trade Center in New York, could still reach millions of people in an urban area. "Potentially, in a big city, you could have an all-advertisement station. An all-real estate station—Joe's Gallery of Homes, followed by Edith's Gallery of Homes, and so on. So that if you were interested in a home within the area, you would just turn to that station. And in rural areas, we can survive on traditional syndicated programming forever." Devaney looked out his office window past a stand of palm trees in the foreground to the smoggy distance, and considered out loud the possibility, five or six years from this day, of kicking HBO right in the ass.

And in a decade? Burbank loomed in the distance.

*   *   *

Holy Mother of God, they were all in there . . . and now here they all came, spilling out of a single Caddie limo like thirty Ringling clowns out of an exploding flivver, detonating at least a megaton of strobe light and sheer, adoring noise: Alan Alda! Mike Farrell! Loretta Swit! William Christopher! Harry Morgan! Jamie Farr! And even . . . David Ogden Stiers! Two hundred fifty million dollars' worth of M*A*S*H cast—on the hoof, in the flesh—swept through a wall of crowd roar and into the Pasadena Civic Auditorium, acknowledging with grins and waves the rolling sea of nonentity beyond the police barriers, which was flecked here and there with the whitecaps of placards bearing such urgent signals as JESUS IS BACK IN TOWN, GET OUT OF EL SALVADOR!, and "CHEERS" NO.1.

The M*A*S*H unit did not, however, sweep too close to the cordon of Nikon-festooned press here to cover Emmy night, who implored, "Alan! Alan! Loretta! Please, over here, Alan! Over here!" Because the M*A*S*H celebs had learned long ago that the safest shots were the ones that came straight out of your publicist's drawer.

A few moments later another limo beached and out undulated Joan Collins, girdled in a suffocating sheath of black sequins as if wrapped to lock in freshness. With no coyness whatever, she teetered as fast as stiletto heels permitted straight for the paparazzi, and swayed there, toasting herself in the heat of their Sunpaks and Vivitars like some hypothermic shipwreck survivor who had been lost on the ocean for months with . . . no PR! They tried to coax a snappy caption out of her but all she gave them was a stoned Mona Lisa smile painted scarlet, while someone in the mob struggled to get a huge sign into the shot that read EMMYS: IDOLATORS.

As Collins preened, Jim Burrows and his wife, followed a little later by Laurie and Peter Grad, floated anonymously up the steps. Betuxed and spectacularly gowned, they had to be *somebody*. But who?

Just inside, backstage, a large room with tables and IBM typewriters had been set up to accommodate the working press and, next door to it, a larger, much nicer room to accommodate the drinking press. Twenty feet of free booze lapped the shores of a giant mesa of food, dominated by an ice-sculpted NBC peacock, buttressed by massive roast hams, shoulders of beef, and enough ancillary goodies for a hundred Friar Tucks. A Jets-Rams game on the lounge's TV, tied at 24 all late in the last quarter, was abruptly obliterated by the beginning of the Emmys, setting off a howl from the reporters, some of whom had been ordering drinks two at a time to save themselves a little legwork.

As the winners began to be announced, each was paraded through the press room, first to an area for deadline photographers, then over to the general press, then to the television cameras, then, last in protocol, to the radio reporters. Outside, the local ABC affiliate was broadcasting the highlights live, in an apparent attempt to blunt the NBC national telecast, which was tape delayed by two hours.

With no delay at all, NBC's shows started steamrollering over the competition. Before the end of the first hour, NBC had won seven out of ten awards. Winners included *Taxi* supporting actor Chris Lloyd and supporting actress Carol Kane, *St. Elsewhere* supporting players James Coco and Doris Roberts, and *Cheers*'s Glen and Les Charles for comedy writing. Then, according to a plan cooked up by Steve Sohmer, NBC cut away to a station in Rapid City, South Dakota, where tuxedoed locals crowed for twenty seconds about the greater glory of the Be There network. This promo met with stony silence in the auditorium.

When the show resumed, MC Joan Rivers strutted out in a new gown—one of seven she modeled during the evening. She said that while she was changing, three stagehands had seen her naked. "One threw up and two turned gay." Describing her dress, she said, "This is something I just got off the rack, which is what Joan Crawford used to say about her daughter." And: "Putting dresses on and taking dresses off, I suddenly have such respect for hookers." The NBC switchboard in Burbank began to jam up.

By the next cutaway, NBC had won ten of fourteen awards. *SCTV* got one for variety show writing and *Hill Street* won drama writing—neither win was a surprise, since both shows had gotten every one of the nominations in their categories. Writer Marshall Herskovitz and his partner, Ed Zwick, who'd written the NBC movie *Special Bulletin*, won the writing award for drama specials, and Herskovitz would later collect the prize for best special. It was an emotional night for Herskovitz, whose father had died just that week. Jim Burrows won the comedy-directing award. The next break took the viewer to Seattle, for another NBC celebration in progress. Scattered hisses in the auditorium mingled with the sound of the promo, and Brandon Tartikoff left his seat to try to cut the commercials from the auditorium monitors.

The slaughter continued as NBC won best miniseries actor, Tommy Lee Jones in *The Executioner's Song*, best comedy with *Cheers*, best drama with *Hill Street*, best dramatic actor with Ed Flanders of *St. Elsewhere*, and best comedy actress with Shelley Long of *Cheers*.

When Ted Danson's category came up, the camera lingered on him and his wife; he could almost feel it touching his face. He had spent the day swimming and playing with his daughter, and had tried not to think about the award. But all day long he'd had a growing sense of optimism. He'd composed a speech in his head.

He lost. *Taxi's* Judd Hirsch won. Hirsch's acceptance speech, a rambling diatribe against man, God, and NBC for canceling the show, met with an icy reception. The glitterati liked their heroes humble.

Outside the auditorium, a tuxedoed journalist shouted into his hand-held mike, over the roar of the departing crowd, "NBC has won virtually two thirds of all the Emmys given out in the major categories. But will the respect of its peers translate into true ratings success? To that question, NBC programming head Brandon Tartikoff replied, 'We'll know in a few weeks.' Sandy Kenyon—CNN—at the Thirty-fifth Annual Emmy Awards in Pasadena."

What Tartikoff was really concerned about at that moment, however, was, why in the name of God did Sohmer have to lay it on so thick? By the end of the evening, the crowd in the auditorium—consisting of practically everyone who counted in the entire industry—was groaning and booing every time NBC cut away to another one of those graceless affiliate brag-fests. Perhaps the cutaways had had to be inserted, since they'd already been arranged and scheduled. But what about that bombastic wrap-up at the end of the show—surely that could have been toned down. At the show's conclusion, an announcer had gone through the complete NBC schedule, night by night, show by show, and then bleated, "Thirty-three Emmys! Wow! *I'll* Be There!"

Tinker was disgruntled, too. He later told journalist Ben Brown, "We all wish we hadn't done it. We haven't been known for playing rough. It's not our style. It looked like we were rubbing other people's noses in it. It was just dumb."

But for Brandon Tartikoff, a comforting fact underlay the embarrassment. Which was: The blunder had been Steve Sohmer's. Mr. Be There, who had come so far, so fast with his pit bull aggressiveness—right up to the doorstep of Tartikoff's office, to be exact—had shot himself in the foot. On that September evening, it was believed throughout the industry, Brandon Tartikoff had been handed at least one more year of job security.

Ted Danson set off for a mandatory party at the Century Plaza, feeling terrible. But when he got there, he was besieged by gushing reporters and noticed that he was being interviewed much more often than the man who had beaten him, Judd Hirsch. The more he talked, the better he felt. Soon it became apparent to him, and to Casey, too, that in some odd fashion the chemistry of Hollywood image making, rather than the Emmy jury, had just declared him the winner.

Allan Katz's dreams looked like they were coming true. Twentieth Century was ratifying *The Hunchback of UCLA* now, step by step. First, Tische and Avnet met with Twentieth feature executives and they gladly approved Danny DeVito as director. Then, a few days later, Katz, DeVito, Tische, and Avnet met with them again to work out the revised story. Here was where Katz figured the ax was bound to fall. "I went in there convinced they were going to pick the movie apart piece by piece. I went in there prepared to answer any possible argument." But when Katz and DeVito finished their presentation, the only objections and questions they got were what they considered very reasonable ones. And the upshot was tremendous: The story was cleared. With any luck *Hunchback* would be rewritten and in production before Valentine's Day 1984.

The crowd was in his pocket—even the HBO cameramen were doubled over—and Harry Anderson was thinking to himself, "Hell, maybe I'm an off-network kind of guy." Harry had on blue eye shadow, rouge, and an industrial earring made out of a black sew-on snap and a watch stem; he was playing the bisexual host of a Home Box Office special being filmed at a Toronto club called Scandal's, a Studio 54 for Canadians. Harry had done a previous show for HBO, called *The Bare Touch of Magic,* which had featured tricks, jokes, and bare nipples, a winner combination on pay TV. The show had been one of the most popular HBO programs ever. It amazed Harry to see how much mileage he could get out of ancient dirty jokes, like "Don't drink water—fish fuck in it," just because they were coming across the usually sanitized twenty-one-inch screen. So when the producers of the Toronto show had told him that the only rule for their program would be to break all the rules, he knew he was home free.

So there he was, going up to a lady in the audience and saying, "Ah! What's this behind your ear?" And pulling out a little wad of bloody scalp.

Then he did the gag of trying to get out of his straitjacket before his assistant could free herself from the chair she was tied to. But this time, the girl he was working with got out of the ropes—and her dress—at the same time, and emerged stark naked.

A male stripper from the nightclub Chippendale's, dressed like Zorro, came on and did his thing, with the additional touch of grabbing women's heads and dive-bombing them into his crotch.

At the end of the show, Harry said, "Hope you all had a good time. And if you didn't, what the hell—most of you are going to go home and screw around with people you don't even care about anyway."

After the show, a little bit drunk, Harry went back to his room. He was pumped full of adrenaline. He lay back on his king-sized bed and rolled a joint. Outside, the night was silent and beginning to chill. He flipped on his TV, found an old black-and-white movie, and contemplated ordering a drink from room service, or maybe a snack. He thought about calling Leslie and Eva. As he started to smoke the joint, the best parts of the show began to drift back into his mind.

Harry thought of a new line he could add to one of the tricks.

The line made him smile, then laugh out loud. He didn't want to be home. He didn't want to start shooting *Night Court.*

"I *love* the road," Harry Anderson thought to himself, reaching for the phone.

The early dark of the fall shadowed Allan Katz's office as he took one last call for the evening. "Hello. . . . Yeah. . . ." He nodded slowly. "Yes. . . . Yes. . . . Well, I find that a little hard to believe. . . . No, because I saw him on Dick Cavett and he said he enjoyed doing theater work very much. . . . Sure, but are you certain it's *his* financial considerations we're talking about? . . . Okay. . . . Okay, I'll talk to you later, bye." Katz hung up. Everything was set to go for his play, *Kaufman and Klein,* save for its leads. With a name star it would be much easier to get the play into a prestigious theater. And getting a big name was hard because a star committing himself to a lengthy run would have to do so for a minimal financial return. By agreeing to direct the production, Jay Sandrich stood to sacrifice tens of thousands of dollars in lost network projects if he didn't juggle his time right. A risk he was willing to take. But Ed Asner was still very much up in the air—his agent had told Katz that he hadn't even received a copy of the script. Which puzzled Katz greatly. Perhaps the messenger had been bushwhacked. And Katz had likewise just got the word that the other big name he was after, Walter Matthau, wasn't interested in doing a play. Any play.

"His agent says Walter doesn't want to do theater, but I'm wondering if it's the agent that doesn't want him to do it, if the agent isn't thinking, 'Ten percent of hardly any money is hardly any money.' If Matthau liked the script, he might not care one way or the other."

The Durning project had also been caught on a minor snag. While ABC seemed quite satisfied with Katz's concept for the show, Twentieth Century now wanted him to write several scenes of a sample episode. Which made Katz uneasy. For his opening episode, he'd planned a script revolving around what Durning's character, Charlie, does when he discovers that a friend of his thirteen-year-old daughter is pregnant. And Katz wasn't eager to rock any unnecessary boats. What if one of the folks over at ABC had a young, unmarried, pregnant daughter right at the moment? When the sale of an entire series depends on the material in a couple of scenes—and you don't have the benefit of exploring the subject matter with the network—you may inadvertently unsell the whole package out of ignorance. And Katz didn't want to lose this one, convinced as he was that the world was ready for an honest adult sitcom minus the usual circus of boffo, zany hooks and heat, a show that simply dealt with the mundane victories, hopes, and desperations of normal people. And at the moment he was already thirty-five pages into the first script, anxious to show himself the viability of the ground he planned to work.

For in twenty-two minutes of television time, Durning had to establish his character, learn that his daughter's friend was pregnant, and register shock that such a thing was happening to a child, astonishment that his own little girl was on the brink of womanhood, and frustration at his inability to communicate with her. The plot had to be resolved with a solution that was not trivial and did not leave the viewer feeling cheated. Plus the whole thing had to be funny. And if he was successful, the network would charge him with the responsibility of performing this stunt twelve more times within the next six to eight months.

"A lotta people think TV is a pot of gold for anyone who wants to take the time to cash in on it," he said. "And in some ways they're right. If you just have talent, you can do fairly well in television. And if you have no talent, but a lot of drive, you can do well, too. If you see bad TV and think to yourself, 'I could have done that,' you're probably right. And I certainly don't buy the idea that you have to be lucky to succeed in this business. No one argues that there are writers in this town getting plenty of work who are just terrible. But if somebody comes up to me and says, 'What you do is pedestrian; you're just a television writer,' or 'TV writers and producers make insane amounts of money for their dubious efforts,' I will say, 'Look, if you're so good, you can make it too. So, come on! Write your script; you'll be able to get someone to read it. Compete with me! See if you have what it takes to write a M*A*S*H or a Rhoda. You may. And if you do, you'll succeed.' This is not a closed business and it doesn't have all the creative people it needs. But I think a lot of the people who think it's so easy are the same kind who make fun of guys like Andy Warhol, saying 'Where's his talent? ' Well all I can offer is 'Okay, you be Andy for a while if it's such a breeze. Buy a canvas, paint whatever you want on it, and then try to sell it for twenty-five thousand dollars. Maybe you'll get it. Who knows?' "

Katz went back to work on his rewrite of *Hunchback*. Originally, he'd planned to only rework 30 to 40 percent of it. Now, over three quarters of the script was being altered, axed, or reinvented. Katz left the office long after his secretary was gone. Outside, under the radiant glare of halogen, his red Ferrari was dusted and dulled by pollution. Time to wash it—after all, it had been a week.

Tony Colvin took off the last Friday in October to go visit his dad in Vail, Colorado—that Republican ghetto in the clouds. *Just Our Luck*'s NAACP problem had quieted down somewhat, although there had been a number of stories in the press, including a negative blurb in *TV Guide*. And the ratings of the first few shows had been lukewarm. But there was nothing Tony could do about either problem. He and Scott had finished their second script, and all the remaining scripts for the first thirteen shows had already been written. So Tony figured he'd catch a few days of clean air before the grind started again.

Scott was talking to two other writers at 4:45 in the afternoon that Friday when executive producer Rick Kellard came in. "I gotta talk to you guys," Kellard said. Scott thought he meant the other two writers, so he started for the door. "No, you, too," Kellard said. Kellard, a tall, well-dressed young guy who looked like actor Michael Sarrazin, came straight to the point. "This is your last day," he said. "Lorimar just told us a few minutes ago that they want to save some money during the hiatus, while ABC decides if it's going to order more shows, and since you guys aren't working on any of the first thirteen, you're expendable. It's nothing against any of you—you've all been doing good work. It's just one of those typical things. Now, I can't really say, 'Go out and get another job,' because I know you still have option time left on your contracts. But I know you've got to make a living.

"You ought to clean your desks out, though, because you never know what'll happen to the offices."

When Kellard left, Scott, who was shocked and upset, talked about this latest turn of events with the other two writers. Neither seemed particularly shook up, and Scott admired their cool. It wasn't until several weeks later that he found out they'd already lined up another pilot deal with Lorimar.

That night, Scott called Tony's wife, to ask if he should tell Tony now or let him enjoy the weekend. She thought Scott might as well wait. Scott then called their manager, Jim Canchola, who told him not to worry, that he'd find out all the details on Monday.

Scott finally called Tony Sunday night. He said, "I've got some bad news. I know I waited a couple of days to tell you, but I didn't want to wreck your weekend. We got laid off."

"What! You're kidding!"

"I wish I was. Kellard told me Friday that we're through until the back nine."

"If there *is* a back nine."

"Correct. I called Jim and he said don't worry."

"That's what he would say. I'm gonna call him." When Tony called Jim Canchola a few minutes later, he said, "What the fuck's going on?"

Canchola sounded hurt, defensive, as if he thought Tony was blaming him. "I don't know what's going on, Tony. But I'll find out. First thing tomorrow. Don't sweat it—it'll work out."

There was, of course, precious little to find out. If the show was renewed, Tony and Scott would have jobs. If it was canceled, they wouldn't. They would know on December 1.

Over the next few weeks, Tony slowly began to implode. At first, he and Scott got together almost every day to work on spec scripts and story ideas. But they ran out of things to do—or out of the energy to do them. Every Wednesday, Tony would get up early and call Paramount to find out the overnights of *Just Our Luck*. He called his former studio because MGM didn't compile its statistics until midmorning, and Tony still had clerical contacts at Paramount—he was still a legend there among the secretaries and janitors. The show held in at about a 19 to 20 share of the audience, not a bad score, considering that it was up against *The A-Team*. And it was obliterating CBS's show in that slot, *The Mississippi*. More importantly, its "comp," the composition of its audience in demographic terms, was a solid "lifeboat" audience—women and children—which meant that advertisers could count on it to reach a particular market segment. *A-Team*, for all its numbers, tended to skew all over the demographic board. So there was just enough good news trickling in to drive Tony insane.

The condition of Tony's apartment slowly deteriorated. His wife, who worked full time, expected Tony to help around the house more, since his days were mostly idle. But Tony wouldn't so much as scrape the mold off the coffee maker. "If I start doing maid's chores," he told himself, "I'm a goner." At least once each day, he had to stop himself from calling an occupational headhunter, who could line him up a job in data processing. He tried to convince himself that he could work DP during the day and write with Scott at night, but he couldn't sell himself the idea. He knew that if he went back into the computer world, he'd never come out.

Scott tried to get out every day to play basketball, but his heart wasn't in it. If he had already made his fortune, a daily double scoop of B-ball would have gone down splendidly, but now, with the whole process of grubstaking his retirement pension still totally ahead of him, basketball just seemed like one more energy drain.

Four days before Thanksgiving, Tony and Scott met at a convenient spot halfway between their homes—the NBC commissary. Totally lacking the grandeur of the executive side of the Paramount commissary, NBC's was little more than a dimestore cafeteria with Formica tables and plastic chairs. At the moment, the Hungry Peacock was festooned with paper pumpkins, Pilgrims, and turkeys, which just made the place all the more depressing.

"Man, I'm going bankrupt," Tony said, slumped against the table, a plastic spoon going around and around in his coffee. "The bills keep coming. We weren't on the show long enough to save anything, and I've depleted my savings down to the bottom in the last couple of years, trying to get into the business, buying picnics for the janitors and whatnot."

"I know the feeling," said Scott. "But listen to this. A film. About the TV business. Like *Network*, only a takeoff on the power of the ratings. We can call it *The Sweeps*."

Tony paused for some time, then slowly began to nod his head. "Not bad. Not bad. Not bad."

"Maybe we can write a treatment," Scott said, "after . . ." He didn't finish.

Tony finished for him. "After our own *personal* sweeps week—ending December first."

About half an hour after Tony and Scott had left, hurrying home in opposite directions to beat the rush hour, Joel Thurm, head of casting at NBC, paced his office, one floor up from the Hungry Peacock, and worried about the mechanics of programming. The new NBC shows had all been aired, and Thurm wasn't too pleased with what he'd seen.

"*Yellow Rose*—the first three episodes were, shall we say, less than interesting. They were slow, and the focus was on outside characters. I wanted to see more of our people. And the two producers just were not compatible. Wilder is *Centennial*—old westerns—and Zinberg is *Hill Street*, very gritty and contemporary. And we've lost Susan Anspach, who had a real problem with Michael Zinberg. Susan is a little, not neurotic—a lot of actresses are a little neurotic—but Susan needs extra attention. After Michael left, she was asked back, but it was too late. Also, the show is aggressively macho—I don't know of any reason why women should watch it.

"But it's salvageable. So is *We Got It Made*. It's one of our highest-rated comedies, but just imagine if it was *good!*

"*Boone* has a problem. In the pilot, I saw Boone as a rebel, a stronger character, and now all of a sudden he's a saint. The show is very, very soft. And I think they should have cheated a little on their music, and not been so authentic—tried a little Stray Cats kind of thing, more contemporary. They softened the daughter—they took sex away from her. I think all those things are mistakes. This show hasn't done Tom Byrd any good, because his character's so soft.

"*Bay City Blues* has the problem of keeping track of who is who—there are so many similar characters. And it's not in a life-

and-death situation, like a cop or medical show. It's also doubt-ful whether women are really watching the show, since it deals with baseball. When *St. Elsewhere* went on the air, there was a lot of loose time, breathing time. They shut themselves down and cast three actors and changed directors. *Bay City* hasn't had the chance to do that.

"*Mr. Smith*. We all should have realized that to have one of those masks that really works, like E.T. or Yoda, it costs several million dollars. It's not easy to get all those expressions, and make them lovable. Maybe they'll add a few more people, so the ape doesn't have to be center stage all the time.

"The last couple of shows of *Jennifer Slept Here* have been very funny. I don't want to say it's a silly show, but it is a silly show. They're still finding their characters and finding out what works. Another show that's having problems with execution is *Manimal*. What got lost there is humor, and the relationship be-tween the male and female leads. If you don't chuckle a little bit while you watch that show, it just isn't going to work. *Roust-ers*, I think, is never going to work. We just all believed that Steve Cannell knew something that everybody else didn't, after he hit so big with *A-Team*. Another one that's not working is *Love and Honor*. There's no center to it—I don't think *Love and Honor* ever knew what *Love and Honor* was."

All of this was taking its toll, Thurm said. The mood at the network, after so much hope, so much optimism, was turning morbid.

"Brandon, I saw him the other day, and he was more drained than I've ever seen him. He looked *tired*. Not the Brandon we know and love."

Out Thurm's eastern window, in the autumn-lengthened evening, a giant saucer of harvest moon, stained pastel in the Los Angeles smog, moved slowly toward the top of the sky. An uncharacteristic cool was in the air. Even in Los Angeles, win-ter has to come sometime.

14
Exeunt

November–December 1983

Finished. Kaput. Over. Done. Goodbye. Sayonara. Adiós. So long. Despite cash outlays of over $30 million for pilot development, despite the efforts of the best producers and creative talents in the industry, despite the promotional brilliance of Steve Sohmer and an overwhelmingly favorable press, within the first month of the new season, seven of the nine new NBC shows were on their way out. Neilsen ratings for the week ending October 23 showed *Jennifer Slept Here* in fifty-sixth position out of seventy, *Boone* in fifty-eighth, *Mr. Smith* in fifty-ninth, *Manimal* in sixty-third, *Yellow Rose* in sixty-sixth, *Rousters* in sixty-ninth, and *For Love and Honor* at the very bottom: seventieth place. *Bay City Blues* would soon come on and sink like the *Titanic*. The November sweeps were a disaster—NBC chugged in five full rating points behind winner CBS, and three points behind ABC.

Three weeks later, the dust having settled even further, NBC's new shows occupied seven of the bottom ten slots. By the first week of December, the network officially decided to cancel *all* of its new prime-time shows, with the exceptions of Fred Silverman's *We Got It Made*—which was doing reasonably well in the ratings—and *Yellow Rose*, which was getting bad ratings but was starting to at least look artistically promising to the network.

At the Polo Lounge, Jeff Sagansky, the young head of NBC dramatic development, remained marginally optimistic. "If you measured the network just Sunday through Thursday," he said, "we would be doing all right. It's Friday and Saturday that have really caused us problems." Problems akin to building the Parthenon over the La Brea tar pits. "There is just no solid ground for us at the end of the week. *Love Boat* and *Dallas* were killers." It was breakfast time; Sagansky ordered shredded wheat and strawberries from a white-coated waiter. "We had hoped *Mr. Smith* would do on Friday nights what *Knight Rider* did last year, but people just didn't buy the special-effects mask that allowed the ape to talk. So, without a foundation, *Jennifer Slept Here* had nothing to stand on. Then you had a show like *Manimal.* Vintage high concept. And the rules there are fairly simple: The show will either catch on early or not at all. It didn't." Sagansky, thirty-two years old, red-haired, angular, impatient, was as dumbstruck as most of the other NBC programmers at the incredible speed at which their schedule had disintegrated.

It had been impossible even to attempt a thorough autopsy, the collapse had been so massive and quick. Theories, though, had already sprouted at NBC's offices. "I think we should have taken an earlier jump with *Yellow Rose*," he said. "If we made a mistake, it was that we didn't maneuver effectively to make sure our Friday and Saturday night shows had every chance of exposure just prior to the time of the traditional fall premieres. We expected people to pay attention to these shows, to at least *sample* them, and they didn't." The early consensus on Steve Bochco's *Bay City Blues* was that, despite the quality of the show, the public wasn't buying soap opera baseball. Women certainly weren't. And it was growing obvious, after *Bay City* and the continuing softness of *St. Elsewhere*, that "large ensemble cast" might have been the three magic words for *Hill Street*, but they could also simply mean a whole bunch of people working at the same place. As for *Boone*, Earl Hamner's show, there was a hindsight realization that doing a family show about the roots of rock 'n' roll was like producing a ballet about a meat-packing plant. To make *Boone* palatable as an eight-o'clock show, too much of the grease, dirty fingernails, and rawness of the era and environment had been shampooed and manicured out, a wrong-headed attempt to blow-dry the past.

The most chilling outcome of the season, though, was that NBC's attempt to be the quality network had flopped. Shows like *Love and Honor, Boone, Bay City Blues,* and *Mr. Smith* had tried to rise a cut above the norm—but the public just didn't . . . watch.

Not only that, NBC's attempts at inanity had also bombed. *Jennifer Slept Here, Manimal,* and *The Rousters* had done just as badly as the higher-quality shows, probably for the same basic reason—they didn't have strong lead-ins. NBC's biggest weakness, after all, was that it had started weak.

The melding of quality and silliness had been particularly ineffective. Trying to walk both sides of the picket fence was apparently not something that a third-place network could get away with.

There was no doubt, even as early as the first month of 1984, that NBC would remain third for at least one more season. It was simply too far behind to catch up.

Owing to clever program-stunting, however, NBC was actually less far behind the competition by midseason than it had been at the same time a year earlier. A handful of hoked-up specials, like *Johnny Carson's Favorite Practical Jokes,* had kept the network from being completely invisible. But the beams and girders of its schedule—its prime-time series—were in appalling disarray.

Now hope born of desperation abounded at NBC. Already the programmers were talking confidently about the possibilities of their new shows, which would begin to pop into the schedule by early January. Steve Cannell had a new two-hunks-on-a-boat action-adventure, *Riptide.* At this time, taking no chances, NBC was throwing it in directly after *The A-Team.* Former *Saturday Night Live* producer Lorne Michaels had *The New Show,* and was hoping a few million of his old *SNL* fans were now ready for prime time. The new comedy *The Duck Factory* was almost ready to go, and *Buffalo Bill* was back.

But as much as anything else on the network, the NBC men were counting on Harry Anderson and *Night Court.* It had finally come to this: The immediate future of the network had come to rest, to no small degree, upon the shoulders of a former con man and street magician.

Eva Anderson, squirming on her mother's lap, said, "Where's Papa?"

"He's a judge," said Leslie, "in Night Court."

"He's not a judge."

"He is now."

Harry came to the rail separating the audience from the actors. "I'm scared shitless," he said. He didn't have on his judge's robe yet, and no one in the audience recognized him. Sitting in the plastic bucket seats, waiting for the first episode of Night Court to start, were a couple of NBC executives, a few of Harry's friends, and a busload of retirees from the Whittier Mobile Home Country Club. Although the room was fairly crowded, network comedy programmer Warren Littlefield was sprawled across a grand total of five seats, each arm hugging one, each leg on a seat in front. On the stage was the paper model of the Chrysler Building that Harry had put together in the spring, while waiting to hear if NBC had picked up his pilot. Jim Burrows, fresh from a Cheers rehearsal, arrived and took a seat.

The show was about a prostitute falling in love with Harry the Judge. The network had wanted to insert the behind-closed-doors thumping of raucous humping—a little offhand intercourse to heighten the concept and 9:30-ize the show, which they were thinking of throwing against the last half of Dynasty. But Reinhold Weege had fought off the suggestion. In the show, Harry ended up extinguishing the hooker's lust by treating her with respect. The story line was still raunchy enough to fill a 9:30 slot, but it was not a cartoon; even the old folks from the mobile home country club liked it. They applauded even when the sign wasn't lit.

Afterward, Burrows said, "It's a good show. But it will take a long time to get started. There's no reason for people to watch it. Just because it's good, that's no reason. People will only watch high concept initially. They want familiarity from TV."

But Burrows was confident that his own low-concept Cheers, which now regularly finished as high as twenty-fifth place, would be back again for a third year. He was reminded that last spring he'd said the third year put a show over the hump. "I said that? Well, now I think the *fourth* year is the major hump." Burrows said hello to his buddy Jay Sandrich, who was directing the episode, and hurried off for home. For him, tomorrow, like every other day of the week, would start early and finish late.

That night, in his home just south of Griffith Park, Harry Anderson smoked his first joint of the week. "I don't have time for pot anymore," he said. "But being straight all week gives me a new perspective on life—it shows me how fucking boring it is."

"Has anybody told Reinie yet where the network is going to put the show?" asked Leslie, finishing the homemade sauces and dressings that she had spent the morning on. She had changed out of the minidress, red tights, and silver high heels she'd worn to the shoot, and was padding around the kitchen in jeans and sneakers.

"I know where it ought to go," said Harry, his eyes starting to burn. "Right after *A-Team*. Imagine riding along in the wake of that thing, like a bike following a semi. I wonder who you have to blow to get that slot."

"You could have your own doll, like Mr. T. The Harry Doll," said Leslie.

"Right! The Mr. A Doll. Then I could be in *Star Trek III*."

"You could get your own cartoon."

"On HBO! Filthy animation for guys who get up on Saturday mornings and smoke dope and watch cartoons.

"Let's go for a ride."

They had two new cars, courtesy of a six-week stream of $10,000 checks from Warner Brothers. Leslie now owned a sensible 1984 Honda Accord, and Harry a $13,000, five-liter, snow-white 1984 Mustang convertible. It was the first new car he'd ever owned. For many years, he'd driven a 1952 International Harvester truck; he'd paid $175 for it and put on a quarter of a million miles, mostly from show to show.

"Did you remember to make the car payment?" said Leslie, strapping herself into the Mustang as they backed out the driveway and headed toward Los Feliz Boulevard, in the direction of downtown Hollywood. Harry didn't respond immediately; the new-smelling, soundless car had hypnotized him, like a drug.

"Yeah. But I hate dealing with the bank. Aesthetically, it sucks."

"We could always sell them."

Harry looked at her as if she were crazy. "You'd let a stranger take *this car*? Where are your *priorities*?"

"What are your priorities? At the moment?"

This time Harry did respond immediately. "Fame! And fortune! Of course."

"Oh, Harry, you don't really care about that."

"I do!" He swung onto Hollywood Boulevard. "Remember that *Taxi* where Latka finally becomes an American?" Leslie shook her head. "You don't remember that one? Latka—Andy Kaufman—was into enterprise. He was making these cookies that didn't really taste very good, but he was so enterprising that everybody was really encouraging him. Then everybody started acting weird from the cookies, because Latka's been putting coca leaves in them—cocaine—only he doesn't know what it is. So Jim—Chris Lloyd—comes in to analyze them. He smells one and says, 'Peru.' He smells it again—'Nineteen seventy-seven!' And again. 'After the rain.'

"Meanwhile, Latka goes home, and he's making cookies, and tasting his batter about every three seconds. He's completely wired, his eyes all bugged out. Finally he's so buzzed he goes into a trance. And Famous Amos—the cookie mogul—appears to him. And Famous Amos, to Latka, is like God. So Latka says, all downhearted, 'Famous Amos, I've gotten so fucked up. I'm so corrupt, all I care about now is fame and fortune. Tell me—what should I do? Whatever you say, I'll do it.'

"Famous Amos says, 'Latka, what I would like to tell you is that fame and fortune, they mean nothing. They are whimsy, and will soon be gone. They are not worth what you must give up for them. I'd like to tell you that. But I can't. Because fame and fortune are great! They're *everything*. *Everything!* So get out there and *go for it!*'

"And Latka became an American."

"Oh, Harry," said Leslie, "you don't really believe all that."

"I do. I do. I want it. And if I can't get it from *Night Court*, I'll go back to the Great America, in New Jersey."

"Where they almost killed you?"

"What are you talking about? They loved me. I never felt so much love in one room. I could play there all the time and become the headliner-proprietor, like Wayne Newton is at the Stardust. I can see the marquee—thirty feet of blazing neon—'Harry Anderson: the Great American.' "

"Harry, you don't want that."

"I do!" Harry Anderson punched his dashboard, and the convertible's top opened the car to the starry, starry, smoggy night.

"It's a strange world," Allan Katz said. "If it weren't for television, me and a lot of other people like me in the industry would probably be institutionalized." His mammoth Alamo-like house towered in succeeding stories behind him, his Ferrari sat all washed and polished in his garage, and he was strapping his skis atop the new Mitsubishi four-wheel-drive jeep he'd just bought. It was Christmas vacation and time to go to Sun Valley. "So I owe television a lot. It provides very well-paying work for people who probably couldn't fit in anywhere else. I mean, can you imagine somebody like Mr. T trying to make a living selling Chevies?

"But it's a bad time for the business," he said, clamping his Rossignols to his ski rack. "Look at the prime-time TV schedule. It's disgraceful. People trying to break into comedy writing sometimes give me scripts to look at. And do you know what they're all written for? *Cheers.* You know why? Because it's the only TV sitcom on now that's got enough substance to really stick in people's imaginations. A few years ago there were shows like *Mary Tyler Moore, Rhoda, M*A*S*H, All in the Family, The Rockford Files,* all on during the same weeks. Now, at best only about twenty percent of the series on today try to deliver something innovative or challenging." Katz counted on his fingers, "*Cheers, Hill Street, St. Elsewhere* . . ." He shrugged. "There's a few more, but . . . what's really galling is that there is so little room in prime time and that it's filled up with such junk.

"You have somebody like Fred Silverman, a man who's run the prime-time schedules of all three networks, and he's got the gall to shoot so low when it comes time to create a show of his own. He comes to NBC and says to Brandon, 'Look, I've got an idea for a show that will knock your socks off. For one thing it's got really nice breasts in it. But that's not all—it's got two grown men who are horny all the time. And attached to those great breasts, by the way, is the heroine, who doesn't have a brain in her head. And the best part,' Fred says, 'is that it's almost exactly like another show. And not just any show, but *the worst show* on TV, *Three's Company.*' That's what probably swung the whole deal."

Katz would spend two weeks in Sun Valley, then return to Los Angeles. He had a lot of work to do. The world was going his way. Twentieth had definitely okayed Danny DeVito to direct *Hunchback*, and ABC was most pleased with the sample scenes from his Charles Durning script. Things were actually going so well that *Kaufman and Klein* had been put off until spring because he no longer had time to shepherd it onto the stage. He had even rejected the offer of a New York group to produce it in Manhattan.

"See, that's the problem with a show like *We Got It Made.* No one at the network is saying this show should be supported because the concept is original or the writing is exceptional or the actors are really bringing a new dimension to the screen. Nobody is supporting it because it's particularly good, but because it's successful. Sort of successful, anyway. And it makes you wonder about the attitudes over there. This isn't like back in the days when NBC was pumping shows like *Hill Street* or *Fame* or *Cheers* because they were superior entertainment. Now, they're looking for success, they just want *something*, anything, to succeed, even if it's lard-headed, 'Oh, no, I lost my bra' comedy stuff."

So what would be Katz's solution?

"Simple. First of all, if I was Grant Tinker, I'd say to the producers, 'From now on, bring projects to the network that you really want to do. Not ones that you necessarily think are the most commercial or whatever, but the projects you'd kill to have made.' Then I'd say, 'From now on, at least twenty-five percent of the development season will be devoted to pilots that are truly innovative. And that still leaves plenty of room for the silly and frivolous stuff, except that from now on, it's going to be *great* silly and frivolous stuff.' I'd say, 'Look, I like gratuitous sex and violence as much as the next guy, but from now on, it's got to be *meaningful* gratuitous sex and violence.' "

And what would it take for Katz himself to put in a stint at a network?

"A lobotomy."

Allan Katz battened down his skis, fired the Mitsubishi's engine, and was gone.

Shortly after Christmas, Ed Asner got a script to read from Allan Katz. It was the Durning project. ABC had finally accepted every aspect of it, but the network's option on Durning ran out at the same time that NBC called Durning, offering to package him in a deal with Norman Lear. There would be no meandering—they would go straight to pilot. Durning took the NBC offer, and Katz sent his script to Asner.

Asner loved it. It was just the kind of no-orangutan comedy he had hoped he'd someday be able to do. And now it looked as if he'd get the chance to do it—the graylist was fading. In fact, he had four other comedy pilots to consider. One was with Eileen Brennan, in which he'd play a boss in a garment factory, and one was from a movie of the week he'd made with Jean Simmons, called *Small Killing*. But the Katz project, which would be directed by Asner's old friend Jay Sandrich, director of most of the *Mary Tyler Moore* shows, was the one he was hungry for. It appeared that the industry once again had an appetite for him also. His views on Central America were being absorbed by the political centrists, and the controversy was receding from the public mind.

Asner had known his persona as the Che Guevara of Hollywood was paling the minute the first SAG election returns had begun to come in. The early numbers said landslide. Asner was sure that an overwhelming victory would be indicative of not just support for his political stands, but—even more conducive to industry acceptance—*indifference* about them. The final tally gave him 73 percent of the vote, a towering victory.

Suddenly, after the election, as the new pilot production season began to gear up, he was once again the kind of actor the networks crave: big, and safe. ABC felt instantly more comfortable with him in Katz's project than they had with Durning.

Asner had even finished second, to Paul Newman, in a *National Enquirer* poll asking, "What celebrity would you most like to see in the White House?" It was an eminently frivolous sampling, of course—but just the kind of "research" that network programmers need to justify their decisions.

And *Lou Grant* was selling. One by one, stations had realized that there was, after all, a market for America's favorite boss.

"All in all," said Asner, getting ready to go to a luncheon meeting, the Katz script still on top of his desk, "I'd have to say that things seem to be coming my way again. I'm not sure I could say I feel secure, because I've never felt particularly secure. I guess it's because I was a minority kid in a WASP atmosphere. At the time of Hitler. That automatically instills in you a sense of wariness. Of fear. Caution. And I was the youngest of five, the lowest man on that totem pole. All of which started me running, a la Sammy Glick, to prove my worth.

"Which I eventually did.

"I just kinda wish it hadn't *taken* so fucking long."

Ted Danson, rehearsing on Paramount's Stage 25 as *Cheers* began to wind down for its long Christmas hiatus, had just about everything he wanted, except for the thing he wanted most.

It looked as if the show was safe for a third year, during which his weekly pay would doubtless escalate toward the headier regions of star-salary.

And things were great at home. He and Casey had been helped by an adoption agency to meet a young woman who was scheduled to give birth in late January, at which point the Dansons would adopt the child. The mother did not feel she was old enough to care for the baby herself.

Also, he had just bought a home in Santa Monica. It was ridiculously expensive, but his salary from the show would absorb the cost.

Physically, he felt better than ever. He was staying away from cigarettes and pot, and was working out for an hour and a half every morning.

But his major goal still escaped him. He had not landed a leading-man role in a big movie for the spring hiatus, and time was running out. After the Emmy Awards, there had been a burst of enthusiasm from film producers—his agent had gotten a number of calls and several scripts had been submitted. But not much had come of it. Danson was still, as far as most movie producers were concerned, an unknown commodity. To them, he was a TV actor. Even worse, a sitcom actor. The question they had was, does this guy have the weight, the sheer manly balls, to anchor a $15 million to $20 million investment?

One role that Danson wanted, for example, in a movie called *Little Treasure*, would have paired him romantically with Margot Kidder—Superman's girlfriend, for God's sake—and also with Burt Lancaster, who even in his relative dotage was powerful enough to blow most other actors right off the screen. Could Ted Danson, who even had trouble hanging in there against Shelley Long on the small screen, carry a movie that featured these two powerhouses?

At an act break in the rehearsal, Danson was given a message by the second AD to call his manager.

He had been offered the role in *Little Treasure*. There was only one problem—filming would start in mid-February, just three weeks after the baby was due, and the shooting would be in the badlands of Mexico.

Danson called Casey, they talked briefly, and Danson called his agent back.

On Valentine's Day of the new year, Ted, Casey, Katie, and the baby would head for Mexico.

From the top of a spindly barstool, with about 10 percent of Corky Hubbert's body weight consisting of beer in his belly, life looked—what word but the California superlative could describe it?—*bitchin*! So bitchin that if things got any tastier Corky might just croak from the sheer bliss of it. Oh, what joy a well-managed career could bring!

For starters, he was getting word that *Not for Publication* director Paul Bartels thought his talent was just plain *mondo bizzarro* big. One of Bartels's assistants had just taken him out and said, "Corky, darling, you simply must start thinking about a stronger *leading-man* orientation. Second or third place in the credit crawl is absolutely not where God intends you to be."

She'd been making noises about a follow-up role in a future film, but Corky had had to cut her short.

Because he was already in the throes of negotiation with another of his all-time favorite directors, Ridley Scott, the guy who had done *Blade Runner,* which had just sold to cable for about enough money to buy Disneyland—oh, to have points in that baby! What Ridley Scott had proposed was $3,000 per week for a couple of months' filming in England. But Corky's manager had to politely decline—he wanted three and a half bills a week. Not so much because Corky was hurting for the extra $500—right now, Corky had cash in every pocket and drawer—but to show the entertainment world that Mr. Hubbert was not just another five-and-dime dwarf. Once known as the midget John Belushi, now the midget Tom Selleck, Corky Hubbert was a hot property and available only to the most serious of bidders.

Which might just turn out to be network television. After all, what could possibly beat successive years of $10,000, $20,000, $35,000 per week? The project in question was a pilot for producer Allan Sacks, who had ABC titillated over a sitcom about Cupid. Starring, of course . . . "Here's the gig," said Corky, draining a shot of Bushmills and waving cash to obtain a beer chaser. "I'm up in heaven, taking a lot of crap from my mom, Venus. So I tell her, 'Hey, I'd rather be a first-rate deity on earth than a second-rate god up here. So, hey, bitch, watch my taillights fade.' Then I hightail it to earth and go into partnership with this really nice couple who run a video dating service. I help 'em out, trying to get people to fall in love with each other. Though sometimes I have to resort to the old bow and arrow"— Cork mimed drawing an arrow from his quiver and pulling back the bowstring—"and, you know, thunk! Right in the old solar plexus."

It was a made-to-order role for Cork. He'd drive the story and build a Q so monstrous he'd be the first choice for every short role until the twenty-third century.

But not even any of this stuff had the turmoil value of what had happened to him during the filming of *Not for Publication.* Halfway through the movie, he'd gotten a case of extreme hormonal instability over this girl who played the mouse in the animal orgy scene. She was a former *Playboy* "Girls from Texas" model, twenty-four years old, who had already published a science fiction novel. He wanted to come on to her, but Christ, who needed the kind of rejection that a *Playboy* sex superstar could hand out? But finally he figured, "What the hell, no dream too large," and spilled out his feelings. And voilà!—there they were, two-backing it from dusk to dawn. Turned out she had a crush on him, too, but was afraid that he was in love with the girl who played the pig.

And now, as their affair was percolating steamily along, she wanted to become business partners with him. She wanted him to co-produce a teenage sexploitation flick. "You know," said Corky, "a few car crashes and a lot of flesh!"

There was just one cloud on the horizon. "The only thing that worries me is that her hobby is psychic reading, and she got me to this séance where she said she realized I'd been married before, *twice*. I haven't even been engaged, but she still wants to hear all about my ex-wives. What the hell am I supposed to say?" But it was the kind of problem he liked having.

Corky swept a dime off the counter, jumped from the precarious height of the barstool, his windbreaker parachuting behind him, and hitched off to a pay phone to call a new guy he wanted to get in touch with, Allan Katz, about appearing in a movie he'd heard about that was bound to be the sleeper hit of the summer, *The Hunchback of UCLA*.

He U-turned and fetched a second dime. He was also going to call his new girlfriend. "Who'da thought," Corky said, "that in the end, I woulda got the girl? And the money? And the fame? The whole enchilada!" He rasped his scratchy laugh, and was still chortling when he punched in his first number. No dream too big, no dreamer too small.

Larry Colton borrowed a suit, dug up his one white shirt from the clothes closet of a former girlfriend, got a tie from his second wife's current husband, then went out and sold $40,000 worth of advertising for *Pillars of Portland* in two days. Everybody wanted to buy, and by the time *Pillars* was ready to air on December 13, CBS affiliate KOIN had nearly sold out almost three dozen ad spots for the show at $1,700 apiece.

Nevertheless, Colton was growing more and more uneasy

with the project. Director Tom Chamberlin had raised hell when Colton asked to be named *Pillars'* executive producer, a post that had been promised to the man who provided the company its editing equipment. At first Colton had been outraged and planned to appropriate the title regardless, even if he had to letter it onto the credit crawl himself. After all, he hadn't raised money for this thing and humped it through every agonizing step of production simply to be dismissed as its writer, had he?

Well, maybe . . .

For, by this time, despite all the hype and hoopla around the show, Colton grew queasier and queasier every time he saw bits and pieces of *Pillars* on a monitor or editing machine. For starters, there were just too goddamned many characters. How the hell were viewers going to keep track of them? *Pillars* made *Hill Street Blues* look like *Waiting for Godot* in terms of the number of people appearing on the screen. Also, what were they doing? Just prior to airing Colton found himself agreeing with Tom Chamberlin that the therapy sequences in *Pillars* were extraneous, and that they should be axed. A move he immediately regretted, for therapy provided the show with at least a fragile central theme, and a semblance of plot.

So *Pillars* became less and less Colton's dream come true and more and more his own personal ticking time bomb. Prior to its airing on December 13, he'd become a regular Gertrude Stein in what passed for Portland's creative circle, his phone ringing off the proverbial hook with requests for interviews and talks. Now the talk he thought might be most appropriate was "How Cocky Writers Who Think That Just Because They're Handy with an IBM Selectric They Can Become Hotshot TV Producers Overnight Are Completely Full of Shit. I Know—I've Been There."

And after viewing the premiere, Colton nearly fell over dead. By his lights, the whole show stunk. The only good parts were provided by the actors, so far as he was concerned. The efforts of an entire year had been for nothing. The show was panned politely in the daily *Oregonian*. "But if it was me," Colton said, "I would have ripped it totally apart." And despite the fact that the phenomenon of *Pillars of Portland* was soon to be celebrated in *The Wall Street Journal* as a possible revolutionary wave of the TV future, for Larry Colton enough was enough. It was time to learn how to do things right. To hell with Portland, Oregon. He rented a car and began a twelve-hundred-mile drive to L.A., where, among other things, his friend Corky Hubbert, raconteur and midget, was keying up for a crack at superstardom. And as Colton sped south down I-5, visions of a new and more real TV world burned at the fore of his fevered mind. It was time to go legit.

On November 30, the last day before ABC was contractually obligated to make up its mind on the extension of *Just Our Luck*, Scott called Tony. "Heard anything?" he said.

"Not a peep. Looks like it's going to go down to the wire."

"Well, if you hear before I do, give me a call, okay?" Tony promised that he would. They were both feeling guardedly optimistic, because the show's most recent ratings had been reasonably good, and also because of the lackluster ratings garnered by *Smokey and the Bandit*, the movie the network had substituted in their slot one night, in order to get a relative test of their strength against *The A-Team*. Tony and Scott also believed the network realized the show had improved significantly since its inception. When Tony and Scott had seen the pilot, they'd thought the show was patently lame, a real ABC lobotomizer— but in recent weeks, several of the episodes had been witty and appealing.

Tony had a hard time sleeping on the night of the thirtieth, and wasn't able to get any work done the next morning. In the afternoon, the phone finally rang. It was the executive producer's secretary. After Tony got off the phone with her, he called Scott.

"We lost the sweeps," he said. "The show's gone."

"Oh. Well. Okay. Well, what do you want to do now?"

"I dunno. What do we *gotta* do now?"

"Get some work, I guess. Hell, I thought we had a real shot. I thought we'd get the pickup."

"I did, too." Tony couldn't think of anything else to say. He felt numb, unreal. It seemed as if Scott felt that way, too. "Talk to you later."

"Sure. Take it easy."

Twenty minutes later, Scott called back. The shock was wearing off and the wound was starting to burn. "Shit!" Scott cried, "the show didn't get picked up. Do you realize that?"

"Yeah, I realize that," said Tony.

"All we needed was the back nine," Scott said. He was seething. "If this was NBC, Tinker and Tartikoff would have let us have it. They don't know just network—they know production. They would have seen we were starting to click." Tony calmed Scott down and got off the line.

Twenty minutes later, Tony called Scott. "Christ!" he yelled, walking back and forth in his kitchen, "do you *realize* what has happened? Do you realize? And Lorimar still owes us for those last two weeks! Think we'll ever see that money? Shit! If we had just . . ."

Scott cut Tony off. "With ifs and buts and candies and nuts, oh, what a merry Christmas it would be," said Scott.

"Don't remind me of Christmas. I'm broke."

For the next three weeks, as the holidays approached, Tony toyed with three projects. One was to line up freelance scripts for other shows. He had ideas for *Benson, Webster,* and *The Jeffersons,* three shows with producers who were familiar with his and Scott's work. But all three shows, by coincidence, were based on black actors, and Tony was reluctant to write for a black-oriented show. The recent actions of the NAACP—which could not possibly have been a positive factor in ABC's decision on *Just Our Luck*—had left a bad taste in Tony's mouth. He couldn't stand the idea of anyone thinking he was working on a show just because he was black.

The second project was to write an outline for a novel entitled *The "Leave It to Beaver" Syndrome.* The book would be about how virtually every living American under age forty had grown up thinking that he or she would eventually become Ward or June Cleaver, and have kids that created merry mischief, and be able to buy a new convertible every year. And if one of the kids screwed up, the wife would just have to say, "Wait until your father gets home," and then the good old guy would straighten everything out with two minutes of laugh-track scolding in the den. "It's an exposé," Tony told Scott. "*Of life.*"

"It says some fuckin' writer who wrote this shit screwed me up. 'Cause I *believed* it!"

Tony never seriously thought he'd finish the book; long form wasn't really his style. He was even having a hard time getting himself to start it. But outlining the novel, he thought, would be a good way to keep his identity as a writer. He was sure that if he could plug away at the book, he'd soon be back in television.

The third project was to update his data-processing résumé. The chore was taking him forever, because he hated the idea of getting back into computers. But he had a family to feed. He had borrowed $1,000 from his father and almost that much from his mother. He was sure he could land a sweet deal in DP.

A few days before Christmas, Tony and his wife went to the *Cheers* Christmas party at Paramount's Stage 25. It was nice to see his old friends, and he was given one of a big pile of giant beach towels that said FROM THE CAST OF CHEERS. Tony enjoyed the party, but he was hoping to do a little business. But it was a no-go. To the *Cheers* people, he could tell, he was still a gofer.

Two days after Christmas, Tony sat at his desk. His wife was off at work, the boys were in school, and he had his résumé on the desk, with a list of people to send it to. He also had a thick stack of blank typing paper that he'd picked up to sketch out the *"Beaver" Syndrome* on. It was time for him to decide whether or not to address some envelopes and get the résumé in the mail, or to just . . . forget about it. He'd been putting off the decision for almost a month.

The phone rang. It was Scott, in Rhode Island for the holidays. He wanted to know if their agent had lined up an appointment for them to pitch a script for *Webster*.

Tony told him the meeting was still floating; *Webster* had already assigned most of its scripts for the back nine and wasn't particularly interested in new freelancers.

"You think all this shit is worth it?" Tony asked.

"Sure it's worth it," Scott said. "We've already done something some people think about doing all their lives and never even try. We wrote a TV show. We proved we could do it. We proved it to Lorimar. We proved it to ABC. We proved it to the producers. And the next time we go in somewhere and say that we can write a TV show, it won't just be a pitch."

When Tony hung up, he pushed the résumé aside—not all the way off his desk, but to one side of it. Then he pulled his typewriter to the heart of his desk and began pounding away.

Epilogue:
The Great
Adventure

May 1984

So much had happened; so little had changed.

"I'm gonna throw up," moaned Tony Colvin, the fingers of one hand sliding slowly down his face, shock settling into his eyes. Tony, at work, wearing a cream-colored polo shirt with GOLDEN WEST RENT-A-CAR embroidered across his thick left pectoral, had just been informed that his old rival at Paramount, writer Michael Weithorn, had recently linked up with the ultra-monied Steven Spielberg on a movie project. "Michael Fucking Weithorn is working with Steven Fucking Spielberg?" he cried, his voice well into the distress octave. "Oh, this really makes me sick." Tony plopped down on his midget, woodlike veneer desk, his rear end covering almost half of it. But Tony had to contain his grief; a customer had just entered to return a Ford Fairmont. The car, which appeared to be steaming under the hood, sat in the parking lot of the combined rental agency—miniwarehouse, a shimmering metallic glare of sun and chrome. "I suspect Goldberg set him up with it?" Tony's suspicions were confirmed. "I thought so. Well, good for Weithorn," Tony said. But his eyes said: "Deliver me, O Lord, from mine enemies."

*　*　*

Gary David Goldberg, prior to becoming rich and powerful, had never had a career. He'd dropped out of Brandeis University in 1964 and worked as a waiter in Greenwich Village. Stringing odd jobs together, he and Diana, the woman he lived with after 1968, traveled around the world. In 1971, they had a daughter and opened a day care center. A little later, Diana moved to San Diego to work on her Ph.D. Goldberg "tagged along"—his words—quite content to take care of the baby, lie back, and watch life slip slowly past.

Diana had to hand in writing for her course work, so Goldberg started to write, too. He didn't feel any burning creative compulsion; he just didn't have anything better to do. But he found that he really loved to write. For as much as fifteen hours each day, he would crank out autobiographical stories. A past president of the Writers Guild visited San Diego and Goldberg met him and showed him his stories. The work was then passed on to an agent in Hollywood, who thought it looked promising and requested a meeting. Goldberg hitchhiked to L.A.

"What are your favorite TV shows?" asked the agent.

"I don't have any," said Goldberg.

"Why not?"

"Can't afford a TV."

That could have been the end of it, except that a motel close to where Goldberg lived was going out of business, and the managers were willing to let go of one of their hulking, crate-sized, black-and-white sets for $25. After Goldberg wrestled one home he watched his first complete TV program in ages. It was *Get Christie Love*. When it was over, Gary turned to Diana and said, "I can do that. I can *definitely* do that."

Every couple of days Goldberg would send his agent a new TV script, and he regularly hitchhiked up to L.A. to meet with the agent. Goldberg wrote a pilot spec script called *Free Clinic*, which was shown to the producers of a program called *The Dumplings*. He got an assignment and wrote a script. The script was shown to Gordon and Lynn Farr, who were producing *The Bob Newhart Show*.

"They singled me out," according to Goldberg, "and said, 'Boy, I'm gonna make you a star.' " And they did. About a year after he'd bought his black-and-white TV, Gary Goldberg was co-producer of *The Tony Randall Show,* with Hugh Wilson, who would soon create *WKRP in Cincinnati.* When *Tony Randall* folded after a two-year run, Goldberg spent two years on the writing staff of *Lou Grant,* then did sixteen episodes of a show he created for MTM, *The Last Resort.* CBS offered to let him develop four pilots. The first one was called *Making the Grade,* a show about high school that aired six times. For that show, he hired a twenty-four-year-old writer named Michael Weithorn, who'd just completed three years of teaching high school. Weithorn barely had enough credits to impress even his own mother—he'd written one episode of *Benson*—but Goldberg liked him, and gambled. The two were a lot alike. Though physically different—Weithorn was blond, muscled, and neat, while Goldberg was shaggy and sloppy—they shared one trait that is not altogether common in Hollywood—they were actually . . . almost . . . humble. Neither thought that he was Jesus Christ, nor even one of the twelve disciples. So when *Making the Grade* was canceled, Goldberg hired Weithorn as story editor on his second CBS pilot, *Family Ties,* which was handed by that network to NBC. And a year later, Weithorn, at the unripe age of twenty-six, was made producer. All the while, Tony and Scott had watched Weithorn's ascent from their offices across the street at Paramount . . . salivating.

Weithorn has been given his break as a producer because Goldberg had less time for the show, having been offered in the summer of '83 the stuff that Hollywood dreams are made of— the chance to make a movie with the reigning heavyweight champ of filmdom, director Steven Spielberg. Spielberg, a TV addict, had seen Goldberg's name at the end of a number of shows he'd liked. So out of the blue, one fine day, Gary Goldberg's phone had rung.

Because Goldberg's daughter, now twelve, was at camp, Goldberg agreed to write the film Spielberg wanted, a love story set in Hollywood, entitled *Reel to Reel.* If she'd been home, Goldberg might have turned Spielberg down in order to spend more time with her. His essential indifference to having a career—still intact—had begun to serve Goldberg: It gave him the upper hand in all of his negotiations. It made him *dangerous.*

By early 1984, Goldberg was working on still another movie

with Spielberg, and he needed help with *Reel to Reel*. So he brought in Michael Weithorn.

By the first week of May 1984, Gary Goldberg had amassed for himself his own production company, several movie and television deals, a separate arrangement to develop pilots for NBC, a house in town, a house at the beach, a small fortune in automobiles, and more just plain stuff than he'd ever imagined a person could own. Grant Tinker, Goldberg's first employer in the MTM days, liked to chide Goldberg about his late-born aquisitiveness. "It's not that you sold out," Tinker often said to him, "it's that you sold out so *completely*." But when Goldberg himself considered his success, it wasn't the sheer enormity of it that fascinated him. What interested Goldberg was this: that this fortune, this fate that had settled on him so arbitrarily, so whimsically, so haphazardly, seemed, in moments of quiet retrospect, to have somehow been utterly, absolutely, inescapably . . . *inevitable*.

To Tony, though, at this particular moment in May, the idea of inevitability seemed absurd. If anything had been burned into his brain over the past year, it was that nothing is inevitable— no fate is sealed, so long as the victim of that fate is willing to work two shifts. Which was exactly what Tony was doing. All day long he rented Toyotas, Rabbits, Datsuns, and Fords for Golden West Rent-a-Car, near the L.A. airport, and all night long he wrote. He had more leisure time than usual—his wife had left him.

Fed up with the money worries, the insecurity, and the lack of a full-time husband, she'd packed up the three kids and walked out. Tony thought she would come back, though, if he lined up something solid in the new season. He'd tried to relieve their marital strain by looking for a job in data processing, after all of his freelance writing projects had failed to bear fruit, but the data-processing industry wanted no part of him. To the staid folks who did the corporate hiring, Tony was Mr. Glitter—they'd made him feel like he was John DeLorean coming to apply for a job as a narc. "They wouldn't take me at face value. They thought I was crazy. They told me they couldn't pay the thousand a week I'd been getting. They clearly resented that I'd left their industry to go to work in . . . in . . .'' Tony sighed. ''. . . Sodom and Gomorrah.''

Scott was also now working outside the industry. Currently, he was employed at a racketball club, passing out towels. An easy gig, by any standards. Not only was it simple, but he was upwardly mobile—he'd just been given a 25-cent-per-hour raise. And there was even talk of letting him try to sell memberships, on commission. Yes—*profit participation*.

Since the money had stopped rolling in, Tony had been forced to refine his lifestyle. Tony, once the proud MC of nightly $100 picnics, was now in ongoing negotiations with the gas and electric companies, doing his best to keep his apartment lit and warm. The phone was gone—if Tony had to make a call, well, wasn't that what quarters were for?

Thus was the life Tony had created for himself in his headlong rush to crack the television industry. And the only way to improve this life, since his alternatives had dwindled so drastically, was to get back into TV. So every night now he was working on his new pilot script. It was about a rent-a-car office. What luck he had to be here! Fortune had smiled upon him once more! Because it turned out that a rent-a-car place was absolutely the perfect setting for a sitcom. You had your regular characters as employees, and a rich broth of guest stars and weirdos coming in to rent cars. You could keep the guest stars in the office as long as you wanted, waiting for their cars, and you could contrive an endless stream of annoyances and solutions. Think of the heat! The hooks! The buttons! Oh, Tony was one lucky dog, all right. Why, just yesterday Tony had driven a guy to the airport whose job it was to pry donations out of university alumni. And what university did he work for? None other than Tony's alma mater, the University of Washington. If a greedy fundraiser trying to score a donation from a down-on-his-luck alum wasn't the foundation for a great plot, then nothing was. Tony was already starting to put the script together. Wouldn't John Houseman make a great fundraiser? My God—the unimaginable luck! Was this not exactly the kind of luck that a future king of television would use to forge his own fate?

And there was even more to be thankful for. For starters, Tony had been in a wreck. His Lancia Beta Coupe, the last vestige of his once-vast fleet of cars, had been plowed into by a guy in a truck. The truck—bless its steel frame heart—had crumpled $2,300 worth of damage into Tony's Lancia, none of which kept the sportser from running. Granted, the mangled, misaligned car now went through a set of retreads about every seven weeks, but it got him to work, and had provided over two grand of grocery money. God certainly must love Tony.

And to finish off his bounty of blessings, there was the IRS. Just a year ago, Tony had been forced to fight off an IRS grab for $19,400 of his money, but now the tables were turned. While he was at *Just Our Luck*, Tony had farmed out his tax management to the same accounting company that serviced the show's stars. These savvy gentlemen, these showbiz smarties, had milked every possible deduction out of Tony's finances. Consequently, he was on the verge of receiving a windfall tax refund of $5,700. Almost six *big bills*. Due to arrive at any moment.

J. Antony Colvin, his alternatives winnowed down to include only complete success—nothing else—was a happy man. And why shouldn't he be?

The check was in the mail.

What happened?

The season, for all intents and purposes, was over, and the NBC programmers had to ask themselves: What happened? But, try as they might, they weren't quite able to pose the question in the mood of cold, sober calculation. No, when the question issued forth, as it did again and again, it came more in the manner of a man coming slowly to his senses after being in a car crash. He knows he's in his car, he knows it's smashed up, he knows the motor is whining, he knows a telephone pole is in his backseat—but . . . *what happened?*

This was Grant Tinker's answer: "Somebody has to lose every September. Let's say it was our turn." Which said, of course, absolutely nothing, and was therefore the most honest of answers. Because in the last analysis, *no one knew* what had happened. They knew only that NBC had logged its worst season in twenty years, and probably its worst season in the history of the network. The season's ratings put NBC in last place, with a 14.9 average rating, distantly behind CBS, with 18.1, and ABC, with

17.2. Going into the new season, just four of NBC's 1983–84 shows stood even a chance of renewal—Harry Anderson's *Night Court*, Norman Lear's *Double Trouble*, Dick Clark's *TV Bloopers*, and Steve Cannell's *Riptide*. The rest of the shows, which had cost about enough money to balance the trade deficit of a small country, were little more than evaporated wisps of memory. How long would it be before America completely forgot about shows like *The Rousters*, or *Legmen*, or even *Boone*? Had they ever even existed?

The NBC shows, as a whole, had been so immediately forgettable that the network had lost every sweeps race of the season, seriously undermining the network's relationship with the affiliates. In the February sweeps, NBC had finished almost three entire rating points out of *second* place. Oh, it was miserable— no doubt about it—if you had the guts to stop and think about it. More so when you considered that no one even knew what had happened.

There was no mass execution of executives at NBC in response to the failed season, and Tartikoff looked solid for at least another year as programming chief. But one head did roll—that of Robert Mulholland, who had ultimate corporate responsibility for the prime-time schedule. Widely rumored as legatee of the job was Steve Sohmer, Mr. Be There, but when all was said and done, Sohmer didn't get the post. He stayed in his present position . . . just down the hall from Tartikoff.

Sohmer was beginning to look like a fortune-teller for his prediction that immense TV audiences could now be lured only by the aura of a Major Televised Event. The top-rated shows of the season—*The Day After*, about nuclear war; *Something About Amelia*, a Ted Danson movie about incest; and *Adam*, about child kidnapping—had all been presented to the nation as important cultural landmarks, and the viewing public had lunged for them like piranhas to chuck steak.

Contriving a major event, though, wasn't easy. Even with the "big events" of the February sweeps—the miniseries *Lace*, *Celebrity*, and *Master of the Game*, as well as the Winter Olympics—the networks had actually drawn 5 percent fewer viewers than the previous year. The exodus of viewers to cable and independent stations appeared unstoppable. And because of it, the market was becoming diluted. HBO had finished the season with a 10 percent loss of audience. Everyone in the industry wanted to know: *How many rats can this cage hold?*

To stop the viewer ebb, the Big Three needed a hot new trend for the 1984–85 season. But they didn't have one. The past season had taught them . . . not much. Other than that kids like rock 'n' roll. This nugget of wisdom had been gleaned by observing the obvious success of MTV. Therefore, all three networks were incorporating some aspect of music videos into their schedules. The general trend was to force a rapid-fire, flashing-image pace into programs, and lay more music over the action—in other words, to make shows look more like videos, or, more accurately, like the progenitors of music videos: television commercials.

The only other discernible trends that would be incorporated in the 1984–85 schedule were the reemergence of action-adventure and a move away from comedy. Actually, comedy was being incorporated into action-adventure. This "lighter action form" was being fueled by, as much as anything else, the inescapable reality that the dramas which the networks were offering—*Riptide, Airwolf, The A-Team*—were virtually impossible for most trainable, nonretarded persons to take with a straight face.

When NBC announced its 1984–85 prime-time schedule, on May 10, it had several of these guns-and-laughs shows on the board, including *Partners in Crime* and *Hot Pursuit*. There were only three new comedies, including *Punky Brewster* and *The Bill Cosby Show*. And there was just one comedy show left from the 1983–84 season, the season that had started with such brilliant optimism and ended in squalid defeat. The sole survivor was *Night Court*. Harry Anderson had kept his day job.

But Harry wasn't able to celebrate on the day of the announcement. He was back on the road. At the time of the schedule's unveiling, Harry Anderson was somewhere over Idaho, in a DC-9, sitting as close as he could to the "black box" that was strategically situated to survive a crash.

The next night, the haunted house of the New Jersey amusement park that Harry had played the previous spring—where the crowd stoned him with coins—burned to the ground.

Eight people died, while the rock band on the stage next door wailed obliviously on. Harry read about it in the papers while he was in Chicago for NBC's annual flatter-the-advertisers party. It turned out, though, that the place was not called the Great American as Harry had thought. Its real name was the Great Adventure.

A hiss: "Hey, midget!"

Corky Hubbert stopped and turned, surveying the tables at El Coyote. He'd just downed an enchilada and two double margaritas and was lighting out when again, a hiss: "Hey, midget!"

From a table a few feet away, Corky ambled back. It was some kid with his date. Who'd already blushed the color of a stop light. Corky introduced himself and asked, "Do I know you?"

The guy didn't say anything.

His date said, "I'm really sorry. He was just joking."

Corky nodded. It had been an arduous week. With a big role in the feature film *Legend*, directed by Ridley Scott, almost in the bag, it looked like ten years of grubbing for bit parts on TV shows and in the movies might finally be over. He was even going corporate. For tax reasons. His manager was forming Cork Co. Corp.

Corky took the girl's hand, smiled, and kissed it. Ever so gently. Then he picked up her water glass and threw its contents in the guy's face.

Pandemonium in El Coyote. The guy bolted up and Corky grabbed him by the collar and pulled him, shoulders first, out of the booth. The kid fell, backpeddling, into a wall. Corky, at four feet eight inches, built like a cross between a butterball and a pit bull, yanked the guy down so that his back formed a question mark. "Midget, huh? Maybe you oughta pick on a paraplegic next! One more word to me, just one more word! and I'll put you in the hospital."

Then Hollywood's fourth-top midget let the shmuck go and headed toward the door. But then, still angry, he wheeled around to the girl, blurted, "I'm sorry, cuz I bet you're really nice," picked up her dinner, an El Coyote Super Tostada with shredded chicken, guacamole, sour cream, lettuce, and olives, and slammed it square into the guy's face. Before easing out the door.

So satisfying! Midget, huh? He couldn't have done it better on *Magnum* or *The Fall Guy*. "Not bad," Corky thought, catching his breath, "—for real life."

Relishing the debacle, Corky headed home with all deliberate speed. Home now usually consisted of a plush apartment near the beach—and for a week it had consisted of nothing less than Hollywood's Chateau Marmont. Corky had shelled out $700 to rent a room for a week at the hotel, where his idol John Belushi had ODed, so he could wine and dine his visiting girlfriend in Gothic splendor. Only trouble was—no girlfriend. His *Playboy* "Girls from Texas" dreamdish, who'd played the mouse in the animal orgy scene in *Not for Publication*, had called at the last minute to ask for a rain check. What could Cork do, even with a seven-hundred-buck room tab to eat—tell the lovely little gumdrop to get her rear in gear, tell her that it was now or never? Not likely. Anyhow, she had a good excuse. Turned out, her roommate was supposed to be getting married to a famous rock star, but was calling the whole thing off and needed emotional support from Corky's squeeze. The rock star, it seemed, loathed sexual congress—but that part was OK, because the roommate was sculpted of ice herself, in spite of the fact that, as far as Corky was concerned, she made "Raquel Welch look like Kermit the Frog." What was not OK, though, was that this frigid lovely had just discovered a tape recording in which the rocker "confessed" not just to murder, but to having drunk his victim's blood as well. If that wasn't horrific enough, she also discovered a tape in which he confessed that not only was he not in love with her, but he didn't even *like* her that much. Which she considered just cause for prenuptial alarm. Even in Dallas, there were civilized limits to a marriage of convenience.

So the roommate went to the superstar's manager, insisting that her betrothed at least see a psychiatrist before they tied the knot. But the manager demurred. "I don't care whose blood he's been drinking, we got an album to do." Exact words.

But all was not lost. Supplanting the Girl from Texas was another girlfriend. Of sorts. An anthropologist-stripper from Portland. She was nice and really cute and extremely enthusiastic about Hollywood—Corky had once taken her to Spago for pizza and every two seconds she'd barked into his ear, *"There's*

Sly Stallone!" or "There's Linda Evans!" or "There's Henry Winkler!" She was great. Somehow, though, it just wasn't the same.

But—screw it!—Corky wasn't a lovesick teenager. He was a career guy. Even though his TV career was on hold because the co-creator of *Mr. Cupid*, the sitcom with Corky in the title role, was sick and unable to come up with a pilot script for the 1984–85 season. Reportedly, the writer's entire digestive canal was completely inflamed.

Still, all and all, was this a great year or what? Corky had come a long way from years gone by, when he'd had to make rent money doing gigs on *The Gong Show* or *The Dating Game* (where he'd wowed the Farrah Fawcett look-alike who asked him, "What is the most important thing in life that your parents couldn't teach you?" by replying, "Trigonometry. I had to learn that in the streets").

And speaking of stabs from the past, what about the time he'd had to resort to an appearance in a porno flick, just to make ends meet? *Ultraflesh.* He'd played it supersafe. His job was comic relief, and he insisted on wearing a fake goatee, a rubber nose, and a sweater over his head, and on keeping all of his clothes on. Not a bad trip, actually. Still, it was tough, improvising dialogue with a woman who had baby batter all over her face.

Back at his place in Santa Monica, Corky stood by the edge of the pool. Palm trees provided swaying islands of shade in the background. Corky dove. His compact manatee-style body sprang to surprising height as he leaped from the pool's edge, before stalling in midair and arching, *ker-splash*, into the water. An instant later he bobbed back to the surface. "What worries me," he said, dogpaddling, "is now my agent and manager keep telling the movie people how great I am and how much money I'm worth because I'm so talented." Corky wiped water out of his eyes. "I'm afraid that if they convince 'em I'm much more fantastic and valuable, I'm going to end up back out on the sidewalk."

But it didn't happen. Within two weeks, Corky had locked up the part—the fifth-highest starring part—in the Ridley Scott movie, budgeted at over $20 million. Corky's salary for about twenty-five weeks of work would enable him, if he so chose, to purchase an only slightly used Volkswagen Rabbit on or about the middle of the week, every week. Or he could lump it all together and buy a large part of a condo complex, which his manager was advising. Whatever his investment strategies, Corky was so well set financially that it wasn't even funny.

Now he sat at Fellini's, on Melrose in Hollywood, bidding farewell to his pal, producer Allan Sacks. Who was in a black shirt. Black coat. Black jeans. Black boots. Sacks was maybe five foot eight or nine, thirty-five or forty. He had an impish pale face and looked like a walking East Village. Co-creator of the hit series *Welcome Back, Kotter*, producer of TV movies like *The Women of West Point*, and would-be producer of *Mr. Cupid*, Sacks had pulled a creative 180 with his latest effort, which was not ready for prime time. The project was a rock feature about Joan Jett, tentatively titled *We're All Crazy Now*, in which Corky had a two-second role as an S&M leather dwarf.

"This could be a blockbuster, Cork," said Sacks of his movie. "But it's got so much satanic imagery in it, I'm, like, afraid I'm doing the devil's work. Scary. I've been thinking about hanging crucifixes all over my apartment, just to keep the place exorcised. Which is embarrassing, because I'm Jewish." A waitress passed and Sacks put a hand up. "Excuse me. Another zombie for my friend and a Coors for me, please."

"It's a great film, though," Corky said. "I mean, like, when the chick comes out wearing a shawl of bloody pig's intestines! Only you could have come up with a move like that, Alan."

"Thanks, Cork."

In a day or so, Corky would be jetting, first class, to London. He would rehearse two weeks, then get a month off at full salary. His departure was bittersweet. For one thing, he'd been rolled. On Sunset Boulevard. Two black guys had jumped out of their Caddy, slammed Corky to the ground, and divested his wallet of $140. Also, it didn't look like he was going to get a part in Allan Katz's *Hunchback of UCLA*. He'd shown Katz his special homemade Betamax promo clips—Corky playing the commander of a Red Chinese submarine intent on blowing up the world, Corky getting murdered in the desert, writhing around on the sand, etc. Katz liked the clips, remarking, "Boy, Cork, it's a good thing none of this stuff is self-indulgent." The problem was, *Hunchback* was full of so much craziness already that even the briefest appearance by Cork could push it right over the edge.

Still, what the hell! Katz would make other movies. Blockbusters! And Corky would be there.

And hadn't he just scored a big one? *The* big one?

Leaving Fellini's, Corky walked down the street, eyeing two hookers. "Imagine trying to build a life out of fifteen-dollar blow jobs. Depressing." Then he looked up at the sky. "*God! I promise! No more fuckups! If you let me make it this time, I'll never screw up again!*"

Then, a more earthly aside. "I came into this town riding on my thumb. And I'm leaving it in a limo. Just like I planned."

His triumph. His ascension. It had all been so sweetly, so perfectly . . . inevitable.

A Note on
Sources

Four books were especially important sources in constructing this account: Sally Bedell's *Up the Tube*, Todd Gitlin's *Inside Prime Time*, Bob Shanks's *The Cool Fire*, and Jib Fowles's *Television Viewers vs. Media Snobs*.

We also drew from literally hundreds of magazine and newspaper accounts of current television, including those appearing in *Broadcasting* magazine, *Channel* magazine, *Cable View* magazine, *TV Guide*, *Playboy*, the *Los Angeles Times*, *Daily Variety*, *Hollywood Reporter*, *The New York Times*, *The Wall Street Journal*, *USA Today*, *Time* and *Newsweek*.

We'd also like to thank several journalists for their assistance and advice, including Steve Oney—whose intimate and excellent tale of NBC programming in *California* magazine, "That Championship Season," was very helpful—Howard Rosenberg of the *Los Angeles Times*, Sally Bedell of *The New York Times*, Ben Brown of *USA Today*, and Richard Hack and Alan Gansberg of *Hollywood Reporter*.

Index